ENVIRONMENTAL SCIENCE, ENGINEERING AND TECHNOLOGY

ENVIRONMENTAL STEWARDSHIP AND ECOLOGICAL PROTECTION

ENVIRONMENTAL SCIENCE, ENGINEERING AND TECHNOLOGY

Additional books in this series can be found on Nova's website under the Series tab.

Additional E-books in this series can be found on Nova's website under the E-books tab.

ENVIRONMENTAL SCIENCE, ENGINEERING AND TECHNOLOGY

ENVIRONMENTAL STEWARDSHIP AND ECOLOGICAL PROTECTION

DANIEL J. MORAN
EDITOR

Nova Science Publishers, Inc.
New York

For permission to use material from this book please contact us:
Telephone 631-231-7269; Fax 631-231-8175
Web Site: http://www.novapublishers.com

LIBRARY OF CONGRESS CATALOGING-IN-PUBLICATION DATA
Environmental stewardship and ecological protection / Daniel J. Moran, editor.
p. cm.
Includes index.
ISBN 978-1-61209-341-3 (hardcover)
1. Environmental protection. 2. Environmental policy. I. Moran, Daniel J.
TD170.E63 2011
333.72--dc22
2010051548

Published by Nova Science Publishers, Inc. † New York

CONTENTS

Preface vii

Chapter 1 Valuing the Protection of Ecological Systems and Services: A Report of the EPA Science Advisory Board 1
United States Environmental Protection Agency

Chapter 2 The Use of Markets to Increase Private Investment in Environmental Stewardship 181
Marc Ribaudo, LeRoy Hansen, Daniel Hellerstein and Catherine Greene

Index 251

PREFACE

There is increasing recognition of the numerous and important services that ecosystems provide to human populations, such as flood protection, water purification, and climate control. Protecting ecological systems and services is part of the EPA's core mission. This new book describes how the EPA can use an "expanded and integrated approach" to ecological valuation to encourage greater collaboration among a wide range of disciplines, including ecologists, economists at each step of the valuation process.

Chapter 1- EPA'S Science Advisory Board (SAB) created the committee on valuing the protection of Ecological Systems and Services (c-vpESS) to offer advice to the Agency on how EPA might better assess the value of protecting ecological systems and services. As used in this chapter, the term "valuation" refers to the process of measuring values associated with a change in an ecosystem, its components, or the services it provides. the SAB charged the committee to:

- Assess EPA's needs for valuation to support decision making.
- Assess the state of the art and science of valuing the protection of ecological systems and services.
- Identify key areas for improving knowledge, methodologies, practice, and research at the Agency.

This chapter provides recommendations to the Agency for improving EPA's current approach to ecological valuation and for supporting new research to strengthen the science base for future valuations.

Chapter 2- U.S. farmers and ranchers produce a wide variety of commodities for food, fuel, and fiber in response to market signals. Farms also contain significant amounts of natural resources that can provide a host of environmental services, including cleaner air and water, flood control, and improved wildlife habitat. Environmental services are often valued by society, but because they are a public good—that is, people can obtain them without paying for them—farmers and ranchers may not benefit financially from producing them. As a result, farmers and ranchers underprovide these services.

In: Environmental Stewardship and Ecological Protection
Editor: Daniel J. Moran
ISBN: 978-1-61209-341-3

Chapter 1

VALUING THE PROTECTION OF ECOLOGICAL SYSTEMS AND SERVICES: A REPORT OF THE EPA SCIENCE ADVISORY BOARD[*]

United States Environmental Protection Agency

ACKNOWLEDGEMENTS

EPA's Science Advisory Board Committee on Valuing the Protection of Ecological Systems and Services would like to acknowledge many individuals who provided their perspectives and insights for the committee's consideration in the development of this chapter.

Many individuals provided perspectives at public meetings of the committee and to members of the committee's Steering Committee. These individuals included: Mr. Geoffrey Anderson, EPA Office of Policy, Economics, and Innovation; Ms. Karen Bandhauer, EPA Region 5; Mr. Devereaux Barnes, EPA Office of Solid Waste and Emergency Response; Ms. Jan Baxter, EPA Region 9; Dr. Richard Bernkopf, U.S. Geological Survey; Dr. Lenore Beyer-Clowe, Openlands Project; Dr. Ned Black, EPA Region 9; Mr. Robert Brenner, EPA Office of Air and Radiation; Dr. Thomas Brown, USDA Forest Service; Dr. Randall Bruins, EPA, Office of Research and Development; Mr. David Chapman, Stratus Consulting; Mr. James DeMocker, EPA, Office of Air and Radiation; Mr. Richard Durbrow, EPA Region 4; Mr. Jesse Elam, northeast illinois planning commission; Mr. William Eyring, center for neighborhood Technologies; Ms. Jerri-Anne Garl, EPA Region 5; Ms. Iris Goodman, EPA Office of Research and Development; Dr. Robin Gregory, Decision Research, Eugene, Oregon; Dr. Sharon Hayes, EPA, Office of Research and Development; Dr. Bruce Herbold, EPA Region 9; Dr. Julie Hewitt, EPA Office of Water; Mr. James Jones, EPA Office of Prevention, Pesticides, and Toxic Substances; Dr. Robert E. Lee II, EPA Office of Pollution Prevention and Toxics; Dr. Richard Linthurst, EPA Office of Research and Development; Dr.

[*] This is an edited, reformatted and augmented edition of a United States Environmental Protection Agency publication, Report EPA-SAB-09-012, dated May 2009.

Albert McGartland, EPA Office of Policy, Economics, and Innovation; Mr. James Middaugh, City of Portland; Dr. Chris Miller, EPA Office of Water; Mr. Michael Monroe, EPA Region 9; Dr. Chris Mulvaney, Chicago Wilderness; Dr. Wayne R. Munns, EPA Office of Research and Development; Mr. David S. Nicholas, EPA Office of Solid Waste and Emergency Response; Ms. Gillian Ockner, David Evans and Associates, Inc.; Dr. Nicole Owens, EPA Office of Policy, Economics, and Innovation; Dr. Stefano Pagiola, World Bank; Ms. Karen Schwinn, EPA Region 9; Dr. Michael Slimak, EPA Office of Research and Development; Dr. Matthew Small; EPA Region 9; Mr. David Smith, EPA Region 9; Dr. Ivar Strand, University of Maryland; Ms. Alexis Strauss, EPA Region 9; Ms. patti Lynne Tyler, EPA Region 8; Mr. John Ungvarsky, EPA Region 9; Mr. James van der Kloot, EPA Region 5; Mr. William Wheeler, EPA Office of Research and Development; Ms. Louise Wise, EPA Office of Policy, Economics, and Innovation; Dr. T.J. Wyatt, EPA Office of Pesticide Programs; and Mr. Steve Young, EPA Office of Environmental Information.

The committee would also like to thank the speakers and participants at the SAB Workshop Science for Valuation of EPA's Ecological Protection Decisions and Programs, December 13-14, 2005: Dr. Mark Bain, Cornell University; Mr. Robert Brenner, EPA Office of Air and Radiation; Ms. Kathleen Callahan, EPA Region 2; Dr. Ann Fisher, Pennsylvania State University; Dr. George Gray, EPA Office of Research and Development; Dr. Brooke Hecht, Center for Humans and Nature; Dr. Bruce Hull, Virginia Institute of Technology; Dr. DeWitt John, Bowdoin College; Dr. Robert Johnston, University of connecticut; Dr. Dennis King, University System of Maryland; Dr. Joseph Meyer, University of Wyoming; Mr. Walter v. Reid, Stanford University; Dr. James Salzman, University of north carolina; Dr. Michael Shapiro, EPA Office of Water; Dr. Elizabeth Strange, Stratus Consulting; and Dr. Paul C. West, The nature conservancy.

The committee especially thanks the experts who provided independent review of the report in draft form. Dr. Roger Kasperson, clark University, Dr. James opaluch, University of Rhode island, and Dr. Duncan patton, Montana State University, reviewed a draft in October 2007. Dr. Trudy Ann Cameron, University of Oregon, Dr. Thomas Dietz, Michigan State University, Dr. Alan Krupnick, Resources for the Future, and Dr. Mark Schwartz, University of California - Davis, reviewed a draft in May 2008. The reviewers provided many constructive comments and suggestions, but were not, however, asked to endorse the conclusions or recommendations in the report. finally, the chair and vice chair of the committee thank former committee members Dr. Domenico Grasso (committee chair 2003-2005) and Dr. Valerie Thomas (committee member 2003-2004), who served on a steering group for the committee during its early years.

EXECUTIVE SUMMARY

EPA'S Science Advisory Board (SAB) created the committee on valuing the protection of Ecological Systems and Services (c-vpESS) to offer advice to the Agency on how EPA might better assess the value of protecting ecological systems and services. As used in this chapter, the term "valuation" refers to the process of measuring values associated with a change in an ecosystem, its components, or the services it provides. the SAB charged the committee to:

- Assess EPA's needs for valuation to support decision making.
- Assess the state of the art and science of valuing the protection of ecological systems and services.
- Identify key areas for improving knowledge, methodologies, practice, and research at the Agency.

This chapter provides recommendations to the Agency for improving EPA's current approach to ecological valuation and for supporting new research to strengthen the science base for future valuations.

General Findings and Advice

EPA'S mission to protect human health and the environment requires the Agency to understand and protect ecosystems and the numerous and varied services they provide. Ecosystems play a vital role in our lives, providing such services as water purification, flood protection, pollination, recreation, aesthetic satisfaction, and the control of diseases, pests, and climate. EPA's regulations, programs, and other actions, as well as the decisions of other agencies with which EPA partners, can affect ecosystem conditions and the flow of ecosystem services at a local, regional, national, or global scale. to date, however, policy analyses have typically focused on only a limited set of ecological factors.

Just as policy makers at EPA and elsewhere need information about how their actions might affect human health in order to make good decisions, they also need information about how ecosystems contribute to society's well-being and how contemplated actions might affect those contributions. Such information can also help inform the public about the need for ecosystem protection, the extent to which specific policy alternatives address that need, and the value of the protection.

Valuation of ecological systems and services is important in national rule makings, where executive orders often require cost-benefit analyses and several statutes require weighing of benefits and costs. Regional EPA offices can find valuation important in setting program priorities and in assisting other governmental and non-governmental organizations in choosing among environmental options and communicating the importance of their actions to the public. Ecological valuation can also help EPA to improve the remediation of hazardous waste sites and make other site-specific decisions.

This chapter describes and illustrates how EPA can use an "expanded and integrated approach" to ecological valuation. The proposed approach is "expanded" in seeking to assess and quantify a broader range of values than EPA has historically addressed and through consideration of a larger suite of valuation methods. the proposed approach is "integrated" in encouraging greater collaboration among a wide range of disciplines, including ecologists, economists, and other social and behavioral scientists, at each step of the valuation process.

Value is not a single, simple concept. people may use many different concepts of value when assessing the protection of ecosystems and their services. for this reason, the committee considered several value concepts. these included measures of value based on people's preferences for alternative goods and services (measures of attitudes or judgments, economic values, community- based values, and constructed preferences) and measures based on biophysical standards of potential public importance (such as biodiversity or energy flows).

To date, EPA has primarily sought to measure economic benefits, as required in many settings by statute or executive order. The report concludes that information based on some other concepts of value may also be a useful input into decisions affecting ecosystems, although members of the committee hold different views regarding the extent to which specific methods and concepts of values should be used in particular policy contexts.

In addition, the Agency's value assessments have often focused on those ecosystem services or components for which EPA has concluded that it could relatively easily measure economic benefits, rather than on those services or components that may ultimately be most important to society. Such a focus can diminish the relevance and impact of a value assessment. this chapter therefore advises the Agency to identify the services and components of likely importance to the public at an early stage of a valuation and then to focus on characterizing, measuring, and assessing the value of the responses of those services and components to EPA's actions.

EPA should seek to measure the values that people hold and would express if they were well informed about the relevant ecological and human well-being factors involved. This chapter therefore advises EPA to explicitly incorporate that information into the valuation process when changes to ecosystems and ecosystem services are involved. Valuation surveys, for example, should provide relevant ecological information to survey respondents. Valuation questions should be framed in terms of services or changes that people understand and can value. Likewise, deliberative processes should convey relevant information to participants. the report also encourages EPA to consider public education efforts where gaps exist between public knowledge and scientific understanding of the contributions of ecological processes.

All steps in the valuation process, beginning with problem formulation and continuing through the characterization, representation, and measurement of values, require information and input from a wide variety of disciplines. instead of ecologists, economists, and other social and behavioral scientists working independently, experts should collaborate throughout the process. Ecological models need to provide usable inputs for valuation, and valuation methods need to incorporate important ecological and biophysical effects.

Of course, EPA conducts ecological valuations within a set of institutional, legal, and practical constraints. these constraints include substantive directives, procedural requirements relating to timing and oversight, and resource limitations (both monetary and personnel). For example, the preparation of regulatory impact analyses (RiAs) for proposed regulations is subject to Office of Management and Budget (oMB) oversight and approval. oMB's circular A-4 on *Regulatory Analysis* makes it clear that RiAs should include an economic analysis of the benefits and costs of proposed regulations conducted in accordance with the methods and procedures of standard welfare economics. At the same time, the circular provides that where EPA cannot quantify a benefit in monetary terms, EPA should still try to measure the effect of the Agency's action in terms of its physical units or, where such quantification is not possible, describe the effect and its value in qualitative terms. Regional and site-specific programs and decisions, which are not subject to the same legal requirements as national rule makings, can offer useful opportunities for testing and implementing a broader suite of valuation methods.

Three Key Recommendations

The committee's principal advice to EPA, as noted above, is to pursue an expanded, integrated approach to assessing the value of the ecological effects of its regulations, programs, and other actions. the report contains three overarching recommendations for achieving this goal. in particular, the report recommends that the Agency:

1. identify early in the valuation process the ecological responses that are likely to be of greatest importance to people, using information about ecological importance, likely human and social consequences, and public concerns. EPA should then focus its valuation efforts on those responses. This will help expand the range of ecological responses that EPA characterizes or quantifies or for which it estimates values.
2. Predict ecological responses in terms that are relevant to valuation. prediction of ecological responses is a key step in valuation efforts. to predict responses in value-relevant terms, EPA should focus on the effects of decisions on ecosystem services or other ecological features that are of direct concern to people. this, in turn, will require the Agency to go beyond merely predicting the biophysical effects of decisions and to map those effects to responses in ecosystem services or components that the public values.
3. Consider the use of a wider range of possible valuation methods, either to provide information about multiple sources and concepts of value or to better capture the full range of contributions stemming from ecosystem protection. in considering the use of different methods, however, care must be taken to ensure that only methods that meet appropriate validity and related criteria are used, and to recognize that different methods may measure different things and thus not be directly additive or comparable. this chapter therefore calls on EPA to develop criteria to evaluate and determine the appropriate use of each method. EPA should also carefully evaluate its use of value information collected at one site in the valuation of policy impacts at a different site (transfers of value information) and more fully characterize and communicate uncertainty for all valuations.

Implementing the Recommendations

The report provides specific advice on how to achieve these overarching recommendations. the report proposes a large number of steps, some of which can be implemented in the short run, but others of which will require investments in research or method development, policy changes, and/or new resources. EPA should begin the process of adopting a more expanded, integrated approach to ecological valuation by prioritizing the steps that it will take to accomplish the report's recommendations, taking into account the relative ease and cost of each potential step.

Implementing Recommendation #1

The first major recommendation, as noted, is to identify from an early stage in the valuation process the ecological responses that contribute to human well-being and are likely to be of greatest importance to people, and then to focus valuation efforts on these responses. to accomplish this, the report recommends that EPA:

- Begin each valuation by developing a conceptual model of the relevant ecosystem and the ecosystem services that it generates. this model should serve as a road map to guide the valuation.
- Involve staff throughout EPA, as well as outside experts in the biophysical and social sciences, in constructing the conceptual model.
- EPA should also seek information about relevant public concerns and needs. EPA can identify public concerns through a variety of methods, drawing on either existing knowledge or interactive processes designed to elicit public input.
- Incorporate new information into the model, in an iterative process, as the value assessment proceeds.

Implementing Recommendation #2

Ecological valuation requires both prediction of ecological responses and an estimation of the value of those responses. to predict ecological responses in value-relevant terms, EPA should focus on the effects of decisions on ecosystem services and should map responses in ecological systems to responses in services or ecosystem components that the public can directly value. Unfortunately, the science needed to do this has been limited, presenting a barrier to effective valuation of ecological systems and services. to better predict ecological responses in value-relevant terms in the future, EPA should:

- Identify and develop measures of ecosystem services that are relevant to and directly useful for valuation. This will require increased interaction within EPA between natural and social scientists. In identifying and assessing the value of services, EPA should describe them in terms that are meaningful and understandable to the public.
- Where possible, use ecological production functions to estimate how effects on the structure and function of ecosystems, resulting from the actions of EPA or partnering agencies, will affect the provision of ecosystem services for which values can then be estimated. development of a broad suite of ecological production functions currently faces numerous challenges and can benefit from new research.
- Where complete ecological production functions do not exist:
  - Examine available ecological indicators that are correlated with changes in ecosystem services to provide information about the effects of governmental actions on those services.
  - Use methods such as meta-analysis that can provide general information about key ecological relationships important in the valuation.
- Support all ecological valuations by ecological models and data sufficient to understand and estimate the likely ecological responses to the major alternatives being considered by decision makers. Analyze and report on the uncertainty involved in biophysical projections.

Implementing Recommendation #3

In characterizing, measuring, or quantifying the value of ecological responses to actions by EPA or other agencies, EPA should consider the use of a broader suite of valuation methods than it has historically employed. As summarized in Table 3 at pages 42-43, this chapter considers the possible use of not only economic methods, but also such alternative methods as measures of attitudes, preferences, and intentions; civic valuation; decision science approaches; ecosystem benefit indicators, biophysical ranking methods; and cost as a proxy for value. A broader suite of methods could allow EPA to better capture the full range of contributions stemming from ecosystem protection and the multiple sources of value derived from ecosystems. non-economic valuation methods may also usefully support and improve economic valuation by helping to identify the ecological responses that people care about, by providing indicators of economic benefits that EPA cannot monetize using economic valuation, and by offering supplemental information outside strict benefit-cost analysis. In this regard, EPA should:

- Pilot and evaluate the use of alternative methods where legally permissible and scientifically appropriate.
- Develop criteria to determine the suitability of alternative methods for use in specific decision contexts. An over-arching criterion should be validity – i.e., how well the method measures the underlying construct that it is intended to measure. Given differences in premises, goals, concerns, and external constraints, appropriate uses will vary among methods and contexts. Different methods are also at different stages of development and validation.

EPA could also improve its ecological valuations by carefully evaluating the transfer of value information and more fully characterizing and communicating uncertainty. in this regard, EPA should:

- Identify relevant criteria for determining the appropriateness of the transfer of value information. these criteria should consider similarities and differences in societal preferences and the nature of the biophysical systems between the study site and the policy site. Using these criteria, EPA analysts and those providing oversight should flag problematic transfers and clarify assumptions and limitations of the study-site results.
- Go beyond simple sensitivity analysis in assessing uncertainty, and make greater use of approaches, such as Monte carlo analysis, that provide more useful and appropriate characterizations of uncertainty in complex contexts such as ecological valuation.
- Provide information to decision makers and the public about the level of uncertainty involved in ecological valuation efforts. EPA should not relegate uncertainty analyses to appendices but should ensure that a summary of uncertainty is given as much prominence as the valuation estimate itself, with careful attention to how recipients are likely to understand the uncertainties. EPA should also explain qualitatively any limitations in the uncertainty analysis.

While EPA should improve its characterization and reporting of uncertainty, the mere existence of uncertainty should not be an excuse for delaying actions where the benefits of immediate action outweigh the value of attempting to further reduce the uncertainty Some uncertainty will always exist.

Context-Specific Recommendations

The report also examines how to implement an expanded and integrated approach to ecological valuation in three specific contexts: national rule makings, regional partnerships, and local site-specific decisions.

National Rule Making

Applying the expanded and integrated valuation approach to national rule making will entail some challenges, but also offers important opportunities for improvement. EPA can implement some, but not all, of the committee's recommendations using the existing knowledge base. The committee also recognizes that EPA must conduct valuations for national rule making in compliance with statutory and executive mandates. Specific recommendations for improving valuations for national rule making in the short run include:

- EPA should develop a conceptual model at the beginning of each valuation, as discussed above, to serve as a guide or road map. to ensure that the model captures the ecological properties and services that are potentially important to people, EPA should incorporate input both from relevant science and about public preferences and concerns.
- The Agency should address site-specific variability in the impact of a rule by producing case studies for important ecosystem types and then aggregating across the studies where information about the distribution of ecosystem types and affected populations is available.
- EPA should not compromise the quality of its valuations by inappropriately transferring information about values. Where the values of ecosystem services are primarily local, the Agency can rely on scientifically-sound value transfers using prior valuations at the local level. However, for services valued more broadly, EPA should draw from studies with broad geographical coverage (in terms of both the changes that are valued and the population whose values are assessed).
- EPA should pilot and evaluate the use of a broader suite of valuation methods to support and improve RIAs. Although OMB Circular A-4 requires RIAs to monetize benefits to the extent possible using economic valuation methods, other methods could be useful in the following ways:
  - Helping to identify early in the process the ecosystem services that are likely to be of concern to the public and that should therefore be the focus of the benefit-cost analysis.
  - Addressing the requirement in Circular A-4 to provide quantitative or qualitative information about the possible magnitude of benefits (and costs) when they cannot be monetized using economic valuation.

  - Providing supplemental information outside the formal benefit-cost analysis about sources and concepts of value that might be of interest to EPA and the public but not reflected in economic values.
- To ensure that RiAs do not inappropriately focus only on impacts that have been monetized, EPA should also report on other ecological impacts in appropriate units where possible, as required by Circular A-4. The Agency should label aggregate monetized economic benefits as "total economic benefits that could be monetized," not as "total benefits."
- EPA should include a separate chapter on uncertainty characterization in each RiA or value assessment.

Regional Partnerships

The committee sees great potential in undertaking a comprehensive and systematic approach to estimating the value of protecting ecosystems and services at a regional scale, in part because of the effectiveness with which EPA regional offices can partner with other agencies and state and local governments. Regional-scale analyses hold great potential to inform decision makers and the public about the value of protecting ecosystems and services, but this potential is at present largely unrealized. the general recommendations of this chapter provide a guide for regional valuations. Regional valuations are a particularly appropriate setting in which to test alternative valuation methods because there are generally fewer legal directives or restrictions regarding the value concepts and methods to be used. the report also includes several recommendations specific to regions, including:

- EPA should encourage its regions to engage in valuation efforts to support decision making both by the regions and by partnering governmental agencies.
- EPA should provide adequate resources to EPA regional staff to develop the expertise needed to undertake comprehensive and systematic studies of the value of protecting ecosystems and services.
- To ensure that regions can learn from valuation efforts by other regions, EPA regional offices should document valuation efforts and share them with other regional offices, EPA's National Center for Environmental Economics, and EPA's Office of Research and development.

Site-Specific Decisions

Incorporation of ecological valuation into local decisions about the remediation and redevelopment of contaminated sites can help enhance the ecosystem services provided by such sites in the long run and thus the sites' contributions to local well-being. The general recommendations of the report provide a useful guide for such site-specific valuations. The report also includes several recommendations of particular relevance to site- specific decisions, including:

- EPA should provide regional offices with the staff and resources needed to effectively incorporate ecological valuation into the remediation and redevelopment of contaminated sites.

- EPA should determine the ecosystem services and values important to the community and affected parties at the beginning of the remediation and redevelopment process.
- EPA should adapt current ecological risk assessment practices to incorporate ecological production functions and predict the effects of remediation and redevelopment options on ecosystem services.
- EPA should communicate information about ecosystem services in discussing options for remediation and redevelopment with the public and affected parties.
- EPA should create formal systems and processes to foster information-sharing about ecological valuations at different sites.

Recommendations for Research and Data Sharing

The report provides several recommendations for EPA's research programs that are designed to provide the ecological information needed for valuation, develop and test valuation methods, and share data. in a number of cases, these recommendations parallel research plans that have been developed by the Office of Research and development and other Agency groups. As an over-arching recommendation, the report advises EPA to more closely coordinate its research programs on the valuation of ecosystem services and to develop links with other governmental agencies and organizations engaged in valuation and valuation research. it advises, at a more general level, fostering greater interaction between natural scientists and social scientists in identifying relevant ecosystem services and developing and implementing processes for measuring them and estimating their value. The report identifies important research areas but does not attempt to rank or prioritize among all of its research recommendations. The committee recommends that EPA develop a research strategy, building on the recommendations in this chapter, that identifies "low-hanging fruit" and prioritizes studies likely to have the largest payoff for their cost in both advancing valuation methods and providing valuation information of importance to EPA in its work.

To develop EPA's ability to determine and quantify ecological responses to governmental decisions, the Agency should:

- Support the development of quantitative ecosystem models and baseline data on ecological stressors and ecosystem service flows that can support valuation efforts at the local, regional, national, and global levels.
- promote efforts to collect data that can be used to parameterize ecological models for site-specific analysis and case studies or that can be transferred or scaled to other contexts.
- carefully plan and actively pursue research to develop and generate ecological production functions for valuation, including Office of Research and development and STAR research on ecological services and support for modeling and methods development. The committee believes that this is a research area of high priority.
- Given the complexity of developing and using complete ecological production functions, continue and accelerate research to develop key indicators for use in

ecological valuation. Such indicators should meet ecological and social science criteria for effectively simplifying and synthesizing underlying complexity and link to an effective monitoring and reporting program.

To develop EPA's capabilities for estimating the value of ecological responses to governmental decisions, EPA should:

- Support new studies and the development of new methodologies that will enhance the future transfer of value information and other means of generalizing ecological value assessments, particularly at the national level. Such research should include national surveys related to ecosystem services with broad (rather than localized) implications so that value estimates might be usable in multiple rule-making contexts. This should also be a priority area for research.
- Invest in research designed to reduce uncertainties associated with ecological valuation through data collection, improvements in measurement, theory building, and theory validation.
- Incorporate the research needs of regional offices for systematic valuation studies in future calls by EPA for extramural ecological valuation research proposals.

To access and share information to enhance the Agency's capabilities for ecological valuation, EPA should:

- Work with other federal agencies and scientific organizations such as the national Science foundation to encourage the sharing of ecological data and the development of more consistent ecological measures that are useful for valuation purposes. A number of governmental organizations, such as the United States department of Agriculture and the Fish & Wildlife Service, are working on biophysical modeling and valuation, and EPA could usefully partner with them.
- Support efforts to develop Web-based databases of existing valuation studies that could be used in transferring value information. the databases should include valuation studies across a range of ecosystems and ecosystem services. the databases should also carefully describe the characteristics and assumptions of each study, in order to increase the likelihood that those studies most comparable to new valuations can be identified for use.
- Support the development of national-level databases of information useful in the development of new valuation studies. Such information should include data on the joint distribution of ecosystem and human population characteristics that are important determinants of the value of ecosystem services.
- Develop processes and information resources so that EPA staff in one region or office of the Agency can learn effectively from valuation efforts being undertaken elsewhere within the Agency.

1. INTRODUCTION

The mission of the Environmental Protection Agency (EPA) is to protect human health and the environment. During its history, EPA has focused much of its decision-making expertise on the first part of this mission, in particular the risks to human health from chemical stressors in the environment. Although protecting human health is the bedrock of EPA's traditional expertise, the broad mission of EPA goes beyond this. EPA's Strategic Plan (U.S. Environmental Protection Agency [EPA], 2006a) explicitly identifies the need to ensure "healthy communities and ecosystems" as one of its five major goals. Agency publications and independent sources document EPA's efforts to protect ecological resources–and its authority for doing so (EPA, 1994; EPA Risk Assessment Forum, 2003; EPA Science Advisory Board, 2000; Hays, 1989; Russell, 1993).

EPA's mission to protect the environment requires that the Agency understand and protect ecological systems. Ecologists use the term "ecosystem" to describe the dynamic complex of plant, animal, and microorganism communities and non-living environment interacting as a system. For example, a forest ecosystem consists of the trees in the forest, all other living organisms, and the non-living environment with which they interact. Ecosystems provide basic life support for human and animal populations and are the source of spiritual, aesthetic, and other human experiences that are valued in many ways by many people.

There has been a growing recognition of the numerous and varied services that ecosystems provide to human populations through a wide range of ecological functions and processes (e.g., Daily, 1997). Ecosystems not only provide goods and services that are directly consumed by society such as food, timber, and water; they also provide services such as flood protection, disease regulation, pollination, and the control of diseases, pests, and climate. There is, too, increasing recognition of the impact of human activities on ecosystems (e.g., Millennium Ecosystem Assessment Board, 2003; Millennium Ecosystem Assessment, 2005). Among the examples of this impact are traditional air and water pollution (such as sulfur dioxide emissions, ground-level ozone, and eutrophication), as well as global warming; changes in the nitrogen cycle; invasive species; aquifer depletion, and land conversions that lead to deforestation or loss of wetlands and biodiversity.

Given the vital role that ecosystems play in our lives, the state of these systems and the flow of services they provide have important human implications. EPA actions, including regulations, rules, programs, and policy decisions, can affect the condition of ecosystems and the flow of ecosystem services. These effects can occur narrowly, at a local or a regional scale, or broadly, at a national or global scale.

Despite the importance of these ecological effects, EPA policy analyses have tended to focus on a limited set of ecological endpoints, such as those specified in tests for pesticide regulation (e.g., effects on the survival, growth, and reproduction of aquatic invertebrates, fish, birds, mammals, and terrestrial and aquatic plants) or specified in laws administered by the Agency (e.g., mortality to fish, birds, plants, and animals) (EPA Risk Assessment Forum, 2003).[1] Given EPA's responsibility to ensure healthy communities and ecosystems, the Agency should consider the full range of effects that its actions will have. Thus, in addition to evaluating impacts on human health and other environmental goals, EPA should evaluate the effects of its actions, wherever relevant, on individual organisms and plant and animal populations, and on the structure and functions of communities and ecosystems. Such evaluations should be comprehensive and integrated.

To promote good decision making, policy makers also require information about how much ecosystems contribute to society's well-being. EPA increasingly recognizes this need. The stated goal of EPA's *Ecological Benefits Assessment Strategic Plan* is to "help improve Agency decision making by enhancing EPA's ability to identify, quantify, and value the ecological benefits of existing and proposed policies" (2006c, p. xv). Information about the value of ecosystems and the associated effects of EPA actions can also help inform the public about the need for ecosystem protection, the extent to which specific policy alternatives address that need, and the value of the protection compared to the costs.

Despite EPA's stated mission and mandates, a gap exists between the need to understand and protect ecological systems and services and EPA's ability to address this need. This chapter is a step toward filling that gap. It describes how an integrated and expanded approach to ecological valuation can help the Agency describe and measure the value of protecting ecological systems and services, thus better meeting its overall mission. The terms ecological valuation or valuing ecological change, as used in this chapter, refer to the process of estimating or assessing the value of a change in an ecosystem, its components, or the services it provides. The values at interest here are those of the public, and this chapter discusses how to appropriately estimate or assess them.

This chapter was prepared by the Committee on Valuing the Protection of Ecological Systems and Services (C-VPESS) of EPA's Science Advisory Board (SAB). The SAB saw a need to complement the Agency's ongoing work by offering advice on how EPA might better value the protection of ecological systems and services and how that information could support decision making to protect ecological resources. Therefore, in 2003, the SAB Staff Office formed C-VPESS,[2] a group of experts in decision science, ecology, economics, engineering, law, philosophy, political science, and psychology, with a particular understanding of ecosystem protection. The committee's charge was to undertake a project to improve the Agency's ability to value ecological systems and services.[3] The SAB set the following goals:

- Assessing Agency needs for valuation to support decision making
- Assessing the state of the art and science of valuing protection of ecological systems and services
- Identifying key areas for improving knowledge, methodologies, practice, and research at EPA

This chapter provides advice for strengthening the Agency's approaches for valuing the protection of ecological systems and services, facilitating the use of these approaches by decision makers, and investing in the research areas needed to bolster the science underlying ecological valuation.[4] It identifies the need for an expanded and integrated approach for valuing EPA's efforts to protect ecological systems and services. The report also recognizes and highlights issues that need to be addressed in using and improving current valuation methods and recommends new research to address these needs. It provides advice to the Administrator, EPA managers, EPA scientists and analysts, and other staff across the Agency concerned with ecological protection. It addresses valuation in a broad set of contexts, including national rule making, regional decision making, and site-specific decisions that protect ecological systems and services.

This chapter appears at a time of lively interest internationally, nationally, and within EPA in valuing the protection of ecological systems and services. Since the establishment of the SAB C-VPESS, a number of major reports have focused on ways to improve the characterization of the important role of ecological resources (Silva and Pagiola, 2003; National Research Council [NRC], 2004; Pagiola, von Ritter et al., 2004; Millennium Ecosystem Assessment, 2005).[5] In addition, the Agency itself has engaged in efforts to improve ecological valuation. The most recent product of these efforts is the *Ecological Benefits Assessment Strategic Plan* noted above (EPA, 2006c). EPA also has sought to strengthen the science supporting ecological valuation through the extramural Science to Achieve Results (STAR) grants program and the Office of Research and Development's ecosystem-services research program (EPA Science Advisory Board, 2008b).

The committee has both learned from and built upon these recent efforts. However, C-VPESS distinguishes its work from many of the earlier efforts in several key ways. First, C-VPESS considers EPA its principal audience. In particular, C-VPESS analyzes ways in which EPA can value its own contributions to the protection of ecological systems and services, so that the Agency can make better decisions in its eco-protection programs. Many of the recent studies, including the Millennium Assessment and National Research Council report, do not consider the specific policy contexts or constraints faced by EPA. Second, most, but not all, of the previous work has concentrated on economic valuation, and monetary valuation in particular. C-VPESS, by contrast, is interdisciplinary and does not focus solely on monetary or economic methods or values.

The report is structured as follows. Chapter 2 provides an overview of the conceptual framework and general approach advocated by the committee. It discusses fundamental concepts as well as the current state of ecological valuation at EPA. Most importantly, it identifies the need for an expanded and integrated approach to ecological valuation at EPA and describes the key features of this approach. Subsequent chapters develop in more detail the basic principles outlined in chapter 2, focusing on implementation. Chapter 3 discusses predicting the effects of EPA actions and decisions on ecological systems and services. Chapter 4 examines a variety of methods for valuing these changes. More detailed descriptions of the valuation methods, developed by members of the C-VPESS, along with a separate discussion of survey issues relevant to ecological valuation, are available on the SAB Web site at http://yosemite.epa.gov/sab/sabproduct.nsf/WebBOARD/C-VPESS_Web_ Methods_ Draft?OpenDocument. Chapter 5 covers cross-cutting issues related to deliberative approaches, uncertainty, and communication. Recognizing that implementation of the process can vary depending on the decision context, chapter 6 discusses implementation in three specific contexts where ecological valuation could play an important role in EPA analysis: national rule making, regional partnerships, and site-specific decisions (looking specifically at cleanup and restoration). Finally, chapter 7 provides a summary of the report's major findings and recommendations.

Everyone at EPA involved in the design or implementation of valuation efforts, or in the establishment of policy for valuations, will find the first five chapters of this chapter relevant and important to their work. Section 6.1 of this chapter, which addresses national rule making, will be of particular interest to EPA officials in Washington, DC. Sections 6.2 and 6.3, which address regional partnerships and site-specific decisions, will be of special interest to regional offices, as well as those national officials responsible for overseeing regional policy. EPA staff who are looking for a detailed discussion of specific methods will find the information on the SAB Web site useful.

2. Conceptual Framework

2.1. An Overview of Key Concepts

2.1.1. The Concept Of Ecosystems

As noted in chapter 1, the term "ecosystem" describes a dynamic complex of plant, animal, and microorganism communities and their non-living environment, interacting as a system. Ecosystems encompass all organisms within a prescribed area, including humans. Ecosystem functions or processes are the characteristic physical, chemical, and biological activities that influence the flows, storage, and transformation of materials and energy within and through ecosystems. These activities include processes that link organisms with their physical environment (e.g., primary productivity and the cycling of nutrients and water) and processes that link organisms with each other, indirectly influencing flows of energy, water, and nutrients (e.g., pollination, predation, and parasitism). These processes in total describe the functioning of ecosystems.

2.1.2. The Concept Of Ecosystem Services

Ecosystem services are the direct or indirect contributions that ecosystems make to the well-being of human populations. Ecosystem processes and functions contribute to the provision of ecosystem services, but they are not synonymous with ecosystem services. Ecosystem processes and functions describe biophysical relationships that exist whether or not humans benefit from them. These relationships generate ecosystem services only if they contribute to human well-being, defined broadly to include both physical well-being and psychological gratification. Thus, ecosystem services cannot be defined independently of human values.

The Millennium Ecosystem Assessment uses the following categorization of ecosystem services:

- Provisioning services – services from products obtained from ecosystems. These products include food, fuel, fiber, biochemicals, genetic resources, and fresh water. Many, but not all, of these products are traded in markets.

- Regulating services – services received from the regulation of ecosystem processes. This category includes services that improve human well-being by regulating the environment in which people live. These services include flood protection, human disease regulation, water purification, air quality maintenance, pollination, pest control, and climate control. These services are generally not marketed but many have clear value to society.
- Cultural services – services that contribute to the cultural, spiritual, and aesthetic dimensions of people's well-being. They also contribute to establishing a sense of place.
- Supporting services – services that maintain basic ecosystem processes and functions such as soil formation, primary productivity, biogeochemistry, and provisioning of habitat. These services affect human well-being indirectly by maintaining processes necessary for provisioning, regulating, and cultural services.

As this categorization suggests, the Millennium Ecosystem Assessment adopts a very broad definition of ecosystem services, limited only by the requirement of a direct or indirect contribution to human well-being.[6] This broad approach recognizes the myriad ways in which ecosystems support human life and contribute to human well-being. Boyd and Banzhaf (2006) propose a narrower definition that focuses only on those services that are end products of nature, i.e., "components of nature, *directly* enjoyed, consumed or used to yield human well-being" (emphasis added). They stress the need to distinguish between intermediate products and final (or end) products and include only final outputs in the definition of ecosystem services, because these affect people most directly and consequently are what people are most likely to understand. In addition, the focus on final products reduces the potential for double-counting, which can arise if both intermediate and final products or services are valued. Under this definition, ecosystem functions and processes, such as nutrient recycling, are not considered services. Although they contribute to the production of ecological end products or outputs, they are not outputs themselves. Likewise, because supporting services contribute to human well-being indirectly rather than directly, they are recognized as being potentially very important but are not included in Boyd and Banzhaf's definition of ecosystem services.

Regardless of the specific definition used, ecosystem services play a key role in the evaluation of policies that affect ecosystems because they reflect contributions of the ecosystem to human well-being. Simply listing the services derived from an ecosystem, using the best available ecological, social, and behavioral sciences, can help ensure appropriate recognition of the full range of potential ecological responses to a given policy and their effects on human well-being. It can also help make the analysis of the role of ecosystems more transparent and accessible. To ensure consideration of the full range of contributions, this chapter uses the term ecosystem services to refer broadly to both intermediate and final/ end services. In specific valuation contexts, however, it may be important to identify whether the service being valued is an intermediate or a final service (see sections 2.1.4, 2.3.3, and 3.3.2 for related discussions).

The committee recognizes that ecosystems can be important not only because of the services they provide to humans directly or indirectly, but also for other reasons, including respect for nature based on moral or spiritual beliefs and commitments. The committee's name includes reference to the protection of both ecosystem services and the ecosystems themselves. Thus, although much of this chapter focuses on ecosystem services, the discussion of ecological protection and valuation applies both to ecosystem services and to ecosystems per se.

2.1.3. Concepts Of Value

People assign or hold all values. All values, regardless of how they are defined, reflect either explicitly or implicitly what the people assigning them care about. In addition, values can be defined only relative to a given individual or group. The value of an ecological change to one individual might be very different than its value to someone else.

Value is not a single, simple concept. People have material, moral, spiritual, aesthetic, and other interests, all of which can affect their thoughts, attitudes, and actions toward nature in general and, more specifically, toward ecosystems and the services they provide. Thus, when people talk about environmental values, the value of nature, or the values of ecological systems and services, they may have different things in mind that can relate to these different sources of value. Furthermore, experts trained in different disciplines (e.g., decision science, ecology, economics, philosophy, psychology) understand the concept of value in different ways. These differences create challenges for ecological valuations that seek to draw from and integrate insights from multiple disciplines.[7]

A fundamental distinction can be made between those things that are valued as ends or goals and those things that are valued as means. To value something as a means is to value it for its usefulness in helping bring about an end or goal that is valued in its own right. Things or actions valued for their usefulness as means are said to have instrumental value. Alternatively, something can be valued for its own sake as an independent end or goal. While a possible goal is maximizing human well-being, one could envision a range of other possible social goals or ends including protecting biodiversity, sustainability, or protecting the health of children. Things valued as ends are sometimes said to have intrinsic value. This term has been used extensively in the philosophical literature but there is not general agreement on its exact definition.[8]

Ecosystems can be valued both as independent ends or goals and as instrumental means to other ends or goals. This chapter therefore uses the term "value" broadly to include both values that stem from contributions to human well-being and values that reflect other considerations, such as social and civil norms (including rights), and moral and spiritual beliefs and commitments.

The broad definition of value used here extends beyond what are sometimes called the benefits derived from ecosystem services. Even the term "benefits," however, means different things in different contexts. In some contexts (e.g., Millennium Ecosystem Assessment Board, 2003; Millennium Ecosystem Assessment, 2005), benefits refers to the contributions of ecosystem services to human well-being. In contrast, the term has a very precise meaning in the context of EPA regulatory impact analyses conducted under guidance from the U.S. Office of Management and Budget (OMB). In that context, benefits are defined by the economic concept of the willingness to pay for a good or service or willingness to accept compensation for its loss.

Given the many ways in which people think about value, the committee discussed a number of different concepts of value. Table 1 lists the various concepts of value considered by the committee, categorized as either preference-based or biophysical. Although people assign or hold all values, preference-based values reflect individuals' preferences across a variety of goods and services, including (but not limited to) ecosystems and their services. In contrast, biophysical values reflect contributions to explicit or implicit biophysical goals or standards determined to be important. The goal or standard might be chosen directly by decision makers or based on the preferences of the public or relevant groups of the public. Separating values into preference based and biophysical categories is not the only way to categorize values, but it has proven useful for the committee in understanding the various concepts of value used by different disciplines and how they are related.[9]

However, inclusion of a value concept on the list in table 1 does not imply a consensus endorsement by committee members of the use of that concept in ecological valuation. The task of distinguishing what is valued from the concept used to define the value is complex, regardless of the disciplinary perspective adopted. It requires clear, meaningful distinctions between the information available, perceptions of that information, and decisions or actions. Each discipline addresses these issues differently; these disciplinary differences are a potential source of confusion and miscommunication. So to make meaningful distinctions between these, the committee had to agree on a set of key assumptions. The resulting categorization of value concepts cannot be evaluated independent of these assumptions. Table 1 is not the only possible categorization. Rather it is the one that allowed the interdisciplinary C-VPESS to develop guidance relevant to EPA for valuing changes in ecosystem services.

Table 1. A classification of concepts of value as applied to ecological systems and their services

Preference-based values
Attitudes or judgments
Economic values
Community-based values
Constructed preferences
Biophysical values
Bio-ecological values
Energy-based values

The concepts of value listed in table 1 differ in a number of important ways. **Attitude or judgment-based values** are based on empirically derived descriptive theories of human attitudes, preferences, and behavior (e.g., Dietz et al., 2005). These values are not necessarily defined in terms of tradeoffs and are not typically constrained by income or prices, especially those that are outside the context of the specified assessment process. Rather, the values are derived from individuals' judgments of relative importance, acceptability, or preferences across the array of changes in goods or services presented in the assessment. Preferences and judgments are often expressed through responses to surveys asking for choices, ratings, or other indicators of importance. The basis for judgments may be individual self-interest, community well-being, or accepted civic, ethical, or moral obligations.

Economic values assume that individuals are rational and have well-defined and stable preferences over alternative outcomes, which are revealed through actual or stated choices (see, for example, Freeman, 2003). Economic values are based on utilitarianism and assume substitutability, i.e., that different combinations of goods and services can lead to equivalent levels of utility for an individual (broadly defined to allow both self-interest and altruism). They are defined in terms of the tradeoffs that individuals are willing to make, given the constraints they face. The economic value of a change in one good (or service) can be defined as the amount of another good that an individual with a given income is willing to give up in order to get the change in the first good. Alternatively, it can be defined as the change in the amount of the second good that would compensate the individual to forego the change in the first good. Economic values can include both use and nonuse values, and they can be applied to both market and non-market goods.[10] The tradeoffs that define economic values need not be defined in monetary terms (willingness to pay or willingness to accept monetary compensation), although typically they are. Expressing economic values in monetary terms allows a direct comparison of the economic values of ecosystem services with the economic values of other services produced through environmental policy changes (e.g., effects on human health) and with the costs of those policies. However, monetary measures of economic values should not be confused with other monetized measures of economic output, such as the contribution of a given sector or resource to gross domestic product (GDP).[11]

Community-based values are based on the assumption that, when consciously making choices about goods that might benefit the broader public, individuals make their choices based on what they think is good for society as a whole rather than what is good for them as individuals. In this case, individuals could place a positive value on a change that would reduce their own individual well-being (e.g., Jacobs, 1997; Costanza and Folke, 1997; Sagoff, 1998). In contrast to economic values, these values may not reflect tradeoffs that individuals are willing to make, given their income. Instead, an individual might express value in terms of the tradeoffs (perhaps, but not necessarily, in the form of monetary payment or compensation) that the person feels society as a whole – rather than an individual – should be willing to make.

Values based on constructed preferences reflect the view that, particularly when confronted with unfamiliar choice problems, individuals do not have well-formed preferences and hence values. This view is based on conclusions that some researchers have drawn from a body of empirical work addressing this issue (e.g., Gregory and Slovic, 1997; Lichtenstein and Slovic, 2006). It implies that simple statements of preferences or willingness to pay may be unstable (e.g., subject to preference reversals).[12] Some have advocated using a structured or deliberative process as a way to help respondents construct their preferences and values (see section 5.1). This chapter refers to values arrived at by these processes as "constructed values." The difference between economic values and constructed values can be likened to the difference between the work of an archeologist and that of an architect (Gregory et al., 1993). Economic values assume preferences exist and simply need to be "discovered" (implying the analyst works as a type of archeologist), while constructed values assume that preferences need to be built through the valuation process (similar to the work of an architect). As a result, the values expressed by individuals (or groups) engaged in a constructed-value process are expected to be influenced by the process itself. Constructed values can reflect both self- interest and community-based values.

Human preferences directly determine all of the concepts of value described above. In contrast, biophysical values do not depend directly on human preferences. Biophysical values reflect the contribution of ecological changes to a pre-specified biophysical goal or standard identified or set prior to measuring the contribution of those changes. This goal or standard can be defined in ecological terms (e.g., biodiversity or species preservation) or based on a biophysical theory of value (e.g., energy theory of value).

Bio-ecological values depend on known or assumed relationships between targeted ecosystem conditions and functions (e.g., biodiversity, biomass, energy transfer, and transformation), ecosystem functions, and the pre-specified biophysical goal or standard (see Grossman and Comer, 2004). Scientists can determine bio-ecological values in several different ways that contribute to the goals. For example, contributions to a biodiversity goal could be based on individual measures such as genetic distance or species richness, or on more comprehensive measures that reflect multiple ecological considerations.

Energy-based values are defined as the direct and indirect energy required to produce a marketed or un-marketed (e.g., ecological) good or service (see Costanza, 2004). In contrast to economic values, energy-based values are not defined in terms of the preference-based tradeoffs that individuals are willing to make, and hence the two concepts of value are conceptually distinct. Nonetheless, researchers who advocate the use of energy-based values have found that in some cases energy cost estimates are similar in magnitude to economic measures of value. Energy-based methods were designed to provide an alternative way to define value that is independent of short-term human preferences. However, some of the components used to construct these values depend on human choices and the preferences that underlie those choices.

The committee considered all of these various concepts of value in its deliberations. To date, EPA analyses have primarily sought to measure economic values, as required by some statutes and executive orders (see section 6.1). However, the committee believes that information based on other concepts of value can also be an important input into Agency decisions affecting ecosystems. Recognizing the significance of multiple concepts of value is an important first step in valuing the protection of ecological systems and services.

2.1.4. The Concept Of Valuation And Different Valuation Methods

Because the committee's charge relates to the value of protecting ecological systems and services, this chapter focuses on valuing ecological changes, rather than on valuing entire ecosystems or the broader question of assessing environmental values that relate to ecosystem protection.[13] Thus, although ecosystems per se and their associated services have value, the term ecological valuation, as used here, refers to the process of measuring the value of a change in an ecosystem., its components, or the services it provides – i.e., it is predicated on a comparison of a given alternative scenario with a baseline scenario. In its simplest form, valuation requires, first, a prediction of a change in the ecosystem or the flow of ecosystem services, and then, the estimation of the value of that change.

An important issue in ecological valuation is the extent to which individuals who express values understand the contributions of related ecological goods and services to human well-being. In many cases, an ecological change may have important implications that are not widely recognized or understood by the general public. This is particularly true for supporting or intermediate services, where the important contributions to human well-being are indirect. For example, Weslawski et al. (2004) indicated that the invertebrate fauna found in soils and sediments are important in remineralization, waste treatment, biological control, gas and climate regulation, and erosion and sedimentation control. However, the general public had no understanding or appreciation of these services (although the public may have an appreciation of the higher-level services or end-point services, such as clean water, aesthetics, and foods that could be derived from the system). Likewise, although individuals might understand the recreational contributions to human well-being associated with a given EPA action to limit nutrient pollution in streams and lakes, they might not recognize or fully appreciate the associated nutrient- cycling or water-quality implications. If asked to value these services, they may express policy preferences or values that reflect incomplete information. Individuals might respond to a survey, make purchases, or otherwise behave as if they place no value on an ecosystem service if they are ignorant of the role of that service in contributing to their well-being or other goals.

There may be occasions where assessments of existing, uninformed attitudes and values held by the public are desired, such as when ascertaining current public understanding of particular ecosystems or services, and designing communications to improve understanding of ecosystems or services or to solicit public support for specific protection policies. In most cases, however, valuation should seek to measure the values that people hold and would express if they were well informed about the relevant ecological and human well-being factors involved. This embodies two principles. First, the ultimate objective of any valuation exercise is to assess the values of the public, not the personal values or preferences of scientists or experts. Basing valuation on the personal preferences of scientists or experts rather than those of the general public would undermine the usual presumptions that, in a democratic society, values held individually and collectively within society should be considered in public policy decisions, and that public involvement is central to democratic governance (e.g., Berelson, 1952; NRC, 1996).

Second, when EPA assesses values, the Agency should provide the public with as much of the relevant science as necessary to make informed judgments about the human/social consequences of the changes they are being asked to value. Lack of public understanding can pose a potentially serious challenge for ecological valuation. This problem can be reduced by explicitly incorporating into the valuation process information about ecological responses to policy options based on the best available science. For example, valuation exercises employing surveys should provide survey respondents with the relevant ecological information and the associated human/social consequences. Likewise, valuation exercises employing deliberative processes should convey relevant information directly to participants in the process.

The lack of public understanding about underlying ecological functions and processes also highlights the importance of framing valuation-related questions in terms of services that people can directly understand and value (see further discussion in section 3.3.2). In many cases, this means asking people to value final or end services that directly affect them rather than asking them to value intermediate services whose effect is less direct. When an EPA action has an important effect on an intermediate service, it would then be incumbent on experts to predict the expected impact of these changes on final services, which could then be valued. In the example of Weslawski et al. (2004) discussed above, this would mean that individuals should not be asked to value a change in invertebrate fauna or the intermediate services they impact (remineralization, waste treatment, etc.). Rather, relevant science should be used to estimate how these changes would ultimately impact final services that individuals understand and appreciate (such as clean water, aesthetics, etc.), and the valuation questions should be framed in terms of these final services.

Even when valuation is informed by the best available science, the valuation process will almost always involve uncertainty. Uncertainty arises in the prediction of changes in ecosystems, in the resulting change in the flow of services, and in estimating the values associated with those changes. The valuation process needs to recognize, assess, and communicate the various sources of uncertainty (see section 5.2 for further discussion).

The valuation process should also recognize that information about different sources of value may be important for decision making, and it should identify appropriate methods to characterize or measure those values. There are a number of valuation methods that can be used to try to estimate or measure values. The methods considered here differ on a number of dimensions.

Perhaps most importantly, different methods can seek to measure different concepts of value, which differ in their theoretical foundations and assumptions. The committee engaged in considerable discussion and debate about the appropriate role of different methods. Although there is not a one-to-one mapping between valuation methods and the concepts of value discussed above, often different views about the appropriate role of alternative valuation methods stem from different views about the nature of value or the appropriate concept of value to apply in a given context. Researchers with different disciplinary backgrounds (e.g., economics, psychology, ecology, decision science) often adopt a particular concept of value and work primarily with and advocate a specific method or set of methods designed to measure that concept.

For example, a fundamental distinction exists between valuation methods that assume individuals have well- defined preferences and those based on the premise that preferences – and hence values – are constructed through the valuation process. As discussed above, the concept of constructed values is based on the premise that, for complex and relatively unfamiliar goods such as ecosystems and some of their associated services, an individual's preferences may not be well-formed and may be subject to intentional or unintentional manipulation or bias, for example by changes in the wording or framing of surveys (see "Survey issues for ecological valuation: Current best practices and recommendations for research " at http://yosemite.epa.gov/Sab/Sabproduct.nsf/WebFiles/SurveyMethods/$File/Survey_methods.pdf). The extent to which this is true has been the subject of scholarly debate both within the committee and outside, and most likely varies with the context (for different sides of this debate, see Becker and Stigler, 1967; Gregory, Lichtenstein and Slovic, 1993; Lichtenstein and Slovic, 2006; and Tourangeau, 2000). If preferences and values regarding ecological systems and services are not well- formed and are instead constructed, they may not be accurately measured or characterized by valuation methods that assume well-formed preferences. For example, some individuals have strongly held values that they find difficult, impossible, or inappropriate to express in monetary units. Requiring these individuals to express such values in monetary equivalents (e.g., in a survey) may compel them to assume a perspective that is unfamiliar or even offensive. Valuation methods based on discourse and deliberation are designed to make explicit and facilitate the construction of preferences in such contexts.

Methods differ along other dimensions as well. For example, they can differ in the type(s) of metrics or outputs produced. In addition, some valuation methods yield a single metric of value, while others yield multiple metrics. Methods that produce a single metric are not necessarily preferable to those that do not. Which approach is more appropriate or useful depends, in general, on the decision context. For example, if the context requires a ranking or choice based on a single criterion (e.g., net benefits), a valuation approach that yields a single (aggregate) metric is needed. In contrast, in a decision context where multiple values are involved (e.g., human health, threatened species, aesthetics, social equity, and other civil obligations) and decision makers themselves are charged with appropriately weighing and balancing competing interests and resolving trade-offs, a multi-attribute approach is preferable. Depending upon the context, this weighing and balancing might be done through political discourse or through a deliberative, decision-aiding process (see the discussions in section 5.3).

Finally, some methods are well developed and have been applied extensively in different contexts; others are still evolving and require further development and testing. However, even for methods that have been used extensively in the past, applying these methods to value changes in ecological systems and services can pose significant challenges beyond those that might exist in other, less complex contexts.

2.2. Ecological Valuation at EPA

As noted in chapter 1, this chapter is focused on ecological valuation within EPA. This necessitates consideration of some issues that might not be considered in more general discussions of ecological valuation. EPA operates in a variety of different decision contexts where valuation might be useful. Although much of the interest in ecological valuation at EPA has focused on valuation needs in national rule making, valuation can also be useful in other decision contexts. Different parts of the Agency need valuation for different purposes and for different audiences. Some contexts closely prescribe how valuations are to be conducted; other contexts are less prescriptive. In addition, EPA faces institutional constraints that influence and limit how it typically conducts valuation in different contexts.

This section of the report describes the committee's understanding of the Agency's needs and constraints related to ecological valuation. It then discusses the committee's understanding of how ecological valuation is typically done at EPA, using an illustrative example. The committee's observations from this example form the basis of its recommendations for an expanded and integrated approach to valuation discussed in the remainder of this chapter.

2.2.1. Policy Contexts At EPA Where Ecological Valuation Can Be Important

As noted, much of the interest in ecological valuation at EPA stems from the need to better value the ecological effects of EPA actions in national rule makings. Two of EPA's governing statutes (the Toxic Substances Control Act and the Federal Insecticide, Fungicide and Rodenticide Act) require economic assessments for national rule making. In addition, Executive Orders 12866 and 13422 have similar requirements for "significant regulatory actions." These economic assessments provide information about whether the aggregate benefits of a policy or regulatory change exceed the costs, which is an important input into policy decisions (Arrow et al., 1996). An Office of Management and Budget circular on "Regulatory Analysis" (OMB Circular A-4) issued in September 2003 identifies key elements of a regulatory analysis for "economically significant rules." Consistent with the principles of welfare economics that underlie benefit-cost analysis, the Circular defines benefits and costs in terms of economic values. Ecological valuation plays a key role in estimating or characterizing these values (see further discussions in sections 2.2.2 and 6.1.2).

EPA's regional offices may also find valuation important in their partnerships with other governments and organizations where the contributions of ecological protection to human well-being are potentially important. Regional offices, for example, may find valuation useful in setting priorities, such as targeting projects for wetland restoration and enhancement, or in identifying critical ecosystems or ecological resources for attention. Valuation may also assist state and local governments, other federal agencies, and non-governmental organizations in deciding how best to protect lands and land uses and in communicating the suitability of the approach chosen.

Valuation can also be useful to EPA in making site-specific decisions, such as those related to the remediation, restoration, and redevelopment of contaminated sites. By providing information about the value of the ecosystem services that could be obtained from site redevelopment, ecological valuation can improve decisions at cleanup sites, including hazardous waste sites listed on the Superfund National Priority List and other cleanup sites (e.g., sites that are the focus of EPA's Brownfields Economic Redevelopment Initiative, Federal Facilities Restoration and Reuse Program, Underground Storage Tank Program, and Research Conservation and Recovery Act).

Although many of the issues and recommendations throughout this chapter apply across decision contexts, specific valuation needs and opportunities vary across these contexts. For this reason, chapter 6 of this chapter discusses the implementation of the report's general recommendations in these three specific decision contexts: national rule making, regional partnerships, and site-specific restoration or redevelopment. Other examples of contexts where ecological valuation may be useful for EPA include:

- Assessing programs as mandated by the Government Performance and Results Act (GPRA) of 1993[14]
- Identifying Supplemental Environmental Projects (EPA Office of Enforcement and Compliance Assurance, 2001) for enforcement cases where projects involve protection of ecological systems and services
- Reviewing Environmental Impact Statements prepared by other federal agencies, under the National Environmental Protection Act
- Issuing permits to protect water quality for those specific states that have not applied for or been approved to run programs on their own and where established state water quality standards allow discretion to consider ecological valuation information.

Although this chapter does not explicitly discuss these other contexts, the approach and selected valuation methods described can be useful in such contexts.

2.2.2. Institutional And Other Issues Affecting Valuation At EPA

EPA must conduct ecological valuation within a set of institutional, legal, and practical constraints. These constraints include procedural requirements relating to timing and oversight, as well as resource limitations (both monetary and personnel). To better understand the implications of these issues for its work, the committee conducted a series of interviews with Agency staff.[15] The interviews focused on the process of developing economic analyses as part of Regulatory Impact Assessments (RIA) for rule making and on the relationship between EPA and the Office of Management and Budget. The interviews also proved beneficial in better understanding strategic planning, performance reviews, regional analysis, and other situations where the Agency needs to assess the value of protecting ecosystems and ecosystem services.

EPA has a formal rule-development process involving several stages, each of which imposes demands on the Agency. Despite the rigidity of the process, Agency analysts assess the value of protecting ecosystems in different ways. Practices vary considerably across program offices, reflecting differences in mission, in-house expertise, and other factors. Program offices have different statutory and strategic missions and have primary responsibility for developing the rules within their mission-specific areas. The organization, financing, and skills of the program offices differ. Although the National Center for Environmental Economics (NCEE) is the Agency's centralized reviewer of economic analysis within the Agency,[16] the primary expertise and development of the rules resides within the program offices.

The timing of the process largely determines the kinds of analytical techniques that are employed. The timing is influenced by court-imposed deadlines on the rule process, as well as Paperwork Reduction Act requirements related to the collection and analysis of new data. By contrast, the scientific community is accustomed to much longer time horizons for their analyses.

Collecting new data poses a significant bureaucratic problem for the Agency. To collect new information from individuals, businesses, and other entities protected under the Paperwork Reduction Act, the Agency must submit an Information Collection Request, which is reviewed within the Agency and by OMB. The Paperwork Reduction Act requires this hurdle and imposes the review responsibility on OMB, adding a significant amount of time to the assessment process. With a time limit of one or two years, at most, to conduct a study, this kind of review significantly limits the scope of analysis the Agency can conduct. Because EPA most often has not been able to collect new information, the Agency has, by necessity, relied heavily on transferring ecological and social values information from previous studies to new analyses.

OMB also acts as an oversight body to review EPA's economic benefit analyses. EPA must justify its claims regarding the economic benefits of its actions, including any analyses of willingness to pay or willingness to accept for ecological protection. As noted above, OMB's Circular A-4 provides explicit guidance for valuation. For a contribution to human welfare or cost that cannot be expressed in monetary terms, the circular instructs Agency staff to "try to measure it in terms of its physical units," or, alternatively, to "describe the benefit or cost qualitatively" (p. 10).[17] Thus, although Circular A-4 does not require that all economic benefits be monetized, it does require, at a minimum, some scientific characterization of those contributions. However, little guidance is provided on how to carry out this task. The circular instead urges regulators to "exercise professional judgment in identifying the importance of non-quantified factors and assess as best you can how they might change the ranking of alternatives based on estimated net benefits" (p. 10).

In conducting benefit assessments, EPA has an incentive to use valuation methods that have been accepted by OMB in the past. This may create a bias toward the status quo and a disincentive to explore innovative approaches, both when monetizing values using economic valuation and when quantifying or characterizing values that are not monetized. The committee recognizes the importance of consistency in the methods used for valuation, but also sees limitations from relying solely on previously accepted methods when innovative or expanded approaches might also be considered.

A related issue involves review of RIAs by external experts. The Agency does not take a standardized approach to RIA review. EPA staff and managers reported that peer review was focused only on "novel" elements of an analysis, meeting the requirements of EPA's peer review policy (EPA, 2006d). This raises the question of how the term novel is defined by the Agency, and perhaps by OMB. More importantly, the novelty standard, ironically, creates another incentive to avoid conducting innovative analyses because the fastest, cheapest option is to avoid review altogether.

Finally, the Agency relies, to varying degrees, on a variety of offices to develop assessments, including individual program offices and NCEE. It is not clear what form of organization is most effective. The Agency's Ecological Benefits Assessment Strategic Plan (2006c) contains suggestions for addressing some of the limitations on ecological valuation resulting from the Agency's internal structure. It advocates the creation of a high-level Agency oversight committee and a staff-level ecological valuation assessment forum. The committee endorses these recommendations.

The Agency will continue to face significant external constraints when considering ecological valuation. The committee recognizes the practical importance of these constraints and advises the Agency to be as comprehensive as possible in its analyses within the limitations imposed by these constraints.

2.2.3. An Illustrative Example Of Economic Benefit Assessment Related To Ecological Protection At EPA

To better understand the current state of ecological valuation at EPA, the committee thoroughly examined one specific case in which assessment of economic benefits was undertaken: the environmental and economic benefits analysis that EPA prepared in support of new regulations for Concentrated Animal Feeding Operations (CAFOs) (EPA, 2002b).[18,19] In communications with the committee, the Agency indicated that this analysis was illustrative in form and general content of other EPA regulatory analyses and assessments of the economic benefits of ecological protection.

EPA proposed the new CAFO rule in December 2000 under the federal Clean Water Act, to replace 25-yearold technology requirements and permit regulations. EPA published the final rule in December 2003. The new CAFO regulations, which cover more than 15,000 large CAFO operations, require the reduction of manure and wastewater pollutants (from both feedlots and land applications of manure) and remove exemptions for stormwater-only discharges.

Because the proposed new CAFO rule constituted a significant regulatory action, Executive Order 12866 required EPA to assess the economic costs and benefits of the rule. An intra-agency team at EPA, including economists and environmental scientists, worked with the U.S. Department of Agriculture on the economic benefit assessment. Before publishing the draft CAFO rule in December 2000, EPA spent two years preparing an initial assessment of the economic costs and benefits of the major options. After releasing the draft rule, EPA spent another year collecting data, taking public comments, and preparing assessments of new options. EPA published its final assessment in 2003. EPA estimates that it spent approximately $1 million in overall contract support to develop the assessment, with approximately $250,000 to $300,000 allocated to water-quality modeling.

EPA identified a wide variety of potential "use" and "non-use" benefits as part of its analysis.[20] Using various economic valuation methods, EPA provided monetary quantifications for seven benefit categories.[21] Approximately 85 percent of the estimated monetary benefits quantified by EPA were attributed to recreational benefits. According to Agency staff, EPA's analysis was driven by what EPA could monetize. EPA focused on those contributions for which data were known to be available for quantification of both the baseline condition and the likely changes stemming from the proposed rule, and for translation of those changes into monetary equivalents.

EPA's final assessment provides only a brief discussion of the contributions to human welfare that it could not monetize. A table in the Executive Summary lists a variety of non-monetized contributions[22] but designated them only as "not monetized." EPA did not quantify these "contributions" in non-monetary terms (e.g., using biophysical metrics) or present a qualitative analysis of their importance. Instead, it represented the aggregate effect of these "substantial additional environmental benefits" simply by attaching a "+B" placeholder to the estimated range of total monetized benefits. Although the Executive Summary gives a brief description of these "non-monetized" benefits, the remainder of the report devotes little attention to them.

The CAFO economic benefits assessment illustrates a number of limitations in the current state of ecological valuation at EPA. First, as noted above, the CAFO analysis did not provide the full characterization of ecological contributions to human welfare using quantitative and qualitative information, as OMB Circular A-4 would appear to require. The report instead focused on a limited set of economic benefits, driven primarily by the ability to monetize these benefits using generally accepted models and existing value measures.[23] These benefits did not include all of the major ecological contributions to human welfare that the new CAFO rule would likely generate.[24] The circular requires that an assessment identify and characterize all of the important benefits of a proposed rule, not simply those that can be monetized. In this case, the monetized benefits alone exceeded the cost of the rule and hence the focus on benefits that could be readily monetized did not affect the outcome of the regulatory review. However, in a different context an economic benefit assessment based only on easily monetized benefits could inadvertently undermine support for a rule that would be justified based on a more inclusive characterization of contributions to human welfare.

Second, the monetary values for many of the economic benefits were estimated through highly leveraged benefits transfers (transferring benefits derived from one or more study sites to a policy site) that often were based on dated studies conducted in contexts quite different from the CAFO rule application.[25] This was undoubtedly driven to a large extent by time, data, and resource constraints, which made it very difficult for the Agency to conduct new surveys or studies and virtually forced the Agency to develop benefit assessments using existing value estimates. Nonetheless, reliance on dated studies in quite different contexts raises questions about the credibility or validity of the benefit estimates. This is particularly true when values are presented as point estimates, without adequate recognition of uncertainty and data quality.

Third, EPA apparently did not develop a comprehensive conceptual model of the rule's potentially significant ecological effects. The report presents a simple conceptual model that traces outputs (a list of pollutants in manure – Exhibit 2-2 in the CAFO report) through pathways (Exhibit 2-1) to environmental and human health effects.[26] This model provided useful guidance, but was not sufficiently comprehensive to assure identification of all possible significant ecological effects. A conceptual model of the relevant ecosystem(s) at the start of a valuation project, as discussed in section 3.1, can help to identify not only important primary effects but also important secondary effects – which frequently may be of greater consequence or value than the primary effects.[27]

Fourth, the CAFO analysis demonstrates the challenges of conducting required economic benefit assessments of ecological protection at the national level.[28] National rule making inevitably requires EPA to generalize away from geographic specifics, in terms of both ecological responses to policy options and associated values. It is, however, possible (and desirable) to use existing and ongoing research at local and regional scales to conduct intensive case studies (e.g., individual watersheds, lakes, streams, estuaries) in support of the national-scale analyses. Systematically performing and documenting comparisons to intensive study sites can indicate the extent to which certain regions or conditions might yield impacts that vary considerably from the central tendency predicted by the national model. Alternatively, with sufficient data about the joint distribution of ecological, socio-economic, and other relevant conditions, case study results can be combined in a "bottom-up" approach to produce a national level analysis (see further discussion in section 6.1.3.1).

Fifth, although EPA invited public comment on the draft CAFO analysis as required by Executive Order 12866, there is no indication in the draft CAFO report that the Agency consulted with the public for help in identifying, assessing, and prioritizing the effects and values addressed in its analysis. Nor is there discussion in the final CAFO analysis of any public comments that might have been received on the draft CAFO analysis. Early public involvement can play a valuable role in helping the Agency to identify all of the systems and services affected by proposed regulations and to determine the regulatory effects that are likely to be of greatest value.

Sixth, EPA did not conduct a peer review of the benefit estimates used in the analysis of the CAFO rule. While the Agency appropriately emphasized peer review in its analysis and report, EPA did not seek peer review in deriving benefit estimates for the CAFO rule. Once again, this shortcoming is undoubtedly a function of time and resource constraints. However, peer review, especially early in the process, could help EPA staff identify relevant and available data, models, and methods to support its valuation efforts. An effective method would be to review not only individual components of an analysis (e.g., watershed modeling, air dispersal, human health, recreation, and aesthetics) but also the overall conceptual model and analytic scheme as well.

Finally, EPA's analysis and report closely adhered to the requirements of Executive Order 12866. Although the Executive Order provided the proximate reason for preparing the analysis and report, the Agency did not have to limit itself to the goals and requirements contained therein. The Executive Order does not preclude EPA from adopting broader goals and hence conducting other analyses in addition to the required benefit-cost analysis. Assessments such as the CAFO study can serve many purposes, including helping to educate policy makers and the public more generally about the economic benefits and other values that stem from EPA regulations. It is important for EPA to recognize this broader purpose.

2.3. An Integrated and Expanded Approach to Ecological Valuation: Key Features

The CAFO example highlights a number of limitations to the current state of ecological valuation at EPA. The committee's analysis points to the need for an expanded, integrated approach to valuing the ecological effects of EPA actions. This approach focuses on the effects of greatest concern to people and on integrating ecological analysis with valuation. The remainder of this chapter describes the approach to ecological valuation developed and endorsed by the committee. The approach should serve as a guide to EPA staff as they conduct RIAs and seek to implement Circular A-4, as well as in decisions on regional and local priorities and activities. Subsequent chapters provide a more detailed discussion of the implementation of the approach.

As noted above (see section 2.1.4), ecological valuation requires both prediction of ecological changes and an estimation of the value of those changes. The committee recommends that, when conducting ecological valuation, the Agency use a valuation process that has three key, interrelated features:

- Early consideration of effects that are socially important
- Prediction of ecological responses in value-relevant terms
- Consideration of the possible use of a wider range of valuation methods to provide information about values

2.3.1. Early Consideration Of Effects That Are Socially Important

The first key component of the proposed approach is the early identification and prediction of the ecological responses that contribute to human well-being and are likely to be of greatest importance to people, whether or not the contributions are easily measured, monetized, or widely recognized by the public. These could include ecosystem responses that people value directly or the resulting responses in the services provided by the ecosystem. The importance of a given response will depend on both the magnitude and biophysical importance of the effect and on the resulting importance to society. Early in the valuation process EPA needs to obtain information about the ecosystem services or characteristics that are of greatest concern, so that efforts to quantify and characterize values can focus on the related ecological response.

Identifying socially relevant effects requires a systematic consideration of the many possible sources of value from ecosystem protection and an identification of the values that may be relevant to the particular policy under consideration. Such a systematic consideration will likely expand the types of services to be characterized, quantified, or explicitly valued. Previous valuation assessments have often focused on what can be measured relatively easily, rather than what is most important to society. This can diminish the relevance, usefulness, and impact of the assessment.

An obvious question is how to assess the likely importance of different ecological responses prior to completion of the valuation process. A main purpose of a thorough valuation study is to provide an assessment of the importance of ecological responses to different policy options. Nonetheless, in the early stages of the process, preliminary indicators of likely importance can serve as screening devices to provide guidance on the types of responses that are likely to be of greatest concern. EPA can obtain relevant information in a variety of ways. These range from in- depth studies of people's mental models and how their preferences are shaped by their conceptualization of ecosystems and ecological services, to more standard survey responses from prior or purpose-specific studies. In addition, early public involvement[29] or the use of focus groups or workshops, composed of representative individuals from the affected population and relevant scientific experts, can help identify ecological responses of concern.

In identifying what matters to people, it is important to bear in mind that people's preferences depend on their understandings of causal processes and relationships and the information at hand. As noted previously, people's expressions of what is important or of the tradeoffs they are willing to make can change with the amount and kind of information provided, as well as the manner in which it is conveyed. Collaborative interaction between analysts and public representatives can help to ensure that respondents have sufficient information when expressing views and preferences. In fact, EPA can use the ecological valuation process as a mechanism for increasing and augmenting public discourse about ecosystem services and how EPA actions affect those services, thereby narrowing the gap between expert and public knowledge of ecological effects.

The committee's approach to valuation envisions consideration of a broader set of ecological effects. However, the committee recognizes that in most cases the purpose of the ecological valuation is to help answer specific questions that the Agency faces. The analyses do not always have to be complete to provide the information needed to answer a particular question. For example, suppose a state agency partnering with EPA must decide whether to allow logging at a particular site and an analysis focused solely on the recreational value of the unharvested site shows that these values alone exceed the net commercial value of logging. The agency can then conclude that logging will lead to a net social loss without valuing other ecological effects of logging. Thus, if the sole purpose of the valuation exercise is to determine whether the logging would generate a net social gain or loss and that determination can be based on a subset of values, then it would be unnecessary to expend a large effort to analyze the full suite of values. Of course, if the recreation value is less than the net commercial value of logging, the agency cannot conclude that logging would lead to a net social gain. In such cases, the analysis can be reframed to provide a lower bound on the magnitude of the ecological benefits from reduced logging that would be necessary to justify the cost.

2.3.2. Predicting Ecological Responses In Value-Relevant Terms

The second major component of the C-VPESS process is to predict ecological responses in terms relevant for valuation. This should begin with a conceptual model, followed by quantification (where possible) using specific ecological and related models. It requires both the prediction of biophysical responses to EPA actions and the mapping of those responses into effects on ecosystem services or features that are of direct concern to people– first conceptually, and then quantitatively. Ideally, this would be done using an ecological production function that is specified and parameterized for the ecosystem and associated services of relevance.

Numerous mathematical models of ecological processes and functions are available. These models cover the spectrum of biological organization and ecological hierarchy (e.g., individual level, population level, community level, ecosystem level, landscape level, and global biosphere). In principle, models can provide quantitative predictions of ecological responses to a given EPA action at different temporal and spatial levels. Some models are appropriate for specific contexts, such as particular species or geographic location, while others are more general.

Ecological models provide a basis for estimating the ecological changes that could result from a given EPA action or policy (e.g., changes in net primary productivity or tree growth) and the associated changes in ecosystems or ecosystem services. However, many have been developed to satisfy specific research objectives and not EPA policy or regulatory objectives. Using these models to assess the contributions of EPA actions to human well-being thus poses challenges.

The first challenge is to link existing models with Agency actions that are intended to control chemical, physical, and biological sources of stress. The valuation framework outlined here requires estimation of the biophysical responses to a specific EPA action. To be used for this purpose, ecological models must be linked to information about stressors. This link is often not a key feature of ecological models developed for research purposes. Existing models may need to be modified or new models developed to address this need.

Ecological models also need to be appropriately parameterized for use in policy analysis. Numerous ecological studies have been conducted at various levels, for example, at Long-Term Ecological Research Sites (Farber et al., 2006). These might provide a starting point for parameterizing policy-relevant models. A key challenge is to determine whether and to what extent parameters estimated from a given study site or population can be transferred for use in evaluating ecological changes at a different location, time, or scale. In many cases, data do not currently exist to parameterize existing models for use in assessing EPA's actions. Such data may need to be developed before the Agency can use these models fully. To the extent that transferable models and parameter estimates exist, a central repository for this information would be extremely valuable.

The final, but perhaps most important, challenge is translating the responses predicted by standard ecological models into responses in terms of ecosystem services or features that can then be valued. If adapted properly, ecological models can connect material outputs to stocks and service flows (assuming that the services have been well-identified). Providing the link between material outputs and services involves several steps. These steps include: identifying service providers; determining the aspects of ecological community structure that influence function; assessing the key environmental factors that influence the provision of services; and measuring the spatial and temporal scales over which services are provided (Kremen, 2005). However, most ecological models currently are not designed with this objective in mind. In particular, they do not predict biophysical responses to stressors in ways that the public can understand or that directly link to human/social consequences that can be valued.

2.3.3. Use Of A Wider Range Of Valuation Methods

Given predicted ecological responses, the value of these responses needs to be characterized and, when possible, measured or quantified. As noted above, a variety of valuation methods exist. To date, economic valuation methods have been the mainstay of ecological valuation at EPA, not only in the context of national rule making (as required by OMB Circular A-4) but also in decision contexts not governed by OMB guidance. A key tenet of the valuation process proposed by the committee is consideration of both economic valuation methods and other valuation methods.

The committee sees two possible roles that use of a broader suite of methods might play. First, the use of an expanded suite of methods could allow EPA analyses to better capture the full range of contributions stemming from ecosystem protection and the multiple sources of value derived from ecosystems. Different valuation methods are designed to assess different sources of value, using different value concepts, and no single method captures them all. Thus, in contexts where the Agency seeks to capture all sources of value and is not constrained in this regard by legislative or executive rules, consideration of a broader suite of methods can contribute to this goal. The specific method(s) to be used would depend upon the underlying sources and concepts of value the Agency seeks to assess, as well as the specific information needs, legal and regulatory requirements (if any), data availability, and methodological limitations it faces. When the Agency can select from a range of methods, there may be scope for piloting and evaluating the use of methods that are relatively novel and in the developmental stage.

Second, even when the Agency is required or chooses to base its assessment on economic values (for example, in the context of national rule making), non-economic valuation methods may be useful in supporting and improving the economic valuation (benefit assessment) in the following ways:

- Non-economic methods could help identify the ecological responses that people care about. For example, surveys, interviews, or focus groups in which individuals indicate the importance of different environmental and other concerns might provide information about the ecological effects of a specific rule that are likely to be viewed as important.

- Some non-economic methods could provide an indicator of an economic benefit that the Agency cannot monetize using economic valuation. For example, metrics that are primarily biophysical or social-economic indicators of impact, such as acres of habitat restored or the number and characteristics of individuals or communities affected, can serve as indicators of at least some contributions of ecosystem protection to human welfare (see further discussion in section 6.1). As noted earlier, OMB Circular A-4 requires that benefits be quantified when they cannot be monetized; some bio-ecological or attitude/judgment-based metrics provide potentially useful forms of quantification in such circumstances. Although they would not provide full information about the magnitude of benefits, they might be expected to correlate with benefits. Thus, when properly chosen, higher levels of a particular biophysical, socio-economic or attitudinal metric would signal higher benefits.
- Non-economic methods could be used to provide supplemental information outside the strict benefit-cost analysis about sources of value that might not be fully captured in benefit measures that come from economic valuation, such as moral or spiritual values. This is consistent with the EPA's call in its Ecological Benefits Assessment Strategic Plan for exploring supplemental approaches to valuation. Even if not part of a formal benefit-cost analysis, information about non-economic values may be useful to both EPA and the public.

Regardless of the specific role played by different methods, the use of a broader suite of methods must adhere to some fundamental principles. First, only valuation methods that meet appropriate validity and related criteria should ultimately be used. Section 4.1 provides a discussion of criteria for assessing validity. The validity of some methods has already been subjected to considerable scrutiny. For methods that are still in the developmental stage, exploration of the method's potential should include an assessment of the validity of the method using a scientifically based set of criteria.

The second principle relates to aggregation across methods. Clearly, values cannot be aggregated across methods that yield value estimates in different units. However, even when units are comparable (e.g., both methods yield monetary estimates of value), aggregation across methods may not be appropriate. Because of their different assumptions, different methods can measure quite different underlying concepts of value and hence yield measures that are not comparable. As a result, simple aggregation across methods is generally not scientifically justified. For example, it would be conceptually inconsistent to add monetary value estimates obtained from an economic valuation method and monetary estimates obtained from a deliberative process in which preferences are constructed, because the two are not based on the same underlying premises. Nonetheless, information about both estimates of value may be of interest to policy makers. In such cases, value estimates should be reported separately rather than aggregated across methods (see further discussion in section 6.1.3.1.). This is consistent with the suggestion above that, in the context of national rule makings where benefit assessments are conducted under Circular A-4, information about non-economic values should be considered separately (as supplemental information) rather than "added to" the economic benefit estimates to obtain a measure of total value.

A third principle relates to the potential for double-counting when multiple methods are used to measure or characterize values. Even when different valuation methods seek to measure the same underlying concept of value (so that aggregation is conceptually justified), adding estimates from different valuation methods could lead to double counting. This can arise either when the value of a change in a given final service is captured by multiple methods, or when both an intermediate service and the final service to which it contributes are valued separately. Clearly identifying which sources of value are captured by a given method (and which are not) will highlight any potential overlap that might exist when multiple methods are used. This could reduce the likelihood of double-counting.

2.4. Steps in Implementing the Proposed Approach

The previous section provides an overview of an integrated and expanded approach to ecological valuation proposed by the committee. The process for implementing the proposed framework would involve the following steps, depicted in figure 1. The six steps are:

1. Formulate the valuation problem and choose policy options to be considered, given the policy context
2. Identify the significant biophysical responses that could result from the different options
3. Identify the responses in the ecosystem and its services that are socially important
4. Predict the responses in the ecosystem and relevant ecosystem services in biophysical terms that link to human/social consequences and hence to values

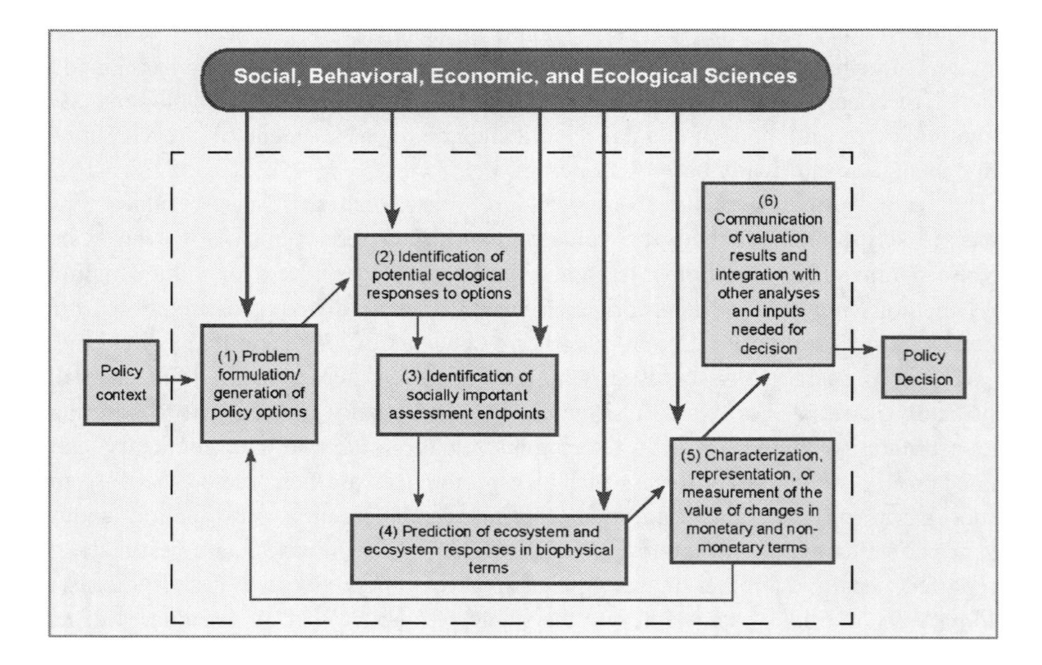

Figure 1. Process for implementing an expanded and integrated approach to ecological valuation

5. Characterize, represent, or measure the value of responses in the ecosystem and its relevant services in monetary or non-monetary terms
6. Communicate results to policy makers for use in policy decisions

Although the steps are depicted sequentially, in actual practice numerous feedbacks should occur with interactions and iterations across steps. For this reason and other reasons (see section 3.1), it is important that the valuation process be based on a conceptual model, developed initially in steps 2 and 3, that can be updated and revised. For example, information about the value of responses in ecosystem services to a given set of policy options might cause a reformulation of the problem or identification of new policy options that could be considered. Also, a projected biophysical effect might suggest human-social values that were not initially considered.

As depicted in figure 1, the implementation of the approach is also contingent upon the specific policy context and intended to provide input for a particular policy decision. As noted above, ecological valuation can play a key role in a number of different decision contexts, including national rule making and regional or local decisions regarding priorities and actions. The valuation problem should be formulated within the specific EPA decision context. Different contexts will generally be governed by different laws, principles, mandates, and public concerns. These contexts can differ not only in the required scale for the analysis (e.g., national vs. local) but possibly also in the type of valuation information that may be needed. For example, in contexts requiring an economic benefit-cost analysis, benefits need to be monetized whenever possible. In contrast, expressing contributions to human welfare in monetary terms might be of little or no relevance to EPA analysts in other contexts. The policy context therefore influences the appropriateness of methods, models, and data.

Figure 1 also highlights the need for information and input from a wide range of disciplines at each step of the process, beginning with problem formulation and the identification of the ecosystem responses that matter and continuing through the valuation of those responses. Instead of ecologists working independently from economists and other social scientists, experts in those disciplines should collaborate throughout. Ecological models need to be developed, modified, or extended to provide usable inputs for value assessments. Likewise, valuation methods and models need to be developed, modified, or extended to address important ecological and biophysical effects that may be underrepresented in value assessments.

Figure 1 suggests a structure that in many ways parallels the Agency's Framework for Ecological Risk Assessment (EPA Risk Assessment Forum, 1992; EPA Risk Assessment Forum, 1998). This framework underlies the ecological risk guidelines developed by EPA to support decision making intended to protect ecological resources (EPA Risk Assessment Forum, 1992). Ecological valuation is a complement to ecological risk assessment. Both processes begin with an EPA decision or policy context requiring information about ecological effects. Next follows a formulation of the problem and an identification of the purpose and objectives of the analysis, as well as the policy options that will be considered. In addition, both ecological risk assessment and ecological valuation involve the prediction and estimation of possible ecological responses to an EPA action or decision. They also both ultimately use this (and related) information in the evaluation of alternative actions or decisions.

Although they are similar, ecological valuation goes beyond ecological risk assessment in an important way. Typically, risk assessments primarily focus on predicting the magnitudes and likelihoods of possible adverse effects on species, populations, and locations, but do not provide information about the societal importance or significance of these effects. In contrast, ecological valuation seeks to characterize the importance to society of predicted ecological effects by providing information on either the value that society places on ecological improvements or the loss it experiences from ecological degradation. By incorporating human values, ecological valuation is closer to risk characterization than risk assessment. Many of the principles that should govern risk characterization outlined in the 1996 National Research Council Report Understanding Risk: Informing Decisions in a Democratic Society pertain to ecological valuation as well. For example, both should be the outcome of an analytical and transparent process that incorporates both scientific information and information from the various interested and affected parties about their concerns and values.

2.5. Conclusions and Recommendations

Ecosystems provide a wide array of services that directly or indirectly support or enhance human populations. People also can value them in their own right for reasons stemming from ethical, spiritual, cultural, or biocentric principles. EPA's broad mission to protect human health and the environment includes the protection of ecosystems.

Many EPA actions affect the state of ecosystems and the services derived from them. To date, ecological valuation at EPA has focused primarily on a limited set of contributions to human well-being from ecological protection. This stems primarily from the difficulty of predicting the responses of ecological systems and services to EPA actions and the difficulty of quantifying, measuring, or characterizing the resulting contributions to human well-being and associated values. The presumption that contributions need to be monetized in order to be carefully characterized also restricts the range of ecological effects that are typically considered in EPA analyses, particularly at the national level.

To implement the key features of an integrated an expanded approach to ecological valuation described in section 2.3 and reiterated in Table 2, the committee recommends that the Agency take the following steps.

- EPA should cover an expanded range of important ecological effects and human considerations using an integrated approach. Such an approach should:
    - Involve, from the beginning and throughout, an interdisciplinary collaboration among natural and social scientists, as well as input about public concerns.
    - Identify early in the process the ecological responses or contributions to human well-being that are likely to be of greatest importance to people and focus valuation efforts on these responses. This would likely expand the range of ecological responses that are valued, recognizing the many sources of value.
    - Predict ecological responses to EPA actions or decisions in value-relevant terms. To do so, the valuation process should highlight the concept of ecosystem services and provide a mapping from responses in ecological systems to

responses in services or ecosystem components that can be directly valued by the public.

- Consider the use of a wider range of possible valuation methods, Methods not currently used by EPA could be used to provide information about multiple sources and concepts of value that might be of interest to the Agency and the public. In addition, they could contribute to assessments based on economic values by (a) helping to identify early in the process the ecosystem services that are likely to be of concern to the public and that should therefore be the focus of the assessment, and (b) addressing the requirement in Circular A-4 to provide quantitative or qualitative information about the possible magnitude of benefits (or costs) when they cannot be monetized using economic valuation.
- Because EPA has limited experience with the use of non-economic valuation methods and some of these methods are still in the developmental stages, the committee believes that it would be wise for the Agency to pilot and evaluate the use of these other methods in different valuation contexts. In the context of national rule making, the Agency should conduct one or two model analyses (perhaps one prospective and one retrospective) of how the use of a wider range of methods could improve benefit assessments in the ways described above. This experience could then guide the Agency's valuation efforts when it conducts subsequent benefit assessments. In addition, the Agency should pilot the use of other valuation methods in local and regional decision contexts, which are less prescriptive and therefore do not need to focus primarily on economic values.
    - As part of this effort, EPA should identify the additional information to be collected and the valuation methods to be used to collect it. After the information is collected and the related valuation completed, EPA should evaluate the contribution of the data collected through the use of new methods to the overall valuation analysis. This evaluation should examine: (a) the properties of the method using a set of explicit criteria (see section 4.1); (b) the contribution of the new information to the decision process; and (c) the potential for using the new information in subsequent analyses (for example, as part of a value transfer).
- EPA should create an institutional structure to facilitate consistent implementation of the proposed valuation approach across the Agency, including the establishment of a high-level oversight body and a staff-level valuation assessment forum, as suggested in the Agency's Ecological Benefits Assessment Strategic Plan (EPA 2006c).

Through the use of the expanded and integrated valuation framework recommended in this chapter, EPA can move toward greater recognition and consideration of the effects that its actions have on ecosystems and the services they provide. This will allow EPA to improve environmental decision making at the national, regional, and site-specific levels and contribute to EPA's overall mission regarding ecosystem protection. EPA can also better use the ecological valuation process to educate the public about the role of ecosystems and the value of ecosystem protection. Through this expanded and integrated approach, different publics can provide EPA with information about how they value ecosystem services.

The remainder of this chapter discusses in more detail how to implement the ideas embodied in the C-VPESS integrated value assessment approach. Some of these ideas can be implemented in the short run, using the existing knowledge base, while others require investments in research and data or method development. Specific recommendations regarding implementation and research needs are included in the chapters that follow.

3. Building a Foundation for Ecological Valuation: Predicting Ecological Responses in Value-Relevant Terms

Chapter 2 presented an overview of an integrated and expanded approach to valuing ecological responses to EPA actions or decisions. This chapter focuses on one part of that approach: predicting ecological responses in value-relevant terms. In every context where the need for valuation arises, information about the magnitude of ecological effects will be a key component of value assessment. No matter what valuation method is used, the valuation process first requires an assessment of the responses of ecosystems and ecosystem services to the relevant EPA action or decision. Even where valuation is not possible, an assessment of these responses can provide valuable information to decision makers and the public. OMB in Circular A-4, for example, provides that, where a benefit cannot be expressed in monetary terms for a major national rule making, EPA "should still try to measure the benefit in its physical units."

This chapter begins with a discussion of the importance of developing an initial conceptual model of the relevant ecosystem and its services that can guide the entire valuation process. Section 3.2 discusses the steps needed to estimate the response of ecosystem and ecosystem services to EPA actions or decisions, including the key importance of ecological production functions. Section 3.3 highlights the challenges that currently exist in trying to implement ecological production functions in specific contexts. These challenges include understanding and modeling the relevant ecology, identifying the relevant ecosystem services, and mapping ecological responses into changes in the relevant ecosystem services. To a large extent, these challenges stem from the site-specificity and underlying complexity of ecosystems. Ecological responses to stressors are often non-linear and discontinuous. Section 3.4 discusses the strategies for evaluating the effects of EPA actions on ecosystem services in the absence of a comprehensive ecological production function. Section 3.5 examines the problem of data availability and conditions where transfer of ecological information might be appropriate. Section 3.6 briefly addresses the importance of new ecological research to support valuation efforts. Finally, section 3.7 summarizes the committee's conclusions and recommendations.

3.1. The Road Map: A Conceptual Model

The key first step in predicting the effects of EPA actions and decisions on ecological systems and services is the formulation of a conceptual model of the relevant ecosystem(s) and its associated services that can guide the valuation effort. The committee recommends that EPA start each ecological valuation by developing such a model. Because the purpose of the model is to guide the valuation process, the model should be context-specific and constructed at a general level. The conceptual model should diagram the predicted relationships among the relevant EPA actions, affected ecosystems, and associated services. The conceptual model is fundamentally a tool to help characterize and predict the ecological and social consequences of the relevant EPA actions and thereby help guide the full valuation process.

Later in the valuation process, EPA will need to use ecological production functions to generate more detailed analyses of key interactions, specific ecological responses to EPA decisions or actions, and resulting consequences to ecosystem services using ecological production functions. As discussed in section 3.3, these analyses will typically require the use of appropriately scaled and parameterized ecological models with a narrower focus. The conceptual model provides a framework for planning for the use of these predictive models at the start of the process and for integrating the more specific analyses into the overall valuation exercise. The goal in the development and use of all models should be to generate information of relevance to the policy making decision facing EPA (Dietz et al., 2003).

The conceptual model should clearly identify the relevant functional levels of the ecosystem, the interrelationships among ecosystem components, and how they contribute to the provision of ecosystem services, either directly or indirectly. Figure 2 provides an example illustrating some aspects of ecosystem services related to nutrient pollution, adapted from Covich et al. (2004).

As figure 2 highlights, the conceptual model should include both information about the underlying ecology and a link to ecological services that are of importance to society. The conceptual model, for example, should include: the impacts of environmental stressors, such as waste disposal, on organisms at different trophic levels; key interactions among species at different levels; and changes at different levels that affect ecological services, such as the food supply, clean water, or recreation.

Not surprisingly, ecologists often focus on underlying ecological relationships (depicted in the lower part of figure 2), and valuation experts tend to focus on the later, value-oriented stages of the process, starting with ecosystem services (shown at the top of the figure). A key principle of this chapter is the need to consider and integrate both aspects of the process. For ecological valuation aimed at improved decision making, a detailed analysis of ecological responses is insufficient unless those responses are mapped to responses in ecosystem services or system components that can be valued. Valuation exercises that do not reflect the key ecological processes and functions are similarly insufficient. Both parts of the valuation process are essential. The development of a conceptual model at the outset of the valuation process can help ensure that the process is guided by this basic principle.

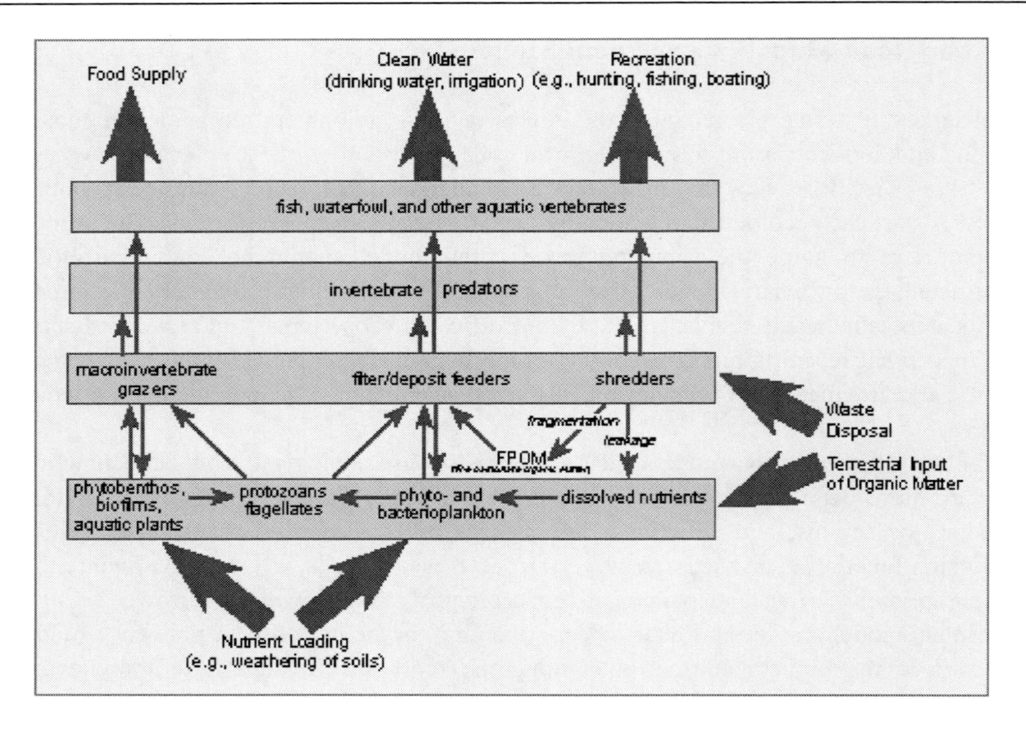

Figure 2. Illustration from Covich et al. (2004) showing relationships of major functional types to ecological services

The development of the conceptual model is a significant task that deserves the attention of EPA staff throughout the Agency, experts in relevant topics from the biophysical and social sciences, and the public. Involving all constituents, including the public, at this stage will enhance transparency, provide the opportunity for more input and better understanding, and ultimately give the process greater legitimacy. Participatory methods such as mediated modeling, described in section 5.3, can play a valuable role in the development of the conceptual model. To promote transparency and understanding, the conceptual model, the process for developing and completing it, and the decisions embedded in it should also be part of the formal record.

The conceptual model should allow for iteration and possible model changes and refinement over time. For example, analysts may initially believe that an action at a local site has local ecological effects, but, on further analysis of the stressors, realize that effects reach to more distant regions downstream or downwind, requiring a change in the conceptual model. Similarly, analysis of the relevant ecological system may show that stressors originally considered insignificant should be added to the conceptual model. As an example, a relatively non-toxic chemical effluent, normally seen as insignificant, might become significant if it is determined that low stream flows or intermittent streams effectively increase the concentration of the chemical to toxic levels during some parts of the year. The need for iterative model changes and refinements is critical and should be part of all valuation efforts.

3.2. The Important Role of Ecological Production Functions in Implementing the Conceptual Model

While the conceptual model serves as a guide for the overall valuation process, the individual components and linkages embodied in that model must be operationalized. The goal is to provide, to the extent possible, quantitative estimates of the responses of ecosystem components or services that can then be valued. Operationalizing the conceptual model requires mapping or describing:

1. How the relevant EPA action will affect the ecosystem
2. How the effects on the ecosystem will, in turn, affect the provision of ecosystem services
3. How people value that ecosystem service response.
   The third step, valuation, is the subject of chapter 4. The remainder of this chapter considers how to implement the first two steps, estimating how the EPA actions will affect the ecosystem, and how the ecosystem response will affect ecosystem services.

The first step requires describing how the EPA action – by reducing or eliminating a stressor or by otherwise protecting or altering an environmental factor – will affect important aspects of ecosystem structure or function. Would a stressor that EPA can eliminate otherwise cause a species to disappear or change in abundance? Would the stressor result in a change in biogeochemistry? For any important effects, EPA should make a quantitative estimate.

The ecological production function is a critical tool for implementing the second step – estimating how the ecological response will affect the provision of ecosystem services. Ecological production functions are similar to the production functions used in economics to define the relationship between inputs (e.g., labor, capital equipment, raw materials) and outputs of goods and services. Ecological production functions describe the relationships between the structure and function of ecosystems, on the one hand, and the provision of various ecosystems services, on the other. These functions capture the biophysical relationships between ecological systems and the services they provide, as well as the inter-related processes and functions, such as sequestration, predation, and nutrient cycling. Coupled with information about how alternative EPA actions or management scenarios will affect the ecological inputs, ecological production functions can be used to predict the effects of the actions or scenarios on ecosystem services.

Ecological production functions could describe the relationship between a broad suite of inputs and ecosystem services. An ecological production function could describe the relationship between inputs for an individual service or, to the extent that two or more services are linked (e.g., produced jointly or in competition), a multiple-output function could capture these linkages.

The analogy between ecological production functions and economic production functions is not perfect. Economic production functions generally involve inputs over which humans have direct control, and the relationship between inputs and outputs is frequently well studied and defined. Ecological production functions, by contrast, involve inputs over which humans have variable and often limited control, and the relationship between inputs and outputs is complex and often very uncertain. Nonetheless, economic production functions provide a useful analogy for the type of relationships and models needed in order to effectively estimate the effect of EPA actions or scenarios on ecosystem services of importance to the public.

Scientists are making progress in understanding and defining ecological production functions for certain ecosystem services. One such service is pollination. Animal pollination is essential for the production globally of about one-third of agricultural crops and the majority of plant species (Kremen and Chaplin, 2007; Kremen et al., 2007). Ecologists have recently built spatially explicit models incorporating land use and its effect on habitat and foraging behavior of pollinators (Kremen et al., 2007). Such models can link changes in ecosystem conditions to the level of pollination of agricultural crops and their yields. Empirical studies using such models have shown the effects of proximity to natural forest on coffee productivity (Ricketts et al., 2004) and the interaction of wild and honey bees on sunflower pollination (Greenleaf and Kremen, 2006).

A second ecosystem service for which considerable progress has been made in developing ecological production functions is carbon sequestration. Agricultural systems, forests, and other ecosystems contain carbon in soil, roots, and above-ground biomass. Rapidly growing markets for carbon sequestration and the potential to generate carbon credits are pushing interest in the accurate assessment of the carbon sequestration potential of agricultural and other managed ecosystems (Willey and Chamaides, 2007). It is possible to quantify aboveground carbon stores fairly accurately in various types of ecosystems such as forests (e.g., Birdsey, 2006; Smith et al., 2006; EPA Office of Atmospheric Programs, 2005), but greater uncertainty remains about stocks of soil carbon that make up the majority of carbon in agricultural and grassland systems (e.g., Antle et al., 2002; EPA Office of Atmospheric Programs, 2005).

Despite this progress, our current understanding of ecological production functions for most ecosystem services remains limited (Balmford et al., 2002; Millennium Ecosystem Assessment, 2005; NRC, 2004). Although many ecological models exist, most do not predict ecosystem service responses. The next section discusses some of the challenges in developing complete ecological production function models for use in ecological valuation.

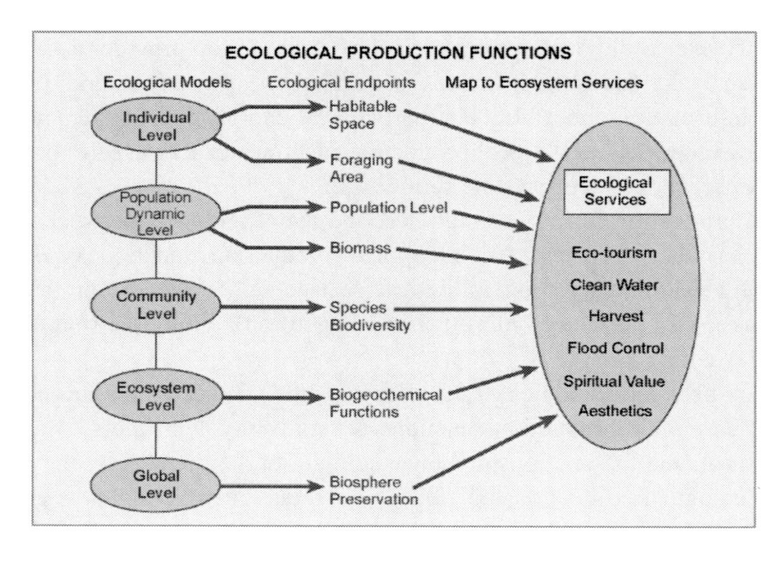

Figure 3. Graphical depiction of ecological production functions

3.3. Challenges in Implementing Ecological Production Functions

Developing and implementing an ecological production function requires:

- Characterizing the ecology of the system
- Identifying the ecosystem services of interest
- Developing a complete mapping from the structure and function of the ecological system to the provision of the relevant ecosystem services

Figure 3 provides a graphical representation of the necessary elements of an ecological production function. On the left side of the figure, ecological models at various organizational levels predict ecological elements or attributes – ecological endpoints – that can be linked to ecosystem services of interest. These ecological models are important components of an ecological production function, but they are not the complete function. An ecological production function requires that the endpoints of these ecological models be mapped or translated into corresponding predictions regarding the ecosystem services of interest to humans.

Each of these three key steps in developing and implementing ecological production functions face challenges that EPA should work to address. This section elaborates on the challenges.

3.3.1. Understanding And Modeling The Underlying Ecology

As noted, the first step in developing an ecological production function is to understand the components, processes, and functioning of the ecosystem that underlie and generate the ecosystem services. Analysts must have a strong understanding of the underlying ecology. Although much is known about ecological systems, current knowledge is still very incomplete, largely because ecosystems are inherently complex, dynamic systems that vary greatly over time and space.

As an example of the complexity of ecological functions, consider the ecological services associated with the activities of soil organisms that might be affected by disposal of waste on that soil. These organisms thrive on organic matter present or added to the soil. By breaking down that organic matter, certain groups of organisms maintain soil structure through their burrowing activities. These, in turn, provide pathways for the movement of water and air. Other kinds of organisms shred the organic material into smaller units that microbes then utilize. The microbes release nutrients in a form that higher plants can use for their growth or in a dissolved form that can move hydrologically from the immediate site into groundwater or a stream. Other groups of specialized microbes may release various nitrogen gases directly to the atmosphere. The nature of soil organisms and the products that they utilize, store, or release all help to regulate the biogeochemistry of the site, as well as the site's hydrology, productivity, and carbon-storage capacity. Predicting the effect of particular actions on ecosystem services such as waste processing and the provision of clean water requires an understanding of these complex ecological relationships.

Complexity also stems from the fact that ecological effects may persist for different periods of time, affecting both the temporal and spatial scales that are relevant for any analysis. The ecological effects from carbon dioxide in the atmosphere, for example, are likely to persist far longer and require a larger temporal and spatial analysis than the effects from acute toxic exposures to hazardous chemicals.

Because of the complexity of most ecosystems, analysts need ecological models to organize information, elicit the interactions among the variables represented in the models, and reveal outcomes under different sets of assumptions or driving variables. Some models are statistical; others are primarily simulation models. Some statistical and theoretical models are relatively small, containing a few equations. Other ecological models are very large, involving hundreds of interacting calculations. Models may be valuable in many of the steps of assessing ecological value including:

- Estimating stress loading
- Estimating the exposure pattern of stress (especially the spatial and temporal implications)
- Identifying ecological elements receiving exposure
- Estimating the exposure-response function of ecological elements
- Estimating the change in stress from potential Agency actions
- Estimating the response of ecosystem services or functions to change in stress

Ecological models can describe ecological systems and ecological relationships that range in scale from local (individual plants) to regional (crop productivity) to national (continental migration of large animals). These models frequently focus on specific ecological characteristics, such as the populations of one or more species or the movement of nutrients through ecosystems, and can cover the full spectrum of biological organization and ecological hierarchy. For instance, a hydrological model might describe possible changes in the timing and amount of water in streams and rivers. A biogeochemical model might predict effects on the levels of various chemical elements in soils, groundwater, and surface waters. A terrestrial carbon cycle model might project changes in plant growth and in carbon sinks or sources. Population and community models might project changes in specific animal and plant populations of concern.

Inevitably, models suffer from limitations. Although many ecological models are well established and used routinely for describing ecological systems, ecological models can only represent the current state of knowledge about the dynamics of an ecological system and generate outputs only as reliable as the data the models use. The dynamism of a system adds to the challenge of modeling, as do the non-linear and discontinuous responses of system components. The model outputs are subject to known, and sometimes unknown, levels of statistical uncertainty. Chapter 5 of this chapter discusses the issue of uncertainty and how EPA should address uncertainty in its valuation efforts (section 5.2.2 examines the specific sources of uncertainty in ecological valuations.) It is important that EPA assess and report on all sources of uncertainty in order to permit a more informed evaluation of and comparison between policy options.

At the moment, the important point to emphasize is that uncertainty pervades the entire valuation process, including the modeling of ecological processes.

Moreover, no ecological model can include all possible interactions. Some ecological models explicitly or implicitly incorporate human dimensions, but most focus primarily on ecological functions. In addition, models capture historical relationships and typically are not able to predict ecosystem patterns for which no modern counterpart exists. For example, if a stressor such as climate change leads species to "reshuffle into novel ecosystems unknown today" for which there is no analog, current models will not predict the effect (Fox, 2007; Dasgupta and Maler, 2004).

Data insufficiency also frequently constrains the applicability, and to some degree formulation, of ecological models. Even when a full theoretical model of an ecosystem exists, applying the model to a specific context of interest will require determining the parameters of the model for that context. However, parameterization is generally difficult because of the complexity of ecological systems and their dependence on an array of site-specific variables. As a result, many ecological models are site specific. The relatively large amounts of site-specific data required to build and parameterize models mean that transferability of the models is limited, either because the model has been developed using spatially constrained data or because inadequate data are available at specific sites with which to drive or parameterize the model. This site-specificity can significantly limit models' applicability to the spatial and temporal complexities required in valuing ecosystem services, especially at regional and national scales.

Ecological models incorporate the best available scientific knowledge of how ecosystems will respond to a given perturbation and the sensitivity of various ecosystem components. The committee therefore recommends that EPA support all of its ecological valuations with ecological models and data sufficient to understand and estimate the likely ecological response to major alternatives being considered by decision makers. Ecological models are essential in representing and analyzing ecological production functions. Guided by the conceptual model described in section 3.1, the Agency should use ecological models to quantify the likely effects of an action on the ecosystem and the resulting effect on ecosystem services.

Given the limitations of many current models, however, the committee also recommends that EPA make the development of effective ecological models one of its research priorities. EPA is already strengthening its approach for developing and using models for decision making. For example, EPA has established the Council for Regulatory Environmental Modeling (CREM), a cross-Agency council of senior managers with the goal of improving the quality, consistency, and transparency of models used by the Agency for environmental decision making. The committee endorses this effort and advises EPA to continue to strengthen its work in this area.

Because many ecological models exist and a variety of models might be used for any particular valuation context (Roughgarden, 1998b), the Agency will often be faced with a choice among one or more predictive models. The appropriate choice of models, and the availability and appropriateness of supporting databases, will depend in part on the scale of analysis (e.g., local vs. national) and the precision of the analysis needed for the relevant policy decision.

The committee recommends that EPA identify clear criteria for selecting ecological models and apply these criteria in a consistent and transparent way. Several existing reports discuss the selection and use of models for environmental decision making and can provide valuable guidance to EPA in the valuation context. In 2005, EPA's Council for Regulatory Environmental Modeling prepared a "Draft Guidance on the Development, Evaluation and Application of Regulatory Environmental Models." In 2006, an EPA

Science Advisory Board panel reviewed the draft report and provided recommendations on revisions (EPA Science Advisory Board, 2006a). Until EPA publishes final guidance, the draft guidance and SAB review can provide EPA with valuable advice in selecting models. A 2007 report of the National Research Council Board on Environmental Studies and Toxicology entitled "Models in Regulatory Environmental Regulatory Decision Making" also provides valuable guidance on selecting appropriate ecological models for use in valuation exercises. The criteria in these reports and the SAB review can guide the Agency both in selecting among models and in setting priorities for future model development.

These reports address environmental modeling in general and do not focus on the use of ecological models for valuation purposes. For valuation purposes, EPA should use the criteria from these reports and choose models that generate outputs either directly in terms of relevant ecosystem services or that are easily translatable into effects on such services. The ultimate goal is to provide a measure of the value of the effects of an action on ecosystem services. The models chosen must advance that goal.

3.3.2. Identifying Ecosystem Services

Another key challenge in implementing ecological production functions is identifying the relevant ecosystem services to be evaluated in any given context. As already emphasized, ecological production functions must ultimately link ecological responses to effects on ecosystem services. This requires that EPA identify the relevant services in a consistent and appropriate way.

Identifying the relevant ecosystem services cannot be done deductively. The relevant services depend on what is important to people in the specific context, once they have been informed about potential ecological effects. The objective is to identify what in nature matters to people and to express this intuitively and in terms that can be commonly understood. Technical expressions or descriptions meaningful only to experts are not sufficient; however, underlying ecological science must inform the identification of relevant services. Identifying relevant services requires a collaborative interaction among ecologists, social scientists, and the public.

The Millennium Ecosystem Assessment (2005) provides a good starting point for identifying potentially relevant ecosystem services by providing an extensive discussion and classification of ecosystem services. In each specific context, however, EPA should also seek input from the general public and from individuals or entities particularly affected by the relevant EPA decision as to what is important. In doing so, EPA can use a variety of sources, such as the valuation methods described in chapter 4 (e.g., surveys, mental models research, or deliberative processes), content analysis of public comments, solicitation of expert opinion and testimony, and summaries of previous decisions in similar circumstances.

Moving toward a common understanding of ecosystem services is important for the success of future valuation efforts. The relative success of EPA efforts to translate air quality problems into human health-related social effects is due in part to the development of agreements about well-defined health outcomes that can be valued. In order to value the health effects of air pollution, it has been necessary to move from describing effects in terms such as oxygen transfer rates in the lung to terms that are more easily understood and valued by the public, such as asthma attacks. Although the search for common health outcomes that can be used for valuation has been difficult, the lesson is clear: If health and social scientists are to productively interact in assessing the value of improved environmental quality, measures of health outcomes that are understandable and meaningful to both groups of scientists are necessary. These outcomes are now understood by disciplines as divergent as pulmonary medicine and urban economics (EPA Science Advisory Board, 2002a). The search for common outcomes that can be valued will be equally important in the ecological realm, where biophysical processes and outcomes can be highly varied and complex.

Some authors have advocated the development of a common list of services to be collectively debated, defined, and used by both ecologists and social scientists across contexts (e.g., Boyd and Banzaf, 2006). Such a list might include:

- Species populations – including those that generate use value, such as harvested species and pollinator species, and those that generate existence values
- Land cover types – such as forests, wetlands, natural land covers and vistas, beaches, open land, and wilderness
- Resource quantities – such as surface water and groundwater availability
- Resource quality – such as air quality, drinking water quality, and soil quality
- Biodiversity

Although only a subset of the services on a common list might be relevant in any particular context, the list would provide some standardization in the definition of ecosystem services across contexts. Advocates argue that development of a common list is the best way to debate and convey a shared mindset, foster the integration of biophysical and social approaches, and provide greater transparency, legitimacy, and public communication about what in nature is being gained and lost. Achieving agreement on a common list might be an important goal, but it is likely to be difficult for complex ecological systems. Converging prematurely on a limited list of services could misdirect valuation efforts and miss important intermediate and end services.

To ensure that the services can be readily and accurately valued, the identification of relevant ecosystem services, either as a common list or for a specific analysis, should follow some basic principles. First, it is important to avoid double counting. All things that matter should be counted, but only once.[30] Second, the ecosystem services should have concrete outcomes that can be clearly expressed in terms that the public can understand. If ecological outcomes are to provide useful input into valuation, they must be described in terms that are meaningful to those whose values are to be assessed.

EPA has launched several initiatives to develop common and useful endpoints for ecological models. These endpoints, however, are typically not themselves ecosystem services. The endpoints instead are often ecological attributes or elements, such as biomass, that serve as inputs to the production of ecosystem services. Although these endpoints often link to the Agency's statutory responsibilities and policy concerns, social scientists typically cannot use them by themselves to value effects on ecosystem services. Looking at figure 3, social scientists need information on the ecosystem services at the right side of the diagram. Most endpoints, shown in the center column of figure 3, are at least one step removed and must still be translated into responses in ecosystem services.

EPA's generic ecological assessment endpoints (GEAEs) (EPA Risk Assessment Forum, 2003) provide a valuable example. The GEAEs are based on legislative, policy, and regulatory mandates. If expanded to include landscape-, regional-, and global-level endpoints (see EPA Risk Assessment Forum, 2003, Table 4.1; Harwell et al., 1999; EPA Science Advisory Board 2002b), they can serve as a first step in characterizing relevant ecological systems and quantifying responses to stressors. Although the GEAEs are a valuable starting point, they also illustrate how far EPA must go in estimating responses in ecosystem services. First, the GEAEs are expressed in technical terms and not in terms of concrete outcomes that the public can understand. These technical terms are certainly appropriate for some regulatory purposes, but most of the public is unlikely to be familiar with them. Therefore, they will have limited use in valuation.

Second, the GEAEs do not necessarily reflect the things in nature that people care about. Although the endpoints reflect policy and regulatory needs (EPA Risk Assessment Forum, 2003, p.5), they depict a narrow range of ecological outcomes, confined to organism, population, and community or ecosystem effects. They do not relate to water availability, aesthetics, or air quality, but rather to kills, gross anomalies, survival, fecundity and growth, extirpation, abundance, production, and taxa richness. These effects are clearly relevant to biological assessment. However, for anglers who care about the abundance of healthy fish in a particular location at a particular time, lost value depends not on the number of kills or anomalies but rather on the abundance of healthy fish.

Another important ecological endpoint initiative is EPA's Environmental Monitoring and Assessment Program (EMAP). Created in the early 1990s, EMAP is a long-term program to assess the status and trends in ecological conditions at regional scales (Hunsaker and Carpenter 1990; Hunsaker, 1993; Lear and Chapman, 1994). Once again, the endpoints developed in EMAP are generally not direct measures of ecosystem services. EMAP does, however, emphasize the importance of developing endpoints that are understandable and useful to decision makers and the public. As EPA has recognized, if an endpoint is to serve as a useful indicator of ecological health, it "must produce results that are clearly understood and accepted by scientists, policy makers, and the public" (Jackson et al., 2000). One study that used focus groups to examine the value of EMAP endpoints as indicators of environmental health similarly concluded that there is a need "to develop language that simultaneously fits within both scientists' and nonscientists' different frames of reference, such that resulting indicators [are] at once technically accurate and understandable" (Schiller et al., 2001). The committee agrees with this conclusion and urges EPA to move further toward this goal.

The Agency is aware of the limitations of current endpoints. The committee emphasizes the limitations for two reasons: to highlight the difference between the Agency's current approach to defining relevant ecological endpoints and the need to identify effects on ecosystem services; and to encourage the Agency to move toward identifying and developing measures of ecosystem services that are relevant and directly useful for valuation.

The identification of relevant ecosystem services will require increased interaction between natural and social scientists within the Agency. The committee urges the Agency to foster this interaction through a dialogue related to the identification and development of measures of ecosystem services. One means of doing this is to encourage greater coordination among the Agency's extramural research programs, including the Decision-Making and Valuation for Environmental Policy grant program and the Office of Research and Development's ecosystem-services research program. A joint research initiative focused on the development of measures of ecosystem services will address a critical policy need and provide a way for the Agency to integrate its ecological and social science expertise in a very concrete fashion.

3.3.3. Mapping From Ecosystem Responses To Changes In Ecosystem Services

Once the underlying ecology is understood and modeled and the relevant ecosystem services are identified, ecological production functions still require a correlation of the ecosystem responses to the relevant ecosystem services. As noted above, although numerous ecological models exist for modeling ecological systems, most of them fall short of estimating effects on ecosystem services. Many of the models have been developed to satisfy research objectives, rather than Agency policy or regulatory objectives. The outputs of these models have not generally been cast in terms of direct concern to people and thus are not useful as inputs to valuation techniques. For example, evapotranspiration rates, rates of carbon turnover, and changes in leaf area are important for ecological understanding, but are not outputs of direct human importance. Some models exist with outputs directly related to human values and include models that predict fish and game populations or forest productivity. These models, however, address only a limited set of ecosystem services.

3.4. Strategies to Provide the Ecological Science to Support Valuation

Although development of a broad suite of ecological production functions faces numerous challenges, EPA can employ several other approaches at this time to gain a better understanding of how ecosystem services respond to its actions. These approaches include using indicators that are correlated with ecosystem services and using meta-analyses. Indicators represent a form of simplification; meta-analysis is based on information aggregation.

3.4.1. Use Of Indicators

As noted above, an ecological production function describes the relationship between ecological inputs and ecosystem services. When a full characterization of this relationship is not available, some indication of the direction and possible magnitude of the changes in the services that would result from an Agency action might still be obtained using indicators. "Indicators," as the term is used here, are measures of key ecosystem properties whose changes are correlated with changes in ecosystem services.[31] In general, an indicator approach involves selecting and measuring key predictive variables rather than defining and implementing a complete ecological production function. Because of the complexity of the interactions between economic and ecological systems, economists frequently take a similar simplified approach that focuses on effects only in the relevant markets, assuming that the effects on the broader market are negligible and can be ignored (Settle et al., 2002).

Indicators can provide useful information about how ecological responses to EPA actions or decisions might affect ecosystem services. If it is known that an indicator is positively or negatively correlated with a specific ecosystem service, predicting the change in the indicator can provide at least a qualitative prediction of the change in the corresponding ecosystem service. Indicators may be important even where models exist that can provide more sophisticated ecological analysis. The use of large, complex ecological models to make numerous or rapid evaluations can be difficult, especially given the quantities of required data and the short time in which assessments generally must be made (Hoagland and Jin, 2006). In these situations, simplification can be far more practical. The use of indicators that simplify and synthesize underlying complexity can have advantages in terms of both generating and effectively communicating information about ecological effects.

Ecologists and environmental scientists have sought to identify indicators of ecosystem condition that might be linked to specific services. Many ecosystem indicators have been proposed (NRC, 2000; EPA, 2002a; EPA, 2007a), and several states have sought to define a relatively small set of indicators of environmental quality. Indicator variables have been established for specific ecosystems such as streams (e.g., Karr, 1993) and for entire countries (e.g., The H. John Heinz III Center for Science, Economics, and the Environment, 2008). The committee acknowledges EPA's work in developing indicators for air, water, and land and for ecosystem condition and encourages the Agency to see where those indicators can be linked to specific services relevant to the valuation of EPA decisions.

There is currently no agreement on a common set of indicators that can be consistently applied and serves the needs of decision makers and researchers in all contexts (Carpenter et al., 2006). However, there are guidelines for specific issues. For example, in evaluating the economic consequences of species invasion, Leung et al. (2005) have developed a framework for rapid assessments based on indicators to guide in prevention and control, simplifying the ecological complexity to a relatively small number of easily estimated parameters.

One potentially useful approach to indicators is to incorporate multiple dimensions into a coherent presentation that describes the status of ecosystems within a region, especially as the ecosystems relate to social values and ecosystem services. For example, "ecosystem report cards," such as those developed for South Florida (Harwell et al., 1999) and for Chicago Wilderness (available at http://www.chicagowilderness. org/pubprod/index.cfm), use an array of indicators designed to provide information about the status and trends associated with the ecological services provided by the ecosystems. The report card identifies seven ecosystem characteristics thought to be important: habitat quality, integrity of the biotic community, ecological processes, water quality, hydrological system, disturbance regime (changes from natural variability), and sediment/soil quality. These characteristics are then related to the goals and objectives for the report card.[32] The outputs are not quantitatively valued or monetized, but rather described by narratives or quantitative/qualitative grades that are scientifically credible and understandable by the public. The report card is designed to:

- Be understandable to multiple audiences
- Address differences in ecosystem responses across time
- Show the status of the ecosystem
- Transparently provide the scientific basis for the assigned grades on the report card

This simplified approach to ecological modeling cannot identify all the possible consequences of EPA actions. The challenge is building ever more complex models that address a wide array of issues over multiple spatial and temporal scales. It may well be that, with accumulated experience, it may be more practical to adopt the simplified approach of selecting a few key indicators or ecological processes that are correlated with specific ecosystem services and can be valued. The committee advises EPA to continue research to develop key indicators for use in ecological valuation. This is likely to be particularly fruitful when those indicators can be used for key repeated rule makings or other repeated decision contexts. Such indicators should meet ecological science and social science criteria for effectively simplifying and synthesizing underlying complexity while still providing scientifically based information about key ecosystem services that can be valued. Use of the chosen indicators should also be accompanied by an effective monitoring and reporting program.

3.4.2. Use Of Meta-Analysis

A second promising approach to providing information about effects on ecosystem services is the use of meta- analysis. Meta-analysis involves collecting data from multiple sources and attempting to draw out consistent patterns and relationships from those data about the links between ecological functions or structures and associated services. For example, Worm et al. (2006) attempted to measure the effects of biodiversity loss on ecosystem services across the global oceans. They combined available data from multiple sources, ranging from small-scale experiments to global fisheries. In these analyses, the impossibility of separating correlation and causation is a severe limitation. But examining data from site-specific studies, coastal regional analyses, and global catch databases allowed these researchers to draw correlative relationships between biodiversity and decreases in commercial fish populations – variables that can be valued and monetized.

In a similar approach, de Zwart et al. (2006) noted that ecological methods for measuring the magnitude of biological degradation in aquatic communities are well established (e.g., Karr, 1981; Karr and Chu, 1999.), but determining probable causes is usually left to a combination of expert opinion, multivariate statistics, and weighing of evidence. As a result, the results are difficult to interpret and communicate, particularly because mixtures of potentially toxic compounds are frequently part of these assessments. To address this issue the authors used a combination of ecological, ecotoxicological, and exposure modeling to provide statistical estimates of probable effects of different natural and anthropogenic stressors on fish. This approach links fish, habitat, and chemistry data collected from hundreds of sites in Ohio streams. It assesses biological conditions at each site and attributes any impairment (e.g., loss of one or more of 117 fish species) to multiple probable causes. When data were aggregated from throughout Ohio, 50 percent of the biological effect was associated with unknown factors and model error; the remaining 50 percent was associated with alterations in stream chemistry and habitat. The technique combines multiple data sets and assessment models to arrive at estimates of the loss of fish species based on broad patterns. Like the Worm et al. (2006) study of the relationship of biodiversity to ocean productivity, this study aggregates data from many sources and uses various models to arrive at estimates that can be easily interpreted and, at least in the case of game fish species, valued and monetized. In a similar context, EPA's Causal Analysis/ Diagnosis Decision Information System (CADDIS) permits scientists to access, share, and use environmental information to evaluate causes of biological effects found in aquatic systems (see http://cf pub.epa.gov/caddis).

3.5. Data Availability

Data availability is a serious problem in the development of ecological production functions. However, data on the structure and function of ecological systems are becoming more available and better organized across the country. Part of the increased availability is simply that Web-based publication now enables authors to make data and analysis readily available to other researchers in electronic format. Also, as government agencies are being held more accountable, these agencies are increasingly making the data they collect and use available to constituents. EPA's National Water-Quality Assessment (NAWQA) program provides useful data on the nation's waterways and aquatic systems in a consistent and comparable fashion (see http://water.usgs.gov/nawqa).

The committee recommends that EPA work with other agencies and scientific organizations, such as the National Science Foundation (NSF), to encourage the sharing of ecological data and the development of more consistent ecological measures that are useful for valuation purposes. EPA should also encourage strong regional initiatives to develop information needed for valuations. Within the ecological research community, the NSF's Long-Term Ecological Research (LTER) program has emphasized organizing and sharing data in easily accessible electronic datasets. Although these data have rarely been collected for valuing ecosystem services, they measure long-term trends and therefore can be particularly valuable in separating short-term fluctuations from longer-term patterns in ecological conditions. Recently, the LTER program has focused on regionalization, in which data are collected from sites surrounding a primary site, providing a regional context for site-based measurements and models.[33]

EPA also can look to the social sciences for useful insights into the building of data-sharing capacity. The social sciences have a lengthy and successful history of sharing data through repositories such as the Inter- University Consortium on Political and Social Research (www.icpsr.umich.edu).

3.5.1. Transferring Ecological Information From One Site To Another

Despite the increasing availability and organization of ecological data, there is rarely enough available information to support many desired analyses. In addition, the costs of collecting extensive data from all the sites in which EPA is considering action would be prohibitive. An important issue is the reliability of transferring ecological information from one site to another or over different spatial or temporal scales. The information can include tools or approaches, data on properties of an ecosystem or its components, and services or contributions to human well-being provided by an ecosystem.

There are no hard and fast rules for when ecological information can be transferred. Confidence in doing so depends on the types of information and the systems in question. Given the complexity of most ecosystems, the richness of interactions, and the propensity for non-linearity, extrapolation of ecological information requires caution. However, certain generalizations are possible. Information is more likely to be transferable when there is greater similarity between ecosystem contexts. Also, aggregate information, such as data on ecosystem properties, is more likely to be transferable than information on particular species or the interactions of particular species. Thus, the ecosystem properties (e.g., leaf area index, primary productivity, or nitrogen-cycling patterns) of an oak-hickory deciduous forest in Tennessee might be transferable to oak-hickory forests in other parts of the eastern United States that are at similar stages of development. To a lesser extent, the information might be transferable to other types of deciduous forests.

Information may be transferable to other spatial or temporal scales if the dynamics over time and space are known for the ecosystems. For instance, if data are available on how the characteristics of an oak-hickory forest change as it develops or goes through cycles of disturbance, data transfers from one point in time to another should be possible. Similarly, if information is available on how the properties of the system vary with spatial environmental variation (e.g., local climate, soil type, or land-use history), the extension of information from one spatial context to another should be possible. EPA and other national and international agencies have sponsored extensive research on the scaling up of data from particular sites to regions (Suter, 2006, chapters 6 and 28; Turner et al., 2007). The results from these analyses are applicable to the transfer of information on ecological properties and services.

To some extent, the same generalizations apply to transferring tools such as models, although success depends on how generally applicable the tool is and how difficult in terms of data requirements it is to parameterize for other situations. For example, forest ecosystem models can often be transferred to other forests using available information from sources such as LTER sites.

3.6. Directions for Ecological Research to Support Valuation

EPA has briefed the committee on its plans to redesign a major part of its intramural and extramural research program to forecast, quantify, and map production of ecosystem services (see briefings to the C-VPESS, EPA Science Advisory Board, 2006b and 2007b). The committee welcomes these efforts as a way to strengthen the foundation for ecological valuation. EPA should evaluate the validity of all models that it develops or uses to assess the reliability of the biophysical changes or responses that they predict.

The committee notes with concern EPA's limited and shrinking resources for ecological research (EPA Science Advisory Board, 2007a). Although the committee has not received any details about Agency plans, it encourages the Agency to carefully focus its research program because the cost of implementing ecological production functions in multiple places on multiple issues may be significant. The committee commends EPA for asking for additional science advice on its Ecological Research Program Strategy and Multi-year Plan and believes this advisory activity should be a priority for an SAB panel of interdisciplinary experts in ecological valuation, drawing on information in this chapter.

3.7. Conclusions and Recommendations

Implementation of the integrated valuation process recommended by this chapter requires the Agency to predict the ecological responses to its actions, identify the relevant ecosystem services of importance to the public, and link the predicted ecological responses to the effect on those services. Estimating the responses of relevant ecosystem services to EPA actions is an essential part of valuation and must be done before the value of those responses can be assessed.

With regard to predicting the responses of ecosystems and ecosystem services, the committee recommends the following:

- EPA should begin each valuation with a conceptual model of the relevant ecosystem and the ecosystem services that it generates. This model should serve as a road map to guide the valuation. EPA should formalize a process for constructing the initial conceptual model, recognizing that the process must be iterative and respond to new information and multiple points of view. The conceptual model should reflect the ultimate goal of valuing the effect of EPA's decision on ecosystems and ecosystem services. The model and its documentation should also clearly describe the reasons for decisions about the spatial and temporal scales of the chosen ecological system, the process used to identify stressors associated with the proposed EPA action, and the methods to be used in estimating the ecological effects. In constructing the conceptual model, the Agency should involve staff throughout EPA, as well as relevant outside experts from the biophysical and social sciences, and seek information about relevant public concerns and needs.

- EPA should identify and develop measures of ecosystem services that are relevant to and directly useful for valuation. This will require increased interaction between natural and social scientists within the Agency. In identifying and evaluating services for any specific valuation effort, EPA should describe them in terms that are meaningful and understandable to the public.
- EPA should seek to use ecological production functions wherever practical to estimate how ecological responses (resulting from different policies or management decisions) will affect the provision of ecosystem services. (EPA Science Advisory Board 2008b).
- All ecological valuations conducted by EPA should be supported by ecological models and data sufficient to understand and estimate the likely ecological responses to major alternatives being considered by decision makers. There are many ecological models. Building on recent efforts within the Agency and elsewhere, EPA should develop criteria or guidelines for model selection that reflect the specific modeling needs of ecological valuation and apply these criteria in a consistent and transparent way.
- Because of the complexity of developing and using complete ecological production functions, the committee advises EPA to continue and accelerate research to develop key indicators for use in ecological valuation. Such indicators should meet ecological and social science criteria for effectively simplifying and synthesizing underlying complexity and be associated with an effective monitoring and reporting program. The Agency can also advance ecological valuations by supporting the use of methods such as meta-analysis that are designed to provide general information about ecological relationships that can be applied in ecological valuation.
- EPA should work with other agencies and scientific organizations, such as the National Science Foundation, to encourage the sharing of ecological data and the development of more consistent ecological measures that are useful for valuation purposes. EPA should similarly encourage strong regional initiatives to develop information needed for valuations. EPA should also promote efforts to develop data that can be used to parameterize ecological models for site-specific analysis and case studies, or that can be transferred or scaled to other contexts.
- EPA should carefully plan and actively pursue research to generate ecological production functions for valuation including research on ecological services and support for modeling and methods development by the Office of Research and Development and the Science to Achieve Results (STAR) program. It is a high

priority to develop ecological models that can be used in valuation efforts and to evaluate the validity of those models.

- Finally, the committee advises the Agency to foster interaction between natural scientists and social scientists in identifying relevant ecosystem services and developing and implementing processes for measuring and valuing them. As part of this effort, EPA should more closely link its research programs on evaluating ecosystem services and valuing ecosystem services

4. METHODS FOR ASSESSING VALUE

In advocating an expanded and integrated approach to valuing the protection of ecological systems and services, the committee urges the Agency to consider, pilot, and evaluate a broader set of valuation methods. This chapter provides an overview of the methods that the committee discussed for possible use in implementing its approach, including methods and approaches for transfer of valuation information.

As noted in chapter 2, the methods considered by the committee vary in the roles that they might play in different decision contexts. For example, as noted previously, benefit assessments for national rule makings must be conducted under the guidance of Office of Management and Budget Circular A-4, which implies that, in that context monetized valuations must be based on appropriate economic methods. Other valuation methods can still provide useful information in this context, but the role of these methods is limited by the need to follow the guidance in the circular (see sections 2.3.3 and 6.1). In other, less-prescribed decision contexts, non-economic valuation methods can play a larger role in analysis (see sections 6.2 and 6.3). Thus, as the Agency considers alternative methods that might be used, it must consider the context of the information needs defined by the particular policy context in which the valuation exercise will be done.

4.1. Criteria for Choosing Valuation Methods

The methods discussed by the committee differ in a number of important respects. These include: the underlying assumptions and concepts of value they seek to measure or characterize; the empirical and analytical techniques used to apply them; their data needs (inputs) and the metrics they generate (outputs); their involvement of the public; the degree to which the method has been developed or utilized; their potential for future use at EPA; and the issues involved in implementing the methods.

Any method used by the Agency must meet relevant scientific standards. Before relying on any given method in a particular valuation process, EPA must determine if there is a scientific basis for the method's use in that context. Methods that are in their early stages of development and application to valuation must be evaluated both for their scientific merit and for their appropriateness in the given context of interest. Methods that are well-developed, have been extensively used for valuation, and have been validated in other contexts should still be evaluated for their suitability in valuing ecosystems and services, because a given context may pose challenges that might not exist in other situations. In addition, when considering what methods to use in specific contexts, EPA should consider the specific policy objectives and whether a given method provides information relevant to that objective. For example, methods that focus on biophysical measures of value may be relevant for objectives defined solely in terms of biophysical criteria but less suitable when policy objectives are defined more broadly in terms of human well-being. In this latter case, methods that allow for consideration of not only ecosystem services but also the many other things that contribute to human well-being (e.g., human health) will be more suitable.

The committee has not developed a full set of criteria for evaluating methods, nor has it applied criteria comprehensively to the methods discussed here. The committee advises EPA to develop criteria and evaluate methods by those criteria prior to use in valuation. This will assist the Agency not only in determining when methods are suitable but also in determining where to invest scarce resources.

Some suggestions for criteria that EPA should consider for inclusion are described briefly in section 4.1.1. In developing criteria for evaluating valuation methods, a distinction should be made between criteria for evaluating the suitability of a particular method in a given context (i.e., evaluating the scientific merit and suitability of the method) and criteria for evaluating the manner in which the method is actually applied (i.e., evaluating the implementation of the method). For example, the question of whether survey methods in general can appropriately be used to estimate or elicit value(s) in a particular context is a different question, requiring different criteria, than the question of whether a specific survey was properly designed and executed to estimate or elicit the intended value(s). If not properly implemented, any method can yield results that are not useful for the intended purpose. For any individual method, EPA can develop criteria to ensure that the method is carefully implemented. Criteria of this type exist for many of the methods described here, and committee members have described criteria for many valuation methods (see valuation method descriptions on the SAB Web site at http://yosemite.epa.gov/sab/sabproduct.nsf/WebBOARD/C-VPESS_Web_Methods_Draft?OpenDocument). The committee recommends that EPA develop a higher-order list of criteria designed to evaluate the suitability of specific methods for a specific valuation context, assuming that any method chosen would be implemented according to best practices.

4.1.1. Suggested Criteria

While not prescribing the specific criteria that EPA should use to evaluate methods before using them in a specific context, the committee offers some suggested criteria. These draw on the literature cited below, as well as the committee's own deliberations.

A primary consideration in evaluating a method should be the extent to which the method seeks to elicit or measure a concept of value that has a consistent and transparent theoretical foundation appropriate for the intended use. Different valuation methods measure different concepts of value. For a method to be appropriate in a valuation context, it must seek to measure a concept of value that is well-defined, theoretically consistent, and relevant for the particular valuation context. For example, a method derived from a biodiversity-based theory of value would not be relevant in a context where biodiversity is not important. Similarly, legal requirements may prescribe a theory of value that must be used in a particular valuation context (most notably, national rule making). Thus, the Agency should consider the theory of value underlying a particular method and its relevance when evaluating the appropriateness of using that method in a specific context.

Assuming a method seeks to elicit or measure a well-defined and relevant concept of value, another over-arching criterion for evaluation is validity – i.e., how well the method measures the underlying construct that it is intended to measure (Gregory et al., 1993; Freeman, 2003; Fischhoff, 1997). Although the underlying construct of value is not directly observable, it can be estimated through the use of valid methods. EPA should use criteria to assess the extent to which a given method is likely to yield a measure, or at least an unbiased estimate, of the underlying construct of value. Examples of criteria that provide information about the validity of a method include:

- Does the method capture the critical features of the relevant population's values, including how deeply they are held? Does it yield value estimates that reflect the intensity of people's preferences or the magnitude of the contribution to a given goal?
- Does the method impose demands on respondents that limit their ability to articulate values in a meaningful way? For example, does the method impose unrealistic cognitive demands on individuals expressing values? Does it allow those individuals to engage in the process that they would normally undertake to identify or formulate and then articulate their values?
- Does the method yield value estimates for individuals that those individuals would, if asked, consent to have used in the proposed way? Fischhoff (2000) suggests that this form of implied informed consent can help to ensure the quality of valuation data generated by a given method and avoid inappropriate use of the resulting value estimates, by ensuring that individuals would "stand behind researchers' interpretation of their responses" (p. 1439). This does not necessarily require that researchers using the method actually seek such consent. Rather, it provides a hypothetical benchmark that can be used in assessing whether a method is capturing what is intended.
- Does the method ensure that measured or elicited values reflect relevant scientific information? A basic premise of the valuation approach proposed by the committee is that a method should elicit or measure values that individuals would hold when well-informed about the relevant science. This does not require that all individuals expressing values know as much as scientific experts in the field, but rather that they understand as much of the science as necessary to make informed judgments about the service(s) they are being asked to value. For example, they should be aware of the magnitude of the changes in ecosystem services or characteristics that would

result from the ecological changes being valued, as well as the implications of those changes for themselves and for others.

- Does the method yield value estimates that are responsive to changes in variables that the relevant theory suggests should be predictors of value, and invariant to changes in variables that are irrelevant to the determination of value? For example, under an economic theory of value, an increase in the quantity of the good or service being valued should result in an increase in the magnitude of expressed values. This form of validity has been termed construct validity (Fischhoff, 1997; Mitchell and Carson, 1989).
- Are the expressions of value resulting from the method stable (i.e., reliable) in the sense that they do not change upon further reflection (Fischhoff, 1997) and are not unduly influenced by irrelevant characteristics of the researcher, process facilitator, or group?
- To what extent does the information elicited from participants in the application of the method (e.g., survey respondents or focus group participants) provide information that can be used to reliably infer something about the values of the targeted group within the relevant population?

These criteria would generally be viewed as necessary for validity, although they are not necessarily sufficient to guarantee it. Methods can also be evaluated on the extent to which the resulting value estimates can be transparently communicated in a useful format to those who will use the value information. Decision makers and the public should be able to understand how the value measures relate to and inform the decision that needs to be made.

4.2. An Expanded Set of Methods

This section provides an overview of, and introduction to, the wide array of methods considered by the committee for possible use in implementing the valuation process proposed in chapter 2. Table 3 provides a listing of these methods, along with an overview of the form of output from each method and the concept(s) of value that it seeks to measure or elicit. Note that, although the concepts of value discussed in chapter 2 are conceptually distinct from valuation methods, methods generally seek to measure specific value concepts, as indicated in Table 3. Hence, specific methods are generally associated with specific concepts of value. However, methods can be complementary, and a given method can sometimes be used to provide information that could be useful in assessing other concepts of value. For example, as discussed in chapter 2, some of the non-economic methods in Table 3 can be useful in supporting and improving economic valuation.

The following discussion of methods is illustrative and introductory rather than comprehensive. The goal is to provide the reader with sufficient information about the methods to allow a preliminary assessment of the role that various methods can play in implementing the proposed valuation process and to direct the interested reader to the relevant scientific literature for further information.

Table 3. Methods Considered by the Committee for Possible Use in Valuation

Method	Form of output/units	Related concepts(s) of value from table 1
Measures of attitudes, preferences, and intentions		
Survey questions eliciting information about attitudes, preferences, and intentions	Attitude scales, preference or importance rankings, behavioral intentions toward depicted environments or conditions	Attitudes and judgments; community-based values
Individual narratives and focus groups	Qualitative summaries and assessments from transcripts	Attitudes and judgments; community-based values
Behavioral observation	Inferences from observations of behavior by individuals interacting with actual or computer-simulated environments	Attitudes and judgments; community-based values
Economic methods		
Market-based methods	Monetary measure of willingness-to-pay (WTP) for ecosystem services that contribute to the provision of marketed goods and services	Economic value
Travel cost	Monetary measure of WTP for ecosystem services that affect decisions to visit different locations	Economic value
Hedonic pricing	Monetary measure of marginal WTP or willingness-to- accept (WTA) as revealed by price for houses or wages paid for jobs with different environmental characteristics	Economic value
Averting behavior	Monetary or other measure of WTP as revealed by responses to opportunities to avoid or reduce damages, for example, through expenditures on protective goods or substitutes	Economic value
Survey questions eliciting stated preferences	Monetary or other measures of WTP or WTA as expressed in survey questions about hypothetical tradeoffs	Economic value
Civic valuation		
Referenda and initiatives	Rankings of alternative options, or monetary or other measure of tradeoffs a community is willing to make, as reflected in community choices	Community-based values; indicator of economic value under some conditions
Citizen valuation juries	Rankings of alternative options, or monetary or other measures of required payment or compensation, based on jury-determined assessments of public values	Community-based values; constructed values
Decision science approaches		
Decision science approaches	Attribute weights that reflect tradeoffs individuals are willing to make across attributes, including ecological attributes, for use in assigning scores to alternative policy options	Constructed values

Table 3. (Continued)

Method	Form of output/units	Related concepts(s) of value from table 1
Ecosystem benefit indicators		
Ecosystem benefit indicators	Quantitative spatially-differentiated metrics or maps related to supply of or demand for ecosystem services	Indicators of economic value and/or community-based values
Biophysical ranking methods		
Conservation value method	Spatially-differentiated index of conservation values across a landscape	Bio-ecological value
Embodied energy analysis	Cost of the total (direct plus indirect) energy required to produce an ecological or economic good or service	Energy-based value
Ecological footprint	Area of an ecosystem (land and/or water) required to support a consumption pattern or population	Bio-ecological value
Cost as a proxy for value		
Replacement cost	Monetary estimate of the cost of replacing an ecosystem service using the next best available alternative	Lower bound on economic value only under limited conditions
Habitat equivalency analysis	Units of habitat (e.g., equivalent acres of habitat) or other compensating changes needed to replace ecosystem services lost through a natural resource injury	Biophysical value; not economic value except under some very limited conditions

The SAB Web site provides supplemental detailed discussions of these methods, including their perceived strengths and weaknesses, that were provided by individual committee members (http://yosemite.epa. gov/sab/sabproduct.nsf/WebBOARD/C-VPESS_Web_Methods_Draft?OpenDocument).[34] In addition, federal agencies have extensively used surveys to elicit value- related information.[35] The SAB Web site provides a separate, detailed discussion of the use of survey methods for ecological valuation. This information (http://yosemite.epa.gov/Sab/Sabproduct.nsf/WebFiles/SurveyMethods/$File/Survey_methods.pdf) is relevant to those economic and other methods discussed below that rely on surveys.

4.2.1. Measures Of Attitudes, Preferences, And Intentions

Social-psychological approaches to assessing the value of ecosystems and ecosystem services employ a number of methods to identify, characterize, and measure the values people hold, express, and advocate with respect to changes in ecological states or their personal and social consequences. These methods elicit value-relevant perceptions and judgments, typically expressed as choices, rankings, or ratings among presented sets of alternative ecosystems protection policies and may include comparisons with potentially competing social and economic goals. Individuals making these judgments may respond on their own behalf or on behalf of others (e.g., society at large or specified subgroups). The basis for judgments can be changes in individual well-being or in civic, ethical, or moral obligations.

Social-psychological value-assessment approaches have relied most strongly on survey methods. For a general discussion of the use of surveys in valuation, see http://yosemite.epa.gov/Sab/Sabproduct.nsf/WebFiles/SurveyMethods/$File/Survey_methods.pdf. **Survey questions eliciting information about attitudes, preferences, and intentions** are most often presented in a verbal format, either in face-to-face or telephone interviews or in printed questionnaires. Assessments of values for ecosystems and ecosystem services can be well-conveyed in perceptual surveys (e.g., assessments based on photographs, computer visualizations, or multimedia representations of targeted ecosystem attributes) and conjoint surveys (e.g., requiring choices among alternatives that systematically combine multiple and potentially competing attributes). Quantitative analyses of survey responses are usually interpreted as ordinal rankings or rough interval-scale measures of differences in assessed values for the alternatives offered. Survey questions about social and psychological constructs may be especially useful when the values at issue are difficult to express or conceive in monetary terms, or where monetary expressions are likely to be viewed as ethically inappropriate.

Further Reading

Adamowicz, W., Boxall, P., Wilhams, M. & Louviere, J. (1998). Stated preference approaches for measuring passive use values: Choice experiments and contingent valuation. *American Journal of Agricultural Economics*, *80*, 64-67.

Dillman, D. A. (1991). The design and administration of mail surveys. *Annual Review of Sociology*, *17*, 225-249.

Dunlap, R. E., Van Liere, K. D., Mertig, A. G. & Jones, R. E. (2000). Measuring endorsement of the New Ecological Paradigm: A revised NEP scale. *Journal of Social Issues*, *56*, 425-442.

Krosnick, J. A. (1999). Survey research. *Annual Review of Psychology*, *50*, 537-67.

Mace, B. L., Bell, P. A. & Loomis, R. J. (1999). Aesthetic, affective, and cognitive effects of noise on natural landscape assessment. *Society & Natural Resources*, *12*, 225-243.

Malm, W., Kelly, K., Molenar, J. & Daniel, T. C. (1981). Human perception of visual air quality: Uniform haze. *Atmospheric Environment*, *15*, 1874-1890.

Ribe, R. G., Armstrong, E. T. & Gobster. P. H. (2002). Scenic vistas and the changing policy landscape: Visualizing and testing the role of visual resources in ecosystem management. *Landscape Journal*, *21*, 42-66.

Schaeffer, N. C. & Presser, S. (2003). The science of asking questions. *Annual Review of Sociology*, *29*, 65-88.

Tourangeau, R. (2004). Survey research and societal change. *Annual Review of Psychology*, *55*, 775-801.

Wilson, T. D., Lisle, D. J., Kraft, D. & Wetzel, C. G. (1989). Preferences as expectation-driven inferences: Effects of affective expectations on affective experience. *Journal of Personality and Social Psychology*, *56*, 45 19-530.

Individual narratives and focus group methods have also been used in values assessments, but these methods are generally more appropriately used as formative tools for the design and testing of formal quantitative surveys. While surveys are typically based on quantitative analyses of responses from large representative samples, individual **narrative methods** – including mental-model analyses, ethnographic analyses, and other relatively unstructured individual interviews – generally employ small samples of informants and analyze responses qualitatively. For example, mental models studies seek to assess how informed people are about the consequences of specific decisions and their decision-relevant beliefs. Mental models studies of risk communication explicitly compare causal beliefs with formal decision models.[36] How people understand relevant causal processes – that is, in this case, their mental models of ecosystems and the services they provide – can be critical to their judgment of the outcomes and effects of environmental programs and can influence their preferences among policy alternatives. Similarly, **focus groups** can be used to elicit information about values and preferences from small groups of relevant members of the public engaging in group discussion led by a facilitator. Rigorous qualitative analyses of transcripts from individual narratives (including mental models studies) or focus groups can expose subtle differences in individual beliefs and perspectives and the inferential bases of participants' expressed values. However, the use of qualitative measures and the uncertainty of any generalizations of results from small respondent samples limit the utility of these methods for formal policy and decision making.

Given the small number of participants, the goal of individual narratives and focus groups is rarely to assess the public's values per se. Rather, these methods seek to identify the types and range of value perspectives, positions, and concerns of individual participants, and to use this information to identify the ecosystem effects that might be particularly important to the public. The open-ended nature of these methods can reveal perspectives and concerns that more structured methods might miss. Thus, these methods can provide useful input early in a valuation process. For example, they are often used in the early stages of designing a formal survey to elicit quantitative value information from a broader representative sample (a "probability sample") of the relevant population.

Further Reading

Bostrom, A., Fischhoff, B. & Morgan, M. G. (2002). Characterizing mental models of hazardous processes: A methodology and an application to radon. *Journal of Social Issues*, *48*, 85-100.

Brandenburg, A. M. & Carroll, M. S. (1995). Your place or mine? The effect of place creation on environmental values and landscape meanings. *Society & Natural Resources*, *8*, 38 1-398.

Gentner, D. & Whitley, E. W. (1997). Mental models of population growth: A preliminary investigation. In *Environment, ethics, and behavior: The psychology of environmental valuation and degradation, ed.* M., Bazerman, D. M., Messick, A. E. Tenbrunsel, & K. WadeBenzoni, 209-233. San Francisco, CA: New Lexington Press.

Kempton, W. (1991). Lay perspectives on global climate change. *Global Environmental Change*, *1*, 183-208.

Merton, R. K., Fiske, M. & Kendall, P. L. (1990). The focused interview: *A manual of problems and procedures*. 2nd ed. London: Collier MacMillan.

Morgan, M. G., Fischhoff, B., Bostrom, A. & Atman. C. J. (2002). *Risk communication: A mental models approach*. Cambridge: Cambridge University Press.

T. Satterfield, & S. (2004). *What's nature worth: Narrative expressions of environmental values*. Salt Lake City: University of Utah Press.

Zaksek, M. & Arvai, J. L. (2004). Toward improved communication about wildland fire: Mental models research to identify information needs for natural resource management. *Risk Analysis*, *24*, 1503-1514.

Recently, researchers have explored the use of **behavioral observation methods** for obtaining information about people's values. These methods elicit values information through observations of behavioral responses by individuals interacting with either actual or computer-simulated environments. Observing how the activities of people change as environmental conditions change can reveal information about the importance of these changes to those people. Researchers can observe changes in actual behavior (e.g., visitation rates) or virtual behavior (e.g., responses in interactive computer simulation games). Behavioral observation methods are consistent with other revealed preference methods (see the following section), but they are still relatively new and untested, particularly in the context of valuing ecosystem services. Nonetheless, they show promise for use in this context.

Further Reading

Bishop, I. D. & Rohrmann, B. (2003). Subjective responses to simulated and real environments: A comparison. *Landscape and Urban Planning*, *65*, 261-267.

Gimblett, H. R., Daniel, T. C., Cherry, S. & Meitner, M. J. (2001). The simulation and visualization of complex human-environment interactions. *Landscape and Urban Planning*, *54*, 63-79.

Wang, B. & Manning, R. E. (2001). Computer simulation modeling for recreation management: A study on carriage road use in Acadia National Park, Maine, USA. *Environmental Management*, *23*, 193-203.

Zacharias, J. (2006). Exploratory spatial behaviour in real and virtual environments. *Landscape and Urban Planning*, *78*, 1-13.

4.2.2. Economic Methods

Economic valuation methods seek to measure the tradeoffs individuals are willing to make for ecological improvements or to avoid ecological degradation, given the constraints they face. An ecological change improving a resource that an individual values will increase that person's utility. The marginal value or economic benefit of that change is defined to be the amount of another good that the individual is willing to give up to enjoy that change (willingness-to-pay) or the amount of compensation that a person would accept in lieu of receiving that change (willingness to accept). Although these tradeoffs are typically expressed in monetary terms, economic methods that express tradeoffs in non-monetary terms (such as conjoint analysis or other choice-based methods) are increasingly being used.

Economic methods can estimate values not only for goods and services for which there are markets but also for non-market goods and services. Economic methods can also value both use and non-use (e.g., existence) values. Thus, economic valuation captures values that extend well beyond commercial or market values. However, economic valuation does not capture non-anthropocentric values (e.g., biocentric values) and values inconsistent with the principle of trrade-offs (such as values based on the concept of intrinsic rights).

There are multiple economic valuation methods that can be used to estimate economic values. These include methods based on observed behavior (market-based and revealed-preference methods) and methods based on information elicited from responses to survey questions about hypothetical tradeoffs (e.g., stated-preference methods). Some of these methods are more applicable to some contexts than to others.

Market-based methods seek to use information about market prices (or market demand) to infer values related to changes in marketed goods and services. For example, when ecological changes lead to a small change in timber or commercial fishing harvests, the market price of timber or fish can be used as a measure of willingness to pay for that marginal change. If the change is large, the current market price alone is not sufficient to determine value. Rather, the demand for timber or fish at various prices must be used to determine willingness to pay for the change. In general, market-based methods can value only those services supplied in well-functioning markets. These methods have been used to assess the welfare effects of a wide variety of public policies.

Further Reading

Barbier, E. B. & Strand, I. (1998). Valuing mangrove-fishery linkages. *Environmental and Resource Economics*, *12*, 151-166.

Boardman, A. E., Greenberg, D. H., Vining, A. R. & Weimer, D. L. (2006). *Cost-benefit analysis: Concepts and practice.* 3rd ed. Upper Saddle River, NJ: Prentice-Hall.

Freeman, A. M. III. (2003). *The measurement of environmental and resource values.* 2nd ed. Washington, DC: Resources for the Future.

Hufbauer, G. & Elliott, K. A. (1994). *Measuring the costs of protection in the US.* Washington, DC: Institute for International Economics.

McConnell, K. E. & Bockstael, N. E. (2005). Valuing the environment as a factor of production. In *Handbook of environmental economics*, ed. K. G. Maler, & J. R. Vincent. Amsterdam: North-Holland.

Winston, C. (1993). Economic deregulation: Days of reckoning for microeconomists. *Journal of Economic Literature.*

Revealed-preference methods exploit the relationship between some forms of individual behavior (e.g., visiting a lake or buying a house) and associated environmental attributes (e.g., of the lake or the house) to estimate value. For example, **travel cost methods** (including applications using random utility models) use information about how much people implicitly or explicitly pay to visit locations with specific environmental attributes including, specific levels of ecosystem services, to infer how much they value changes in those attributes. **Hedonic pricing** uses information about how much people pay for houses or other directly-purchased items with specific environmental attributes (e.g., visibility, proximity to amenities or disamenities) to infer how much they value changes in those attributes. It also may use information about the wages people would be willing to accept for jobs with differing mortality or morbidity risk levels to infer how much they value changes in those risks. In contrast, **averting-behavior methods** use observations on how much people spend to avoid adverse effects, including environmental effects to infer how much they value or are willing to pay for the improvements those expenditures yield.

Further Reading

General

[1] Bockstael, N. B. & McConnell, K. E. (2007). Environmental and resource valuation with revealed preferences: A theoretical guide to empirical models (The economics of non-market goods and resources) New York: Springer.

Travel costs

[1] Phaneuf, D. J. & Smith, V. K. (2005). Recreation demand models. In *Handbook of environmental economics*, vol. 2, ed. K. Mäler, & J. Vincent. Amsterdam: North-Holland.

[2] Randall, A. (1994). A difficulty with the travel cost method. *Land Economics*, *70*, 88-96.

[3] Smith, V. K. & Kaoru, Y. (1990). Signals or noise? Explaining the variation in recreational benefit estimates. *American Journal of Agricultural Economics*, *72*, 419-433.

[4] Walsh, R. G., Johnson, D. M. & McKean, J. R. (1992). Benefit transfer of outdoor recreation demand studies, 1968-1988. *Water Resources Research*, *28*, 707-713.

Hedonic pricing

[1] Cropper, M. L., Deck, L. & McConnell, K. E. (1988). On the choice of functional forms for hedonic price functions. *Review of Economics and Statistics*, *70*, 668-75.

[2] Mahan, B. L., Polasky, S. & Adams, R. M. (2000). Valuing urban wetlands: A property price approach. *Land Economics*, *76*, 100-113.

[3] Palmquist, R. B. (2005). Hedonic models. In *Handbook of environmental economics*, vol. 2, ed. K. Mäler, & J. Vincent. Amsterdam: North- Holland.

[4] Smith, V. K., Poulos, C. & Kim, H. (2002). Treating open space as an urban amenity. *Resource and Energy Economics*, *24*, 107-129.

Averting behavior

[1] Dickie, M. (2003). Defensive behavior and damage cost methods. In *A primer on non-market valuation*, ed. P. A., Champ, K. J. Boyle, & T.C. Brown, Dordrecht: Kluwer Academic Press.

[2] Smith, V. K. (1991). Household production functions and environmental benefit estimation. In *Measuring the demand for environmental quality*, ed. J. B. Braden, & C. D. Kolstad, Amsterdam: North-Holland.

In contrast to revealed-preference methods, **stated-preference methods** infer values or economic benefits from responses to survey questions about hypothetical tradeoffs. As with social-psychological methods, stated-preference methods often use focus groups to improve survey designs. In some cases, survey questions directly elicit information about willingness to pay or accept, while under some survey designs (e.g., conjoint or contingent behavior designs) monetary measures of benefits are not expressed directly. Rather, quantitative analysis of the tradeoffs implied by survey responses is needed to derive economic benefit measures. Although the use of stated-preference methods for environmental valuation has been controversial, there is considerable evidence that the hypothetical responses in these surveys provide useful evidence regarding values (see related detailed discussion on the use of survey methods for ecological valuation on the SAB Web site at http://yosemite.epa.gov/Sab/Sabproduct.nsf/WebFiles/SurveyMethods/$File/Survey_methods.pdf).

Further Reading

Arrow, K., et al. (1993). Report of the NOAA Panel on Contingent Valuation, *Federal Register*, *58*, *10* (Jan. 15, 1993), *4601-4614.*

Banzhaf, S., Burtraw, D., Evans, D. & Krupnick, A. (2004). *Valuation of natural resource improvements in the Adirondacks*. Washington, DC: Resources for the Future.

I. J. Bateman, & K.G. Willis, eds. (1999). Valuing environmental preferences: Theory and practice *of the contingent valuation method in the US, EU, and developing countries.* Oxford: Oxford University Press.

Carson, R. T. & Hanemann, W. M. (2005), *Contingent Valuation in Handbook of Environmental Economics*, *Vol. II*, ed. K. Göran-Mäler, & J. R. Vincent Amsterdam: North Holland, 821-936.

P., Champ, K. J. Boyle, & T. C. Brown, eds. (2003). *A primer on non-market valuation.* Dordrecht: Klumer Academic.

Freeman, A. M. III. (2003). *The measurement of environmental and resource values,* 2nd ed. Washington, DC: Resources for the Future.

Haab, T. C. & McConnell, K. E. (2002). *Valuing environmental and natural resources.* Cheltenham, UK: Edward Elgar.

B. J. Kanninen, ed. (2007). *Valuing environmental amenities using stated choice studies.* Dordrecht: Springer.

R. J., Kopp, W. W. Pommerehne, & N. Schwarz, eds. (1997). Determining the value of non-marketed goods: *Economic, psychological, and policy relevant aspects of contingent valuation methods*. Boston: Kluwer Academic Publishers.

Murphy, J. J., Allen, P. G., Stevens, T. H. & Weatherhead. D. (2005). A meta-analysis of hypothetical bias in stated preference valuation. *Environmental and Resource Economics*, *30*, 313-325.

Smith, V. K. (1997). Pricing what is priceless: A status report on non-market valuation of environmental resources. In *The International Yearbook of Environmental and Resource Economics,* eds. H. Folmer, & T. Tietenberg, Cheltenham, U.K.: Edward Elgar, 156-204.

4.2.3. Civic Valuation

Civic valuation seeks to measure the values that people place on changes in ecosystems or ecosystem services when explicitly considering or acting in their role as citizens. These valuation methods often seek to value changes that would benefit or harm the community at large. They purposefully seek to assess the full value that groups attach to any increase in community well-being attributable to changes in the relevant ecosystems and services.

Civic valuation, like economic valuation, can elicit information about values either through revealed behavior or through stated valuations. One source of information based on revealed behavior is votes on public referenda and initiatives involving the provision of environmental goods and services (e.g., purchases of open space). Another source is community decisions to accept compensation for permitting environmental damage (e.g., by hosting noxious facilities). Where revealed values are difficult or impossible to obtain, citizen valuation juries or other representative groups can be charged with determining the value they would place on changes in particular ecological systems or services when acting on behalf of, or as a representative of, the citizens of the relevant community.

Referenda or initiatives can provide information about how members of the voting population value a particular governmental action involving the environment. Analysis of referenda or initiatives can reveal whether the majority of the voting population feels that a given environmental improvement is worth what it will cost the relevant government body, given a particular means of financing the associated expenditure (and hence, an anticipated cost to the individual who is voting). In casting their votes, individuals may consider not only what they personally would gain or lose but also what the community as a whole stands to gain or lose if the proposal is adopted. Similarly, analyses of public votes about whether to accept an environmental degradation (e.g., through hosting a noxious facility) seek to determine if the majority of the voting population in that community feels that the environmental services that would be lost are worth less than the contributions to well-being the community would realize (e.g., in the form of tax revenues, jobs, or monetary compensation).

These approaches provide information about the policy preferences of the median voter and, under certain conditions, provide information about the mean valuations of those who participate in the voting process. To the extent that voters consider their own budget constraints when voting, these valuations reflect economic values, i.e., willingness to pay or willingness to accept. As with all economic values, the revealed economic value reflects both personal benefits and costs, as well as any altruistic motivation (public regardedness) individual voters have when casting their votes.

Further Reading

Banzhaf, S., Oates, W., Sanchirico, J. N., Simpson, D. & Walsh, R. (2006). Voting for conservation: What is the American electorate revealing? *Resources 16*. Washington, DC: Resources for the Future. http://ww.rff.org/rff/ news/features/loader.cfm?url=/comm. onspot/ security/getfile.cfm&pageid=22017.

Butler, D. & Ranney, A. eds. (1978). *Referendums*. Washington DC: American Enterprise Institute.

Cronin, T. E. (1989). *Direct democracy: The politics of referendum, initiative and recall.* Cambridge: Harvard University Press.

Deacon, R. & Shapiro, P. (1975). Private preference for collective goods revealed through voting on referenda. *American Economic Review*, *65*, 793.

Kahn, M. E. & Matsusaka, J. G. (1997). Demand for environmental goods: Evidence from voting patterns on California initiatives. *Journal of Law and Economics*, *40*, 137-173.

List, J. & Shogren, J. (2002). Calibration of willingness-to-accept. *Journal of Environmental Economics and Management*, *43*, 219-233.

Lupia, A. (1992). Busy voters, agenda control, and the power of information. *American Political Science Review*, *86*, 390-399.

Magleby, D. (1984). *Direct legislation: Voting on ballot propositions in the United States.* Baltimore, MD: Johns Hopkins University Press.

Murphy, J. J., Allen, P. G., et al. (2003). *A meta-analysis of hypothetical bias in stated preference valuation*. Amherst: *Department of Resource Economics*, University of Massachusetts.

Sagoff, M. (2004). *Price, principle and the environment.* Cambridge: Cambridge University Press.

Schläpfer, F., Roschewitz, A. & Hanley, N. (2004). Validation of stated preferences for public goods: A comparison of contingent valuation survey response and voting behavior. *Ecological Economics*, *51*, 1-16.

Shabman, L. & Stephenson, K. (1996). Searching for the correct benefit estimate: Empirical evidence for an alternative perspective. *Land Economics*, *72*, 433-49.

Vossler, C. A., Kerkvliet, J., Polasky, S. & Gainutdinova, O. (2003). Externally validating contingent valuation: An open-space survey and referendum in Corvallis, Oregon. *Journal of Economic Behavior and Organization*, *51*, 261-277.

In contrast to initiatives and referenda, **citizen valuation juries** measure stated rather than revealed value. They also incorporate elements of the deliberative valuation process (see chapter 5). The jury is given extensive information and, after a lengthy discussion, usually asked to agree on a common value or make a group decision. To date, citizen juries have typically been asked to develop a ranking of alternative options for achieving a given goal. Although citizen juries have been used in other contexts, experience using citizen juries for ecological valuation is very limited. Nonetheless, in principle, a jury could be asked to generate a value for how much the public would, or should, be willing to pay for a possible environmental improvement, or, conversely, willing to accept for an environmental degradation. In contrast to estimates of willingness to pay derived from economic valuation methods, the estimates from citizen juries would not reflect the budget constraints of the individual participants and would reflect community-based values rather than economic values. To the extent that a citizen jury engages in group deliberation, resulting value estimates also would reflect constructed values.

Further Reading

Aldred, J. & Jacobs, M. (2000). Citizens and wetlands: Evaluating the Ely citizens' jury. *Ecological Economics*, *34*, 217-232.

Alvarez-Farizo, B. & Hanley, N. (2006). Improving the process of valuing non-market benefits: Combining citizens' juries with choice modelling. *Land Economics*, *82*, 465-478.

Blamey, R. K., et al. (2000). *Citizens' juries and environmental value assessment*. Canberra: *Research School of Social Sciences*, Australian National University.

Brown, T. C., et al. (1995). The values jury to aid natural resource decisions. *Land Economics*, *71*, 250-260.

Gregory, R. & Wellman, K. (2001). Bringing stakeholder values into environmental policy choices: A community-based estuary case study. *Ecological Economics*, *39*, 37-52.

Kenyon, W. & Hanley, N. (2001). *Economic and participatory approaches to environmental evaluation*. Glasgow, U.K., Economics Department, University of Glasgow.

Kenyon, W. & Nevin, C. (2001). The use of economic and participatory approaches to assess forest development: A case study in the Ettrick Valley. *Forest Policy and Economics*, *3*, 69-80.

Macmillan, D. C., et al. (2002). Valuing the non-market benefits of wild goose conservation: A comparison of interview and group-based approaches. *Ecological Economics*, *43*, 49-59.

McDaniels, T. L., et al. (2003). Decision structuring to alleviate embedding in environmental valuation. *Ecological Economics*, *46*, 33-46.

O'Neill, J. & Spash, C. L. (2000). Appendix: Policy research brief: conceptions of value in environmental decision-making. *Environmental Values*, *9*, 521-536.

4.2.4. Decision Science Methods

Decision science valuation methods derive information about people's values through a deliberative process that helps individuals understand and assess tradeoffs among multiple attributes. The ultimate goal is to have an individual or group assign scores to alternatives (e.g., different projects) that can then be used to choose among those alternatives, recognizing that those alternatives will differ along a number of relevant dimensions or attributes. Generally, one alternative will score higher along some dimensions but not others, suggesting that tradeoffs must be made when choosing among alternatives.

Decision science valuation methods are typically embedded in a decision-aiding process. As part of the process, an expert facilitator helps the individual or group decompose the choice problem by identifying and operationalizing objectives as well as relevant attributes. For example, people may feel that the value of a project to protect an estuary depends on attributes such as the estuary's ability to provide nutrient exchange and nursery habitat for anadromous fish, the opportunities it provides for recreation, and the cost of the project. The facilitator leads the individual or group through a process by which they assign weights to each of the attributes. A variety of approaches to assigning weights have been used, including assigning importance points, eliciting ratio weights, determining swing weights, and pricing out attributes. These weights reflect the tradeoffs that the individual or group is willing to make across attributes, and hence reveal information about values.

Once the attribute weights are determined, an aggregating function (or utility function) is used to combine the weights and attribute levels into a score (or measure of multi-attribute utility) for each alternative. Ranking alternative projects or options based on these scores can provide information about which option (and hence which combination of attributes) is viewed as more valuable.

Further Reading

Arvai, J. L. & Gregory, R. (2003). Testing alternative decision approaches for identifying cleanup priorities at contaminated sites. *Environmental Science & Technology*, *37*, 1469-1476.

Clemen, R. T. (1996). *Making hard decisions: An introduction to decision analysis*. Boston: PWS- Kent Publishing.

Gregory, R., Arvai, J. L. & McDaniels, T. (2001). Value-focused thinking for environmental risk consultations. *Research in Social Problems and Public Policy*, *9*, 249-275.

Gregory, R., Lichtenstein, S. & Slovic, P. (1993). Valuing environmental resources: A constructive approach. *Journal of Risk and Uncertainty*, *7*, 177-197.

Hammond, J. S., Keeney, R. L. & Raiffa, H. (1998). Even swaps: A rational method for making trade-offs. *Harvard Business Review*, *76*, 137- 138, 143-148, 150.

Hammond, J., Keeney, R. L. & Raiffa, H. (1999). *Smart choices: A practical guide to making better decisions*. Cambridge: Harvard Business School Press.

Keeney, R. L. (1992). *Value-focused thinking.* A *path to creative decision making.* Cambridge: Harvard University Press.

Keeney, R. L. & Raiffa, H. (1993). *Decisions with multiple objectives: Preferences and value tradeoffs*. Cambridge: Cambridge University Press.

Payne, J. W., Bettman, J. R. & Johnson, E. J. (1992). Behavioral decision research: A constructive processing perspective. *Annual Review of Psychology*, *43*, 87-132.

Slovic, P. (1995). The construction of preference. *American Psychologist*, *50*, 364-371.

4.2.5. Ecosystem Benefit Indicators

Ecosystem benefit indicators offer quantitative metrics that are generally correlated with ecological contributions to human well-being and hence can serve as indicators for these contributions in a specific setting. They use geo-spatial data to provide information related to the demand for, supply (or scarcity) of, and complements to particular ecosystem services across a given landscape, based on social and biophysical features that influence – positively or negatively – the contributions of ecosystem services to human well-being. Examples of these indicators include the percentage of a watershed in a particular land use or of a particular land type, the number of users of a service (e.g., water or recreation) within a given area, and the distance to the nearest vulnerable human community.

Ecological benefit indicators can serve as important quantitative inputs to valuation methods as diverse as citizen juries and economic valuation methods. Ecosystem benefit indicators provide a way to illustrate factors influencing ecological contributions to human welfare in a specific setting. The method can be applied to any ecosystem service where the spatial delivery of services is related to the social landscape in which the service is enjoyed. However, although the resulting indicators can be correlated with other value measures, such as economic values, they do not themselves provide measures of value.

Further Reading

Boyd, J. (2004). What's nature worth? Using indicators to open the black box of ecological valuation. *Resources*.

Boyd, J., King, D. & Wainger, L. (2001). Compensation for lost ecosystem services: The need for benefit-based transfer ratios and restoration criteria. *Stanford Environmental Law Journal, 20.*

Boyd, J. & Wainger, L. (2002). Landscape indicators of ecosystem service benefits. *American Journal of Agricultural Economics, 84.*

Wainger, L., King, D., Salzman, J. & Boyd, J. (2001). Wetland value indicators for scoring mitigation trades. *Stanford Environmental Law Journal, 20.*

4.2.6. Biophysical Ranking Methods

In some contexts, policy makers or analysts are interested in values based on quantification of biophysical indicators. Possible indicators include measures of biodiversity, biomass production, carbon sequestration, or energy and materials use.[37] Quantification of ecological changes in biophysical terms allows these changes to be ranked based on individual or aggregate indicators for use in evaluating policy options based on biophysical criteria previously determined to be relevant to human/social well-being.

Use of a biophysical ranking does not explicitly incorporate human preferences. Rather, it reflects either a non-anthropocentric theory of value (based, for example, on energy flows) or a presumption that the indicators provide a proxy for human value or social preference. This latter presumption is predicated on the belief that the healthy functioning and sustainability of ecosystems is fundamentally important to the well-being of human societies and all living things, and that the contributions to human well-being of any change in ecosystems can be assessed in terms of the calculated effects on ecosystems. Opinion is mixed – among both committee members and the broader scholarly community – on whether it is an asset or a drawback that these ranking methods are not tied directly to human preferences.

The committee discussed two types of biophysical rankings. The first is a ranking method based on conservation value. The **conservation value method** develops a spatially-differentiated index of conservation value across a landscape based on an assessment of rarity, persistence, threat, and other landscape attributes, reflecting the contribution of these attributes to sustained ecosystem diversity and integrity. These values can be used to prioritize land for acquisition, conservation, or other purposes, given relevant biophysical goals. Based on geographic information system (GIS) technology, the method can combine information about a variety of ecosystem characteristics and services across a given landscape and overlay ecological information with other spatial data. Conservation values have been used in various contexts by federal agencies (e.g., the U.S. Forest Service, Fish and Wildlife Service, National Park Service, and Bureau of Land Management), non-governmental organizations (e.g., The Nature Conservancy and NatureServe), and by regional and local planning agencies.

Further Reading

Brown, N., Master, L., Faber-Langendoen, D., Comer, P., Maybury, K., Robles, M., Nichols, J. & Wigley, T. B. (2004). Managing elements of biodiversity in sustainable forestry programs: *Status and utility of NatureServe's information resources to forest managers*. National Council for Air and Stream Improvement Technical Bulletin Number 0885.

Grossman, D. H. & Comer, P. J. (2004). *Setting priorities for biodiversity conservation in Puerto Rico*. NatureServe Technical Report.

Riordan, R. & Barker, K. (2003). *Cultivating biodiversity in Napa*. Geospatial Solutions November.

Stoms, D. M., Comer, P. J., Crist, P. J. & D. H. Grossman, (2005). Choosing surrogates for biodiversity conservation in complex planning environments. *Journal of Conservation Planning*, 1.

The second group of biophysical methods that the committee discussed quantify the flows of energy and materials through complex ecological systems, economic systems, or both. Ecologists have used these methods to identify the resources or resource-equivalents needed to produce a product or service, using a systems or life-cycle ("cradle to grave") approach. For example, **embodied energy analysis** measures the total energy, direct and indirect, required to produce a good or service. Similarly, **ecological footprint analysis** measures the area of an ecosystem (e.g., the amount of land and/or water) required to support a certain level and type of consumption by an individual or population.[38]

In addition to using these methods to measure required inputs, some ecologists have advocated using the cost estimates for embodied energy as a measure of value, based on an energy (or other biophysical input) theory of value. Although conceptually distinct, they have found that these estimates can be of similar magnitude to value estimates based on economic valuation methods.

Further Reading

Embodied energy

Ayres, R. U. (1978). Application of physical principles to economics. In *Resources, environment, and economics: Applications of the materials/energy balance principle*, ed. R.U. Ayres, 37-7 1. New York: Wiley.

Boulding, K. E. (1966). The economics of the coming spaceship Earth. In *Environmental quality in a growing economy*, ed. H. Jarrett, 3-14. Baltimore, MD: Resources for the Future/ Johns Hopkins University Press.

Cleveland, C. J. (1987). Biophysical economics: Historical perspective and current research trends. *Ecological Modelling*, *38*, 47-74.

Cleveland, C. J., Costanza, R., Hall, C. A. S. & Kaufmann, R. (1984). Energy and the U.S. economy: A biophysical perspective. *Science*, *225*, 890-897.

Costanza, R. (1980). Embodied energy and economic valuation. *Science*, *210*, 1219-1224.

Costanza, R. (2004). Value theory and energy. In *Encyclopedia of energy*, *vol 6*, ed. C. Cleveland. 337-346. Amsterdam: Elsevier.

Costanza, R., Farber, S. C. & Maxwell, J. 1989. The valuation and management of wetland ecosystems. *Ecological Economics*, *1*, 335-361.

Hall, C. A. S., Cleveland, C. J. & Kaufmann, K. (1992). *Energy and resource quality: The ecology of the economic process*. New York: Wiley.

Hannon, B., Costanza, R. & Herendeen, R. A. (1986). Measures of energy cost and value in ecosystems. *The Journal of Environmental Economics and Management*, *13*, 391-401.

Kaufmann, R. K. (1992). A biophysical analysis of the energy/real GDP ratio: Implications for substitution and technical change. *Ecological Economics*, *6*, 35-56.

Ruth, M. (1995). Information, order and knowledge in economic and ecological systems: Implications from material and energy use. *Ecological Economics*, *13*, 99-114.

Ecological footprint analysis Costanza, R., ed. (2000). Forum: The ecological footprint. *Ecological Economics*, *32*, 341-394.

Global Footprint Network. (2008). [cited 2008]. Available from http://www.footprintnetwork.org/en/index.php/GFN/

Herendeen, R. (2000). Ecological footprint is a vivid indicator of indirect effects. *Ecological Economics*, *32*, 357-358

Rees, W. E. (2000). Eco-footprint analysis: Merits and brickbats. *Ecological Economics*, *32*, 371- 374

Simmons, C., Lewis, K. & Moore, J. (2000). Two feet – two approaches: A component-based model of ecological footprinting. *Ecological Economics*, *32*, 375-380.

Wackernagel, M., Onisto, L., Bello, P., Linares, A. C., Falfan, I. S. L., Garcia, J. M., Guerrero, A. I. S. & Guerrero, M. G. S. (1999). National natural capital accounting with the ecological footprint concept. *Ecological Economics*, *29*, 375-390.

Wackernagel, M., Schultz, N. B., Deumling, D., Linares, A. C., Jenkins, M., Kapos, V., Monfreda, C., Loh, J., Myers, N., Norgaard, R. B. & Randers, J. (2002). Tracking the ecological overshoot of the human economy. *Proceedings of the National Academy of Sciences*, USA *99*, 9266-9271.

4.2.7. Methods Using Cost As A Proxy For Value

A fundamental principle in economics is the distinction between benefits and costs. In the context of ecosystem services, economic benefits reflect what is gained by increasing the amount of a given service relative to some baseline, while costs reflect what must be given up in order to achieve that increase. Costs can provide information about benefits or value only under specific and limited conditions. Nonetheless, several methods based on costs have been used in the valuation of ecosystem services.

One such method is **replacement cost**. Under this method, the value of a given ecosystem service is viewed as the cost of replacing that service by some alternative means. For example, some studies have valued clean drinking water provided by watershed protection by using the cost savings from not having to build a water filtration plant to provide the clean water (NRC, 2000 and 2004; Sagoff 2005). This type of cost savings can offer a lower-bound estimate of the value of an ecosystem service, but only under limited conditions (Bockstael et al., 2000). First, there must be multiple ways to produce an equivalent amount and quality of the ecosystem service. In the above example, the same quantity and quality of clean water must be provided by both the watershed protection and the filtration plant. Second, the value of the ecosystem service must be greater than or equal to the cost of producing the service via this alternative means, so that society would be better off paying for replacement rather than choosing to forego the ecosystem service. In the example, the value of the clean water provided must exceed the cost of providing it via the filtration plant. When these two conditions are met, it is valid to use the cost of providing the equivalent services via the alternative as a lower-bound estimate of the economic value of the ecosystem service.

Further Reading

Bartik, T. J. (1988). Evaluating the benefits of non-marginal reductions in pollution using information on defensive expenditures." *Journal of Environmental Economics and Management*, *15*, 111-22.

Bockstael, N. E., Freeman, A. M., et al. (2000). On measuring economic values for nature. *Environmental Science and Technology*, *34*, 1384-1389.

Chichilnisky, G. & Heal, G. (1998). Economic returns from the biosphere. *Nature*, *391*, 629-630.

National Research Council. (2000). *Watershed management for potable water supply: Assessing the New York City strategy*. Washington, DC: National Academies Press.

National Research Council. (2004). *Valuing ecosystem services: Toward better environmental decision-making*. Washington, DC: National Academies Press.

Sagoff, M. (2005). *The Catskills parable*. PERC Report. Bozeman, MT: Political Economy Research Center.

Shabman, L. A. & Batie, S. S. (1978). The economic value of coastal wetlands: A critique. *Coastal Zone Management Journal*, *4*, 231-237.

Another cost-related concept is **habitat equivalency analysis** (HEA), which has been used in Natural Resource Damage Assessments under the Comprehensive Environmental Response, Compensation and Liability Act and the Oil Pollution Act. HEA seeks to determine the restoration projects that would provide ecosystem or other related services (including capital investments such as boat docks) sufficient to compensate for a loss from a natural-resource injury (e.g., a hazardous waste release or spill). In principle, to determine whether a set of projects provides sufficient compensation for a loss, HEA should determine the tradeoffs required to make the public whole using utility equivalents of the associated losses and gains – i.e., it should use a value-to-value approach (see Roach and Wade, 2006;

Jones and Pease, 1997). However, in practice HEA is often based on a service-to-service approach specified in biophysical equivalents (e.g., acres) rather than utility equivalents (value). Restoring habitat far from where people live and recreate, however, may not create value equivalent to nearby lost habitat, even if the replacement habitat is of the same size.

Although HEA can provide dollar estimates of the cost of providing replacement services or projects, these estimates do not necessarily satisfy the two conditions noted above that are necessary for replacement cost to provide a lower bound on value. For example, the value of the ecosystem or other services provided by the restoration projects may not exceed the cost of providing those services. Even if it does, several other assumptions are needed to ensure that HEA will provide an actual estimate of the economic value of the lost ecosystem services and these assumptions will often not be met in practice. These include fixed proportions between services and values, as well as unit values that are constant over time and space (Dunford et al., 2004).

Because costs and benefits are two distinctly different concepts, the committee urges caution in the adoption of any methods using costs as a proxy for value. The above conditions for valid use must be satisfied. Analyses of costs should not be interpreted as measures of benefits unless these conditions are met. Nonetheless, when appropriately applied, methods such as replacement cost and HEA may be useful to EPA in policy contexts where there are multiple ways of providing an ecosystem service.

Further Reading

Dunford, R. W., Ginn, T. C. & Desvousges, W. H. (2004). The use of habitat equivalency analysis in natural resource damage assessment. *Ecological Economics*, *48*, 49-70.

King, D. M. (1997). *Comparing ecosystem services and values: With illustrations for performing habitat equivalency analysis*. Service Paper Number 1. Washington, DC: National Oceanic and Atmospheric Administration.

U.S. National Oceanic and Atmospheric Administration. (1995). *Habitat equivalency analysis: An overview.* Policy and Technical Paper Series No. 95-1 (revised 2000). Washington, DC: National Oceanic and Atmospheric Administration.

U.S. National Oceanic and Atmospheric Administration. (1999). *Discounting and the treatment of uncertainty in natural resource damage assessment.* Technical Paper 99-1. Washington, DC: National Oceanic and Atmospheric Administration, Damage Assessment and Restoration Program.

U.S. National Oceanic and Atmospheric Administration. (2001). *Damage assessment and restoration plan and environmental assessment for the Point Comfort/Lavaca Bay NPL site recreational fishing service losses.* Washington, DC: National Oceanic and Atmospheric Administration.

U.S. National Oceanic and Atmospheric Administration. Coastal Service Center Web site. Habitat Equivalency Analysis. http://www.darrp.noaa.gov/library/pdf/heaoverv.pdf

The price of **tradable emissions permits** under cap-and-trade systems will almost never meet the requirements for using cost as a proxy for value. The price of an emission permit in a well-functioning market will reflect the incremental cost of pollution abatement. This price does not reflect the value of pollution reduction unless one of two conditions is met: a) the number of permits is set optimally, so that the incremental cost of pollution equals the incremental benefit of pollution reduction; or b) there are significant purchases of permits for purposes of retiring rather than using the permit, which indicates the willingness-to-pay for pollution reduction by the purchaser. Absent these exceptions, the price of tradable emissions permits should not be used as a proxy for value.

4.3. Transferring Value Information

This section examines the transfer of value information from one policy context to another. For example, values assessed for a change in ecosystem services in one setting (reflecting a combination of biophysical and socio-economic conditions) might be used to estimate values in a different setting, as illustrated by the CAFO example described in section 2.2.3. Value transfers, especially in the form of benefits transfer, have been important to EPA valuation efforts, but there are a number of concerns and conditions that have important implications for the validity of such transfers and that have typically not been consistently or adequately addressed. The following discussion identifies and addresses concepts and methods that are important for achieving valid transfer of any measures or data about ecological values, with specific attention to the transfer of economic benefits. Parallel conceptual and methodological issues apply to all values transfers, and very similar issues are involved in the transfer of ecological data and information between different policy contexts (see the discussion in chapter 3.5.1).

4.3.1. Transfer Of Information About Economic Benefits

Economists often use information about economic benefits derived from a previous valuation study to assign values to changes in another context. This process or method is known as benefits transfer. As an example, suppose that a hedonic property value study used data from the sales of residential homes in Chicago (the study site) to estimate the incremental change in housing prices associated with variations in the air quality conditions near these homes. Given a variety of theoretical and statistical assumptions, measures adapted from the estimates of these price equations can be used to estimate the marginal value of small improvements in air quality in another city, such as New York or Los Angeles (the policy site). The adjustments necessary to use benefit information from a previous study in a new context depend on a number of factors, including the needs of proposed policy application, the available information about the policy site, and the comparability of preferences and supply conditions at the study and policy sites.

In light of the time and money needed to generate original value estimates, EPA relies heavily on benefits transfer. In fact, benefits transfer is the primary method EPA uses to develop the measures of economic trade-offs used in its policy evaluations. Most regulatory impact assessments and policy evaluations rely on adaptation of information from the existing literature. Recent examples of policy evaluations that used benefits transfer methods include EPA's *Economic and Benefits Analysis for the Final Section 316(b) Phase III Existing Facilities Rule, June 1, 2006* (EPA, 2006b), EPA's *Final Report to Congress on Benefits and*

Costs of the Clean Air Act, 1990 to 2010. (EPA, 1999), and the economic benefit- cost analysis of the CAFO regulations.

EPA's heavy reliance on benefits transfer raises a significant issue regarding its validity: under what conditions can the findings derived from existing studies be used to estimate values in new contexts? Inappropriate benefits transfer often is a weak link in valuation studies. A number of environmental economists and other policy analysts have devoted considerable attention to benefits transfer (e.g., Wilson and Hoehn 2006).

The evaluations of benefits transfer in the literature have been mixed. For example, Brouwer (2000) concludes that "no study has yet been able to show under which conditions environmental value transfer is valid" (p. 140). Similarly, Muthke and Holm-Mueller (2004) urge analysts to "forego the international benefit transfer" and remark that "national benefit transfer seems to be possible if margins of error around 50% are deemed to be acceptable" (p. 334). On the other hand, Shrestha and Loomis (2003) conclude that, "Overall, the results suggest that national BTF can be a potentially useful benefit transfer function for recreation benefit estimation at a new policy site" (pp. 94-95).

Because benefits transfer constitutes a wide collection of methods that arise from the specific needs of each policy application, broad conclusions regarding validity are not meaningful. Rather, assessment of the validity of the approach requires case-by-case evaluation of the assumptions used in the specific application of interest and must consider the similarities and dissimilarities between the study site and the policy site(s). For this reason, overall the committee believes that general conclusions regarding the validity of the application of these methods are not possible. However, some applications of benefits transfer by EPA have been valid, while others have not.

4.3.2. Transfer Methods

As noted, benefits transfer refers to a collection of methods rather than a single approach. Values derived from one or more study sites can be transferred to a policy site in three ways. The first is the transfer of a unit value. A unit-value transfer usually interprets an estimate of the tradeoff people make for a change in environmental services as locally constant for each unit of change in the environmental service. For the policy site, the relevant and available values for these factors are used to estimate an adjusted measure for the unit value based on the specific conditions in the policy site (see Brouwer and Bateman, 2005 for an example in the health context). As noted above, the required adjustments will depend on a number of factors.

A second approach is the function transfer approach, which replaces the unit value with a summary function describing the results of a single study or a set of studies. For example, a primary analysis of the value of air-quality improvements might be based on a contingent valuation survey of individuals' willingness to pay to avoid specific episodes of ill health, such as a minor symptom-day (e.g., a day with mildly red watering itchy eyes) or one day of persistent nausea and headache with occasional vomiting (e.g., Ready et al., 2004). A value function in this context relates willingness to pay to respondent characteristics and other factors that are likely to influence it, such as income, health status, demographic attributes, and the availability of health insurance. This value function is then used to estimate willingness-to-pay for populations with different characteristics. Alternatively, the original study might estimate a demand function or discrete choice model based on an underlying random utility model describing revealed preference choices. The demand function or discrete choice model is transferred and then used to estimate economic benefits at the policy site. In this case, the function being transferred is an estimated behavioral model rather than a value function.

Meta-analyses, which statistically combine results from numerous studies, can also involve a type of function transfer. Meta-analyses can be undertaken when there is accumulated evidence on measures of economic tradeoffs for a common set of changes in resources or amenities, provided that the benefit concept that is measured and the resource change that is valued are consistent across the studies that are combined in the analysis (Smith and Pattanayak, 2002). One area with a large number of applications is water quality relevant to recreation (e.g., Johnston et al., 2003; Smith and Kaoru, 1990a, 1990b). EPA recently used this approach in its assessment for the Phase III component of the 316(b) rules.

Some meta-analyses combine unit values to produce a weighted average unit value. While this might be appropriate in some valuation contexts (EPA Science Advisory Board, 2008a), in the context of ecological valuation it can be problematic because it ignores the site-specific variation in the value of ecosystem that stems from heterogeneity in both ecosystem and population characteristics. Alternatively, meta-analyses can combine studies to estimate a meta-regression function, which can be used to identify both site and population characteristics as well as methodological characteristics that influence benefit estimates. Such a function has the potential to be used for benefits transfer and allows an adjustment for characteristics of the policy site, if based on a structural approach that ensures that basic consistency properties are satisfied in order for the results to yield reliable benefit estimates (Bergstrom and Taylor, 2006). These approaches to benefit transfer have not yet been widely used. They need to be evaluated before it would be possible to describe a set of practices for applications, for example, in national rule making (see further discussion in section 6.1).

A third approach to benefits transfer is preference calibration. It uses information from the study site to identify the parameters that describe underlying preferences, with the objective of then using the resulting preference relationship to estimate benefits at the policy site (see Smith et al., 2002). With calibration, not all relevant parameters (in this case relating to preferences) are estimated directly from the data. Rather, some are calculated or inferred from available estimates of other parameters and assumed or observed relationships and constraints. When the parameters can be calibrated or estimated from the existing literature, the transfer uses the calibrated preference function, together with the conditions at the policy site, to measure the tradeoff for the change associated with the policy application.

4.3.3. Challenges Regarding Benefits Transfer

Several challenges arise when using benefits transfer. The first stems from possible differences between the study and policy sites. Regardless of the type of transfer method used, economic benefits or economic value functions derived from a particular ecosystem study site will not necessarily be relevant for a different policy site. How people value the preservation or alteration of an ecosystem depends on two dimensions: their preferences and the nature of the biophysical system. Differences in both biophysical characteristics and human values and preferences dictate that great care must be taken in deciding whether the valuation of benefits in one context can be validly used in another context.

Similarities or differences in preferences are likely to depend on how close the populations in the two cases are along social and economic dimensions that influence marginal willingness to pay. For example, income levels or age profiles are sometimes relevant, as in many cases of valuing recreational opportunities. The particular cultural characteristics of the community also may be relevant. For example, in locations where salmon are seen as iconic species reflecting the entire ecosystem (e.g., Seattle), people are likely to value more highly both salmon and water quality important for preserving the salmon.

When only information on willingness to pay per unit of improvement is available, the analyst must be sensitive to the types of differences that would render a transfer inappropriate. If all the differences between the study site and the policy site are such that one is likely to have a higher value per unit of improvement than the other, the value at the study site can provide either a floor or ceiling for the value at the policy site. When the information from the study site is in functional terms (e.g., willingness to pay as a function of income levels or age), social-economic differences between the study site and the policy site can be accommodated if these specifications are valid.

Although it may be possible to adjust for differences in social-economic characteristics of the populations, the capacity to adjust for biophysical differences is typically more limited. For example, even if the affected populations have identical characteristics (or adjustments can be made for their differences), the value of improving the water quality of one small lake in Minnesota is likely to be quite different from improving water quality in a small lake in Texas, because the effects on the overall provision of ecosystem services are likely to be quite different and not captured by a single relationship.

The challenge of transferring benefit estimates is exacerbated by the fact that often few economic benefit studies are available for use. One consequence is that analysts sometimes rely on benefits estimates that are too old to be reliable for new applications. For example, the regulatory impact assessment conducted for the concentrated animal feeding operations (CAFO) rule based its willingness-to-pay estimates for improved water quality on indices taken from a contingent valuation survey conducted by Carson and Mitchell in November 1983, 20 years before EPA's final CAFO rule was published (see Carson and Mitchell, 1993). In addition, due to lack of suitable previous studies, analysts sometimes inappropriately use values or functions derived from studies designed for purposes other than those relevant for the policy site. For example, the Carson and Mitchell study used in the CAFO rule was not intended to apply to specific rivers or lakes. Moreover, the water quality index used by Carson and Mitchell was highly simplified and not intended to reflect changes in ecosystem services beyond those related to fishing, boating, and swimming.

An additional challenge stems from the difficulty of finding the most appropriate unit values to carry over from the study site to the policy site. In the example below, illustrating willingness to pay for an improved fishing catch rate, several different metrics of value are possible, and the different metrics will have very different implications for the valuation at the policy site. The choice of unit values also has to be appropriate to the scale and context. For example, willingness to pay for increased wilderness areas in a study site may have been expressed in terms of dollars per absolute increase in area (e.g., $100 per taxpayer annually for a 100-acre increase in area, or $1 per acre). This unit value may be reasonable for a small, heavily populated municipality, but far too high for a municipality with substantially more existing wilderness area.

4.3.4. Improving Transfers Of Value Information

The discussion above points to the need for additional research to develop and improve methods and data for use in value transfers. While the discussion in section 4.3.3. focuses on the specific case of benefits transfer, similar challenges arise with other transfers of information about values. A number of strategies or processes can help improve these transfers.

One strategy that can help address the challenges of determining whether and how to conduct a value transfer is the use of a screening process. This procedural approach assumes that a deliberate effort to examine the similarities and differences between study sites and the policy site, by both EPA analysts and those overseeing their work, will help flag problematic transfers and clarify the assumptions and limitations of the study-site results. Several procedures can be considered. One is to contact experts familiar enough with both the previous and current contexts to determine whether to proceed with the proposed transfer. These experts can consider input from a variety of sources, including the public, and then apply the criteria that they regard as relevant, even if the set of criteria is not explicit. Experts knowledgeable in both the study case and the policy case can suggest the most appropriate functional forms and unit values (e.g., Desvousges, Johnson, and Banzhaf, 1998). Experts may also be able to suggest other existing valuations that would be better candidates for transfer of value information.

Another procedure is to make a detailed examination of the appropriateness of the study case a regular part of EPA review of analyses using value transfers. Such oversight of the use of case studies would require analysts to clarify the assumptions, purposes, and units of the study-site analysis so that EPA reviewers can judge the appropriateness of the transfer. Analysts must also be fully transparent regarding the origin and context, including the date, of the original valuation.

More thorough cataloguing of existing valuation studies, with careful descriptions of the characteristics and assumptions of each, would be helpful in increasing the likelihood that the most comparable existing valuations will be identified. This is a compelling rationale for developing databases of valuation studies. The establishment and development of a Web-based platform for data and models focusing on valuation estimates would be very worthwhile. Comparable to the Web sites developed and maintained for other large- scale social science research surveys such as the Panel Study on Income Dynamics (PSID) and the Health and Retirement Study (HRS), such a platform could expand the ability of Agency analysts to search for the most appropriate study cases and to supplement these records with related data for transfers. Some efforts along these lines are currently underway. These include the Environmental Valuation Reference Inventory (EVRI), which was developed by Environment Canada in conjunction with other agencies including EPA (see http://www.evri.ca/), and a database currently being developed for recreational use values (see http://www. cof.orst.edu/cof/fr/research/ruvd/Recreation_Letter. html). However, a more systematic effort across a wide range of ecosystems services is needed (see Loomis and Rosenberger, 2006).

WILLINGNESS TO PAY FOR AN IMPROVED CATCH RATE: THE CHALLENGE OF CHOOSING A UNIT VALUE FOR ECONOMIC BENEFITS TRANSFER

Suppose estimates from the literature imply that the average value of the willingness to pay for a 10% improvement in the catch rate (i.e., fish caught per unit of effort) for a sport fishing trip is $5 per trip. This estimate could be from a study describing specific types of fishing trips by a sample of individuals or it could be an average of several studies.

One approach for developing a unit value transfer would divide $5 by 10% to generate a unit value of $0.50 for each 1% improvement. This strategy implicitly assumes the benefit measure is not influenced by the level of quality – i.e., to be constant for each proportionate improvement.

Another approach would take the same information on average tradeoffs and calculate a unit value using the level of the quality variable, in this case a catch rate that itself embeds another economic decision variable – the effort a recreational fisher devotes to fishing. In this example, the quality or number of fish caught per hour of effort must be known. Suppose that in the study providing the estimated economic benefit, the average number of fish caught with an hour of effort before the improvement was 2. Thus a 10% improvement means that the typical recreationist would catch 0.2 more fish with an hour's effort, implying a unit value of $5 for every additional 0.2 fish caught per hour of effort, or (assuming a linear relationship in terms of the catch rate rather than the proportionate change in this quality measure) $25 for every additional fish caught per hour of effort.

Finally, the unit value could be expressed in terms of improved fishing trips. Suppose the average recreational trip involves 5 hours of fishing over the course of a day. Then the improvement of 0.2 fish per hour implies an average of one more fish caught during a trip. These additional data might be used to imply that the improvement makes typical trips yield incremental economic benefits of $5 per trip (the value of catching 0.2 additional fish per hour for a period of five hours).

Table 4. Table of Alternative Unit Value Transfers

Assumption	Unit Value	Interpretation of Policy	Aggregate Value
Constant unit value for a 1% improvement	$0.50 per 1% improvement	5% improvement per trip	$0.50 x 5 x 3 x 2000 = $15,000
Constant unit value for an extra fish caught per hour of effort	$25 per additional fish per hour	Added fish caught	$25 x .05 x 1 x 3 x 2000 = $7,500
Constant value for an improved trip	$5 per trip	Improved fishing trips	$5 x 3 x 2000 = $30,000

There are other ways this estimate could be interpreted. These examples are not intended to be the only "correct" ones or the best. They illustrate that the information on the baseline conditions, the measurement of quality, and the measurement and terms of use all can affect how a given set of estimates is used in a benefits transfer.

For the study site, all three interpretations are simply arithmetic transformations of the data describing the context for the choices that yield the tradeoff estimates. However, the same conclusions do not hold when they are transferred to a different situation. Suppose the policy site involves a case where we wish to evaluate the effects of reducing the entrainment of fish in power plant cooling towers. Assume further it is known from technical analysis that this regulation would lead to 5% improvement in fishing success along rivers affected by a rule reducing fish entrainment. Table 4 shows the alternative unit value transfers if these areas have 2000 fishers, each taking about 3 trips per season and currently they catch 1 fish per hour.

Clearly these examples deliberately leave out some important information. Trips may be different – longer, requiring more travel time, or involving different features such as different species or related activities. These added features are aspects omitted in the example. These estimates also do not allow for the possibility that fishing success induces current recreationists to take more trips or that people who never took trips may start taking them after the improvement. Under each of these possible outcomes, the sources for error in the transfer compound. Even without such details, these simple examples illustrate how the aggregate economic benefit measures can differ by a factor of four.

In addition to development and maintenance of a comprehensive database of existing valuation studies, more original valuation studies across a wider range of ecosystem services are needed to increase the Agency's capacity to conduct transfers. The committee urges the Agency to support research of this type. This research will be most useful if conducted with the explicit intention of developing value estimates that EPA can use for subsequent transfers. Such an intention can influence how the original valuation studies are conducted and documented. For example, Loomis and Rosenberger (2006) suggest a number of ways of designing original studies to facilitate benefits transfer, such as the use of objective, quantitative measures of quality changes within realistic ranges and the consistent and full reporting of project details. These same criteria would be appropriate for any value transfer.

4.4. Conclusions and Recommendations

The valuation approach proposed in this chapter calls for EPA to consider the use of a broader suite of methods than EPA has typically employed in the past for valuing ecosystems and their services. There is a variety of methods that could be used and the committee urges EPA to pilot and evaluate the use of alternative methods, where legally permissible and scientifically appropriate. Some of the methods considered by the committee have been used extensively in specific decision contexts (e.g., the use of economic methods in national rule making or the use of surveys, as described in endnote 35), while others are still relatively new and in the developmental stages. The methods also differ in a number of important ways, including the underlying assumptions, the concepts and sources of value they seek to characterize, the empirical and analytical techniques used to apply them, their data needs (inputs), and the metrics they generate (outputs), and the extent to which they involve the public. For these reasons, the potential for use by EPA in ecological valuation will be different for the different methods and in different contexts. The Committee advises EPA to:

- Only use methods that are scientifically based and appropriate for the particular decision context at hand.
- Develop a set of criteria to use in evaluating methods to determine their suitability for use in specific decision contexts. This is an important first step in implementing the valuation approach proposed in this chapter
- Explicitly identify relevant criteria to be used in determining whether a contemplated values transfer is appropriate for use in a specific ecological valuation context. Both EPA analysts and those providing oversight of their work must take into account the differences between study site and policy site to flag problematic transfers and clarify the assumptions and limitations of the study site results.
- Support efforts to develop Web-based databases of existing valuation studies across a range of ecosystem services, with careful descriptions of the characteristics and assumptions of each, to assist in increasing the likelihood that the most comparable existing valuations will be identified.

Conduct additional original research on valuation that is designed to be used in subsequent value transfers

5. Cross-Cutting Issues: Deliberative Processes, Uncertainty, and Communications

This chapter addresses three topics important to multiple stages of ecological valuation: analysis of uncertainties related to ecological valuation; communication of ecological valuation information; and the role of deliberative processes.

5.1. Deliberative Processes

Deliberative processes, in which analysts, decision makers, and/or members of the public meet in facilitated interactions, can be useful in estimating and valuing the potential effect of EPA actions on ecosystems and their related services. Such processes can assist at several steps of an assessment, ranging from developing conceptual models and determining the ecosystem services on which the Agency should focus its assessment to valuing those services. For example, where the public is not familiar with key ecosystem services, deliberative processes can provide the public with expert information that may better enable them to identify what services are important to them. Similarly, where the public is not accustomed to valuing particular ecosystem services, deliberative processes may again help members of the public estimate the value that they would place on those services. Deliberative processes also can increase public understanding and acceptance of a valuation effort and, where appropriate, permit the public to play a more active role in shaping and analyzing options.

Two specific types of deliberative processes of potential use to EPA in particular valuation efforts are mediated modeling and constructed value processes. In mediated modeling, analysts work with members of the public to develop a model representing a particular environmental system of interest, ranging from watersheds or local ecosystems to large regions or even the globe (for example, Higgins et al., 1997; Cowling and Costanza, 1997; van den Belt, 2004). Members of the public participate in all stages of the modeling process, from initial problem scoping to model development, implementation, and use. The resulting model can be used for multiple purposes, including determining the ecosystem services that are potentially important to the public and evaluating alternative scenarios or options of interest. If the model is to be used to consider tradeoffs, the model must incorporate values drawn from methods described in chapter 4. Because of public involvement in the modeling process, the model and any results derived from it are likely to enjoy buy-in and reflect group consensus.[39]

Constructed value processes can help in both estimating values and, in some cases, making policy decisions. A central premise of constructed value processes is that people's preferences and values for complex, unfamiliar goods, such as many ecosystem services, are multi-dimensional and that people sometimes construct their preferences and values for such goods during the process of elicitation. This premise contrasts with the premise underlying some valuation methods, most notably economic valuation methods, that assume preferences are given and that values or contributions to well-being can be measured using a single metric such as willingness to pay or accept.

Constructed value processes can be used either as part of a valuation process or directly in decision making. In both situations, constructed value processes involve a number of steps, including identifying objectives, defining the attributes to be used to judge progress toward the objectives, specifying the set of management options, and measuring changes in relevant attributes under the options (Gregory et al., 1993; Gregory et al., 2001; Gregory and Wellman, 2001). Objectives are diverse and often multi-dimensional. Examples include maintaining some requisite level of ecological services, protecting endangered or threatened species, producing particular resources, increasing tourism or recreational opportunities, and supplying a sense of pride or awe (Gregory et al., 2001). The final output is either a judgment about the current state of the system relative to an alternative state (if the context is evaluative) or the selection or identification of a preferred management option (if the context is decision making). Constructed value processes draw on inputs from a variety of disciplines, including economics, ecology, psychology, and sociology. A discussion of the use of decision science approaches for ecological valuation appears in section 4.2.6.

These deliberative processes, if done in a careful way and supported by appropriate resources, can provide useful input for valuation by identifying what people care about.[40]

Deliberative processes can be especially useful for providing input in valuation situations where the public may not be fully informed about ecosystem services. Such processes involving science, agency, and members of the public can be helpful for getting an idea of what an informed public might value. To adequately address and incorporate relevant science, however, it is important that such deliberative processes receive sufficient financial and staff resources (SAB, 2001).

5.2. Analysis and Representation of Uncertainties in Ecological Valuation

5.2.1. Introduction

All aspects of ecological valuation efforts – from the estimation of ecological impacts to valuation – are subject to uncertainty, regardless of the methods used. Assessment of this uncertainty allows for a more informed evaluation of proposed policies and of comparisons among alternative policy options. For each option, decision makers should have sufficient information regarding what is known about the distribution of possible outcomes and associated values in order to take uncertainty into account when they make their policy choices. Identifying key uncertainties can also provide potentially important insights regarding the design of research strategies that can reduce uncertainty in future analyses.

When addressing uncertainty in ecological valuation, four key questions arise: First, what are the major sources of uncertainty and what types of uncertainty are likely to arise when using alternative valuation methods? Second, what methods are available to characterize uncertainty in ecological valuations? Third, how should information regarding uncertainty be communicated to decision makers? Fourth, what types of new research – data collection, improvements in measurement, theory building, theory validation, and others – can reduce uncertainty for particular sources in specific applications? Section 5.2.2 briefly describes the major sources of uncertainty in the valuation of ecosystems and ecosystem services. The overview of specific valuation methods available at http://yosemite.epa.gov/sab/sabproduct.nsf/WebBOARD/CVPESS_Web_Methods_Draft?OpenDocument discusses the uncertainty arising from the use of individual methods. Section 5.2.3 then discusses two approaches to characterizing uncertainty regarding ecological values: Monte Carlo analysis and expert elicitation. Section 5.2.4 addresses the communication of uncertainty information. Section 5.2.5 discusses how EPA can use uncertainty analysis to set research priorities.

Historically, efforts to address uncertainty in ecological valuations and in all economic benefit assessments that are part of regulatory impact analyses have been limited. Providing greater information about uncertainty is consistent with the need for transparency and can improve decision making. In the context of regulatory impact analyses, Office of Management and Budget Circular A-4 explicitly calls for analysis and presentation of important uncertainties. To assess the level of confidence to attribute to projections used in a valuation, decision makers must know the analyst's judgment of the uncertainty of the valuation and its component steps, as well as the assumptions underlying the valuation analysis.

5.2.2. Sources Of Uncertainty In Ecological Valuations

As discussed in chapters 3 and 4, ecological valuation entails several analytic steps, each potentially subject to uncertainty. These steps include predicting ecological impacts of the relevant Agency decision or action, predicting the effects of these impacts on ecosystem services, and valuing the consequences of these effects.[41] Uncertainties in each stage of the analysis are of potential importance, and there is no reason – on the basis of theory alone – to judge one to be more important than the other. Rather, the relative magnitude of the uncertainty involved in each step is fundamentally an empirical question.

At each stage, uncertainty can arise from several sources.[42] First, some of the physical processes might be inherently random or stochastic. Second, there can be uncertainty about which of several alternative models of the process best captures its essential features. Finally, there are uncertainties involved in the statistical estimation of the parameters of the models used in the analysis.

At the biophysical level, for example, any characterization of current or past ecological conditions will have numerous interrelated uncertainties. Any effort to project future conditions, with or without some postulated management action, will magnify and compound these uncertainties. Ecosystems are complex, dynamic over space and time, and subject to the effects of stochastic events (such as weather disturbances, drought, insect outbreaks, and fires). Also, our knowledge of these systems is incomplete and uncertain. Errors in projections of the future states of ecosystems are thus unavoidable and constitute a significant and fundamental source of uncertainty in any ecological valuation.

All social, economic, or political forecasts are also based on implicit or explicit theories of how the world works, either represented by the mental models of the forecasters or by the models underlying the formal and explicit methods used in econometric modeling, systems dynamics modeling, and other forms of modeling. Theories and their expressions as models are unavoidably incomplete and may simply be incorrect in their assumptions and specifications. These sources of uncertainty compound the uncertainty surrounding the relevant biophysical relationships and responses to policy changes.

Uncertainty also arises in determining the value of the predicted ecological responses. The nature and interpretation of this uncertainty is different for different concepts of value. For example, for concepts of value that assume well-defined preferences (such as economic value), uncertainty can arise in the estimation of actual ("true") values, for a number of reasons. Valuation methods are subject to data and theory limitations, and they unavoidably rely on assumptions that introduce uncertainty. In addition, analysts are often required to apply estimated values to contexts that differ from those in which the values were developed, and appropriate adjustments might not have been made in transferring estimates to different contexts. These types of uncertainty in estimating true values can be assessed using the methods for probabilistic uncertainty analysis discussed below. Alternatively, for concepts of value that are based on the premise that preferences must be constructed (such as constructed values), uncertainty reflects the lack of a "true" underlying value. This type of uncertainty differs qualitatively from the uncertainty that arises in estimating true but unknown values. As a result, it is not meaningful to characterize it using a probability distribution function. Rather, in uncertainty analysis, the value should be viewed as a parameter that can be varied to determine the implications of different values for the ranking of alternative policy options (see Morgan and Henrion, 1990).

In identifying the types of uncertainty most likely to be of concern for individual valuation approaches in specific contexts, two issues are relevant: the sensitivity of the approach to the potential sources of uncertainty listed above, and the magnitude of uncertainty thereby generated. The consequence of data limitations can be assessed by determining the variation in results implied by variations in data. Vulnerability to theoretical limitations is more difficult to assess, but can be gauged in some cases by comparing predictions based on alternative models.

5.2.3: Approaches To Assessing Uncertainty

Probabilistic uncertainty analysis, by its very nature, is complex, particularly in the context of ecological valuation. The simplest and probably most common approach to evaluating uncertainties is some form of sensitivity analysis, which typically varies one parameter or model assumption at a time and calculates point estimates for each of the different parameter values or assumptions. The results provide a range of estimates of the "true" value, including lower and upper bounds. No effort is made to assign probabilities to the calculated values or estimate the shape of the distribution of values within the range.

Although sensitivity analysis may be sufficient for some simple problems, its use in the context of ecological valuation is likely to give an incomplete and potentially misleading picture of the true uncertainty associated with the value estimates. Due to the number of sources of uncertainty in many ecological valuations, sensitivity analysis is unlikely to account for the implications of all the sources of uncertainty. In addition, sensitivity analysis becomes unwieldy when the outcomes relevant to the value assessment themselves consist of multiple interrelated variables. For example, it is extremely difficult at the biophysical level to calculate the uncertainty in projecting outcomes from a complex ecological system composed of multiple interacting variables subject to the influence of external stochastic events.

Given the limitations of simple sensitivity analysis, other approaches to characterizing uncertainty have been developed. These include Monte Carlo analysis and the use of expert elicitation. These approaches can provide a more useful and appropriate characterization of uncertainty in complex contexts such as ecological valuation.

Monte Carlo analysis is an approach to characterizing uncertainty that allows simultaneous consideration of multiple sources of uncertainty in complex systems. It requires the development of a model to predict the system's outputs from information about inputs (including parameter values). The underlying inputs that are uncertain are assigned probability distributions. A computer algorithm is then used to draw randomly from all of these distributions simultaneously (rather than one at a time, as in sensitivity analysis) and to predict outputs that would result if the inputs took these values. By repeating this process many times, the analyst can generate probability distributions for outputs that are conditional on the distributions for the inputs.

Developments in computer performance and software have substantially reduced the effort required to conduct calculations for a Monte Carlo analysis once input uncertainties have been characterized. Widely available software allows the execution of Monte Carlo analysis in common spreadsheet programs on a desktop computer. In developing probability distributions for uncertain inputs, uncertainty from statistical variation can also often be characterized with little additional effort relative to that needed to develop point estimates. Much of the needed data already will have been collected for the development of point estimates (although characterizing other sources of uncertainty in inputs can require more effort).

In contrast to sensitivity analysis, Monte Carlo analysis provides information on the likelihood of particular values within a range, which is essential to any meaningful interpretation of that range. Without such an understanding, the presentation of a range of possible outcomes may lead to inappropriate conclusions. For example, a reader may assume that all values within the range are equally likely to be the ultimate outcome, even though this is rarely the case. Others may assume that the distribution of possible values is symmetric. This, also, is often not the case.

Because of its ability to characterize uncertainty in a more meaningful way, Monte Carlo analysis has become common in a variety of fields, including engineering, finance, and a number of scientific disciplines. It has been useful in policy contexts. EPA recognized as early as 1997 that it can be an important element of risk assessments (EPA, 1997). Circular A-4, in calling for the analysis and presentation of uncertainty information as part of regulatory analyses, also notes the potential use of Monte Carlo analysis. However, efforts to quantify uncertainties through Monte Carlo analyses rarely have been undertaken in ecological valuations. More often, uncertainty has been addressed qualitatively or through sensitivity analysis.

The reliable application of Monte Carlo methods requires the specification not only of variances on key variables, but also of covariances across the variables. Without appropriate covariances, the method is unreliable and can lead to biased results. Positive covariance increases the spread of results, while negative covariances decrease the spread.

Where Monte Carlo analysis can be reliably used in the estimation of ecological values, the analysis is unlikely to address all sources of uncertainty. Thus, the results will likely understate the range of possible outcomes that could result from the relevant public policy. Nonetheless, the ranges produced will still provide more reliable information about the implications of known uncertainties than simple sensitivity analysis. In turn, these ranges can better inform judgments by policy makers as to the overall implications of uncertainty for their decisions. The committee therefore urges EPA to move toward greater use of Monte Carlo analysis, where feasible, as a means of characterizing the uncertainties associated with estimating the value of ecological protection.

A variety of expert elicitation methods can also provide indications of the amount and nature of uncertainty associated with estimates of specific values or predictions regarding the impacts of a given activity or change (e.g., Morgan and Henrion, 1990; Cleaves, 1994). In its simplest form, an expert elicitation is a single expert's assessment of the uncertainty of an estimate, forecast, or valuation, whether it is based on implicit judgment or a more explicit approach like the Monte Carlo technique. Policy makers can elicit more information from the expert, such as the assumptions underlying his or her analysis or the bases for uncertainty, to better understand the reliability of the expert's input and the nature of the uncertainty.

Although an elicitation can rely on a single expert, the bulk of expert elicitation methods involve multiple experts, which allows for a comparison of their judgments and an assessment of any disagreements. If the experts are of equal credibility, so that no judgment can be discarded in favor of another, the range of disagreement reflects uncertainty. If top scientists strongly diverge in their estimates, forecasts, or valuations, the existence of a high level of uncertainty is irrefutable. This relationship, however, is asymmetrical because narrow disagreement does not necessarily reflect certainty. The experts may all be equally wrong, a somewhat common occurrence given that experts often pay attention to the same information and operate within the same paradigm for any given issue (Ascher and Overholt, 1983). When experts interact before providing their final conclusions (e.g., by exchanging estimates and adapting them to what they learn from one another), errors due to incompleteness can be reduced. For example, biologists may benefit from the kind of information that atmospheric chemists can provide, and vice versa. While such interactions run the risk of "groupthink" – the unjustified convergence of estimates due to psychological or social pressures to come closer to agreement (Janis, 1982) – structured group processes can help reduce the risk.

For many expert elicitation methods, translation into probabilities is difficult. Simple compilations of estimates (e.g., contemporaneous estimates of species populations) from different experts can generate a table with the range of estimates. However, these compilations are unable to convey the degree of uncertainty that each expert would attribute to his or her estimate. Including confidence intervals can provide this information.

The SAB has been asked to review a draft Agency white paper on expert elicitation and provide advice on the utility of using expert elicitation to support EPA regulatory and non-regulatory analyses and decision-making, including potentially in ecological valuations. Although EPA has historically focused expert elicitations on human health issues, the approach may be useful for ecological valuation as well. The committee suggests that EPA consider using expert elicitation to obtain estimates of parameters and their uncertainty for use in Monte Carlo analysis, if suitable information about the relevant range for the parameter values is not available based on observation (e.g., field work or experiments).

5.2.4. Communicating Uncertainties In Ecological Valuations

It is important not only to analyze the sources and size of uncertainty involved in a valuation but also to effectively communicate that uncertainty to both decision makers and the public – and in a manner that does not overwhelm the recipient and cause them to disregard or misinterpret the information. If improvements in the analysis of uncertainty do not go hand-in-hand with improvements in the communication of that uncertainty, the added information can end up confounding rather than facilitating good decision making (Krupnick et al., 2006).

In the past, point estimates have been given far greater prominence in public documents such as regulatory impact assessments and other government valuations than discussions of the uncertainty associated with them. Uncertainty assessments are often relegated to appendices and discussed in a manner that makes it difficult for readers to discern their significance. This result may be inevitable, given that single-point estimates can be communicated more easily than lengthy qualitative assessments of uncertainty or a series of sensitivity analyses. The ability of Monte Carlo analysis to produce quantitative probability distributions, however, provides a means of summarizing uncertainty that can be communicated nearly as concisely as point estimates. If a summary of uncertainty is not given prominence relative to an estimate itself, decision makers will lose both the context for interpreting the estimate and opportunities to learn from the uncertainty.

Some resistance to the use of formal uncertainty assessments such as through Monte Carlo analysis, and to the prominent presentation of the results, may be due to the perception that such analysis requires greater expert judgment and therefore renders the results more speculative.[43] Also, some might argue that, given the inevitably incomplete nature of any uncertainty analysis, prominently presenting its results could incorrectly lead readers to conclude that the results of an ecological valuation are more certain than they actually are. Both concerns are generally unfounded. As described above, developing characterizations of uncertainty, such as for inputs in a Monte Carlo analysis, often simply involves making explicit and transparent expert judgments that already must be made to develop point estimates for those inputs. To the extent that an uncertainty analysis is incomplete in its characterization of uncertainty, that fact can be communicated qualitatively.

EPA should also consider the use of consistent language and graphical approaches in reporting on uncertainty in its valuations. Common usage, as well as graphical presentation of quantitative information regarding uncertainties, can help improve communication and understanding of relevant uncertainties for both policy makers and interested members of the public (Moss and Schneider, 2000). Organizations such as the Intergovernmental Panel on Climate Change (IPCC) and the Millennium Ecosystem Assessment have consciously adopted clear and consistent language, as well as graphical approaches, in reporting on uncertainty.

5.2.5. Using Uncertainty Assessment To Guide Research Initiatives

Over time, additional research related to data collection, improvements in measurement, theory building, and theory validation can reduce the uncertainties associated with ecological valuation. For example, research can improve our understanding of the relationships governing complex ecological systems and thereby reduce the uncertainty associated with predicting the biophysical impacts of alternative policy options. Even stochastic uncertainty can sometimes be addressed by initiating research that focuses on factors previously treated as exogenous to the theories and models. For example, an earthquake-risk model based on historical frequency will have considerable random variation if detailed analysis of fault-line dynamics is excluded; bringing fault-line behavior into the analysis can lead to reductions in such uncertainty (Budnitz et al., 1997).

Assessments of the magnitude and sources of uncertainty can help to establish research priorities and to inform judgments about whether policy changes should be delayed until research reduces the degree of uncertainty associated with possible changes. Enhanced uncertainty analyses can provide decision makers with information needed to make better decisions. Determining whether the major source of uncertainty comes from weak data, weak theory, randomness, or inadequate methods can help guide the allocation of scarce research funds. Some data needs will simply be too expensive to fulfill, and some methods have intrinsic limitations that no amount of refinement will fully overcome. Uncertainty analysis can provide insight into whether near-term progress in reducing uncertainty is likely, based on the sources of uncertainty and the feasibility of addressing these limitations promptly. However, it is important to avoid the pitfall of delaying an action simply because some uncertainty remains where the benefits of immediate action outweigh the value of attempting to further reduce the uncertainty. Some uncertainty always will remain.

5.3. Communication of Ecological Valuation Information

The success of an integrated and expanded approach to ecological valuation depends in part on how EPA obtains information about public concerns during the valuation process and then communicates the resulting ecological valuation information to decision makers and the public (Fischhoff, 2009). Although the committee has not extensively discussed the communication challenges presented by ecological valuation, it believes that generally accepted practices for communication of technical information apply to the valuation context. Section 5.3.1 discusses general practices of particular relevance to valuation. Section 5.3.2 addresses the special communication challenges that arise for ecological valuation.

Three essential functions of communication in valuing the protection of ecological systems and services are:

- Communication among and between technical experts and the public within the valuation process itself
- Communication of valuation information by analysts to decision makers
- Communication of the results of the valuation and decision making processes to interested and affected members of the public.

Although these communication functions may appear to be separate steps, they overlap. The success of the overall valuation process and any communication step within it, for example, depends on understanding how decision makers use valuation information. Spokespersons must understand how different public groups and experts frame valuation issues before they can effectively communicate the results of a formal valuation analysis.

5.3.1. Applying General Communication Principles To Ecological Valuation

Effective communication should be designed for the relevant audience of the valuation information. The potential pool of interested parties include decision makers, interested and affected members of the public, and experts in social, behavioral, and economic sciences and ecological sciences. A broad public audience is likely to be interested in better understanding the value of protecting ecological systems and services. Also important is an intermediate audience of analysts, who serve as important mediators for valuation information through their analyses and activities. This latter audience needs to access not only value estimates but also technical details and models. To support decisions effectively, communications must be designed to address a recipient's goals and prior knowledge and beliefs, taking into account the effects of context and presentation (Morgan et al., 2002). The committee recommends that EPA formally evaluate the communication needs of the users of valuations and adapt valuation communications to those needs.

As discussed earlier, an effective communication strategy also requires interactive deliberation and iteration (NRC, 1996). Effective communication of values requires systematic interactions with interested parties, where the interaction will differ depending on the technical expertise and focus of the parties. In general, interactive processes are critical for improving understanding, although reports (such as EPA's *Draft Report on the Environment*) are also important, especially in the context of assessment.

Basic guidelines for risk and technical communication are generally applicable to communicating ecological values. Linear graphs, for example, are likely to convey trends more effectively than tables of numbers (Shah and Miyake, 2005), and text that incorporates headers and other reader-friendly attributes will be more effective than text that does not (Schriver, 1989). In developing effective communication approaches for ecological valuation,

EPA can look to guidelines developed for risk and technical communication. Two useful examples of such guidelines are the communication principles in EPA's *Risk Characterization Handbook* (EPA, 2000d) and the guidelines for effective web sites (Spyridakis, 2000). The principles in the *Risk Characterization Handbook* include transparency, clarity, consistency, and reasonableness. Spyridakis, in turn, provides guidance in five categories: content, organization, style, credibility, and communicating with international audiences. Spyridakis provides a concise table for communicating information via Web sites and provides generally accepted guidance useful for communication of valuation information, including: (a) selecting content that takes into account the reader's prior knowledge; (b) grouping information in such a way that it facilitates storing that information in memory hierarchically; (c) stating ideas concisely; and (d) citing sources appropriately, and keeping information up to date.

As in the case of any type of communications, it is difficult to predict the effects of communicating ecological valuations. Good communications practice requires formative evaluation of the communications as part of the design process. Testing messages after the fact will enable assessments of effectiveness, leading to continued improvement in communications (e.g., Scriven, 1967; Rossi et al., 2003). The committee recommends that EPA evaluate its communication of ecological valuations to assess the effects of the communication and to learn how to improve upon Agency communication practices.

5.3.2. Special Communication Challenges Related To Ecological Valuation

Although application of these general communication principles will improve communications of ecological valuations, special challenges arise in this context.

First, communicating the value of protecting ecological systems and services requires conveying not only value information (in terms of metrics such as monetized values and rating scales), but also information about the nature, status, and changes to the ecological systems and services to which the value information applies. The EPA Science Advisory Board review of EPA's *Draft Report on the Environment* (EPA Science Advisory Board, 2005) and other reports (e.g., Schiller et al., 2001; Carpenter et al., 1999; Janssen and Carpenter, 1999) emphasize that people need to understand the underlying causal processes in order to understand how ecological changes affect the things they value, such as ecosystem services.

The causal processes can be conveyed using such visual tools as mapped ecological information, photographs, graphs, and tables of ecological indicators. To the extent that such visual outputs – especially outputs from integrated geographic information systems using best cartographic principles and practices (Brenner, 1993) – can be interactive, the outputs will facilitate sensitivity analysis that can address audience questions about scale and aggregation and may be more effective as communication tools. The EPA Science Advisory Board has proposed this kind of framework for reporting on the condition of ecological resources. EPA's *Draft Report on the Environment* (EPA, 2002a) and Regional Environmental Monitoring and Assessment Program reports illustrate a range of representational approaches.[44]

Second, the many uses and definitions of the term "value" complicate the communication of ecological values. The broad usage of the term in this chapter includes all the concepts of value described in Table 1. Context and framing can strongly influence how people rank, rate, and estimate values (Hitlin and Piliavin, 2004; Horowitz and McConnell, 2002), as well as how they interpret value-related information (e.g. Lichtenstein and Slovic, 2006).

As discussed elsewhere in this chapter, value measures are required or useful in a variety of regulatory and non- regulatory contexts, ranging from national rule making, to site-specific decision making and prioritization of environmental actions, to educational outreach in regional partnerships. In some cases monetization is required, whereas in others (e.g., educational outreach by regional partnerships), narratives and visual representations of values may play a more important role. Little direct evidence exists about how people perceive alternative value measures.

One mechanism for mitigating disconnects when reporting ecological values in different metrics is to employ an iterative, interactive approach to eliciting, studying, and communicating values and tradeoffs, where values are represented in multiple ways. Verbal quantifiers (e.g., "many" or "very likely"), for example, may make technical information more accessible, but the wide variability with which these terms are interpreted (Budescu and Wallsten, 1995) makes it critical to make the underlying numerical information readily available. Appropriate use of graphical and visual approaches, including geographic information systems, can aid interpretation of quantitative information. Visualization can facilitate new insights (MacEachren, 1995).

Third, much remains to be learned about how particular representations of non-monetized and monetized values, such as narrative, numerical attitude ratings, graphics, or other visual information, influence decision makers or the public in either value elicitations or presentations of cost-benefit analyses. Survey and decision research suggest that perceptions – and expressions – of values depend on format as well as context and specific content.[45] For example, including graphics in otherwise equivalent sets of information demonstrably influences expressed values (Chua, Yates and Shah, 2006; Stone et al, 2003). Asking people for ecological value in dollars can produce different results than using other response scales (e.g., Schlapfer, Schmitt and Roschewitz, 2008).

Finally, in many circumstances, interactive communication of ecological valuation information is likely to be more effective than static displays. Interactive communication allows users to manipulate the data or representations of the data, such as with sliders on interactive simulations. Interactive visualization has the potential to allow users to tailor displays to reflect their individual differences and questions. Even with exactly the same presentation, people's understandings of content vary because of differences in educational or cultural background, and different intellectual abilities. Interactive exploration tools give the audience a chance to investigate freely the part in which they are interested or about which they have questions.

As Strecher, Greenwood, Wang, and Dumont (1999) argue, the advantage of interactivity include that it supports: active (rather than passive) audience participation; tailoring information for individual users; assisting the assessment process; and visualizing risks under different scenarios (allowing users to ask "what if" questions). Interactivity is a good solution if the complexity of the visualization has the potential to overwhelm users (Cliburn, Feddema, Miller, and Slocum, 2002). Interactive visualization nonetheless poses challenges as well. 3-D visualization, which has become increasingly popular in visualization practice (Encarnacao et al., 1994), both necessitates interactivity and at the same time challenges it because of the sheer computational power required.

5.4. Conclusions and Recommendations

Deliberative processes can play an important role in the valuation process and the committee makes the following recommendations regarding their use:

- EPA should consider using carefully conducted deliberative processes to provide information about what people care about.
- Particular attention should be paid to deliberative processes where the public may not be fully informed about ecosystem services. Deliberative processes involving scientists, agency personnel, and members of the public can be helpful for getting an idea of what an informed public might value.
- EPA should ensure that deliberative processes receive the financial and staff resources needed to adequately address and incorporate relevant science and best practices.

A recent report of the National Research Council also examines public participation in environmental assessments and is a useful source for additional recommendations on the potential role of deliberative processes in the valuation process (NRC, 2008).

Providing information to decision makers and the public about the level of uncertainty involved in ecological valuation efforts is critical for the informed evaluation of proposed policies and alternative policy options. The committee makes the following recommendations to ensure the effective analysis and representation of uncertainties in ecological valuations:

- In assessing uncertainty, EPA should go beyond simple sensitivity analysis and make greater use of approaches, such as Monte Carlo analysis, that provide more useful and appropriate characterization of uncertainty for the complex contexts of ecological valuation. Sensitivity analysis is unlikely to account for all sources of uncertainty in ecological valuation and can become unwieldy when value outcomes consist of multiple interrelated variables. EPA should also consider using expert elicitation to obtain estimates of parameters and their uncertainty for use in Monte Carlo analysis, if suitable information about the relevant range for the parameter values is not available based on observation.
- The Agency should not relegate uncertainty analyses to appendices but should ensure that a summary of uncertainty is given as much prominence as the valuation estimate itself. EPA should also explain qualitatively any limitations in the uncertainty analysis. EPA should also explain limitations in the valuation exercise due to uncertainties.
- EPA should invest in additional research designed to reduce the uncertainties associated with ecological valuation through data collection, improvements in measurement, theory building, and theory validation. Assessments of the magnitude and sources of uncertainty can help to establish research priorities inform judgments about whether policy changes should be delayed until research reduces the degree of uncertainty associated with possible changes. The Agency, however, should not delay a necessary action simply because some uncertainty remains, because uncertainty always will remain.

The success of ecological valuations also depends on how EPA obtains information about public concerns during the valuation process and then communicates the resulting ecological valuation information to decision makers and the public. To promote effective communications, the committee recommends the following steps:

- EPA should evaluate the users of valuation information and their needs and adopt communications that are responsive to those needs.
- In communicating ecological valuation information, the Agency should follow basic guidelines for risk and technical communication. EPA's *Risk Characterization Handbook* (EPA, 2000d) provides one set of useful guidelines, including transparency, clarity, consistency, and reasonableness.
- EPA should evaluate its communication of ecological valuations to assess its effects and to learn how to improve upon its practices.
- To the extent feasible, the Agency should communicate not only value information but also information about the nature, status, and changes to the ecological systems and services to which the value information applies. Visual tools such as mapped ecological information, photographs, graphs, and tables of ecological indicators can be very useful in conveying causal processes.

Where appropriate, the Agency should employ an iterative, interactive approach to communicating values.

6. Applying the Approach in Three EPA Decision Contexts

This chapter discusses implementing the C-VPESS ecological valuation approach in three specific EPA decision contexts: national rule making, regional partnerships, and site-specific decision making. The committee believes that improved ecological valuation in each context can contribute to improved policy analysis and decisions. The committee examined a number of illustrative examples for each decision context and used these examples to inform its views about application of the approach advocated in this chapter.

The discussions below elaborate on the three key features of the valuation approach advocated in this chapter as they relate to the specific decision contexts:

- Identifying and focusing early in the process on the impacts that are likely to be most important to people
- Predicting ecological changes in value relevant terms
- Using multiple methods in the valuation process.

The discussions are meant to be illustrative rather than comprehensive and the exclusion of a particular method from discussion in a specific context is not intended to suggest inappropriateness. Note that the general principles and concepts used in the discussions below are described in more detail elsewhere in this chapter (see, for example, chapter 4 and descriptions of valuation methods and survey issues and best practices, links available on the SAB Web site at http://yosemite. EPA Web_Methods_Draft?OpenDocument).

6.1. Valuation for National Rule Making

6.1.1. Introduction

This section examines the application of the expanded, integrated approach to ecological valuation in the context of national rule making. Executive Orders and implementation guides often specifically require assessment and analysis of the benefits and costs of national rules to follow prescribed economic methods. Thus, this section is focused on the application of valuation using economic methods. In addition, it discusses the role that the other methods described in chapter 4 can play in this context. As background for this discussion, the committee examined three examples of previous Agency economic benefit assessments:

- The Agency's assessment for the final effluent guidelines for the aquaculture industry (EPA, 2004a)
- The Agency's assessment for the 2002 rule making regarding concentrated animal feeding operations (CAFOs) (EPA, 2002b; chapter 2 also discusses this benefit assessment)
- The prospective analysis of the benefits of the Clean Air Act Amendments of 1990 (EPA, 1999).[46]

Brief descriptions of the three benefit analyses are presented later in this section. These examples provide insights reflected in the discussion and recommendations throughout this section.

6.1.2. Valuation In The National Rule Making Context

As noted previously, valuation by EPA in the national rule making context is typically subject to constraints imposed by statute, executive order, and/ or guidance from the Office of Management and Budget (OMB). Most of the environmental laws administered by the Agency require that regulations such as environmental quality standards and emissions standards be based on criteria other than economic benefits and costs. In some cases, the legislation explicitly precludes consideration of costs or benefits in the standard-setting process. For example, under the Clean Air Act, primary ambient air quality standards for criteria air pollutants must be set to protect human health with an adequate margin of safety. Even where a law, such as the Safe Drinking Water Act, allows consideration of benefits and costs, adherence to a strict "benefits must exceed costs" criterion is not required.

However, even when national EPA rules are not determined by a strict benefit-cost criterion, assessments of the benefits and costs of EPA actions, conducted under prescribed procedures, can be important for a number of reasons. First, Executive Order 12866 (as amended by Executive Order 13422), requires federal agencies to "assess both the costs and the benefits of the intended regulations, and ... propose or adopt a regulation only upon a reasoned determination that the benefits of the intended regulation justify its costs" (Executive Order 12866, October 4, 1993). These assessments are commonly referred to as regulatory impact assessments (RIAs). They generally evaluate, in economic terms, the form and stringency of the rules that are established to meet some other objective, such as protection of human health.

Second, in some cases, an assessment of economic benefits and costs can be mandated by law. For example, Section 812 of the Clean Air Act Amendments of 1990 requires the Agency to develop periodic reports to Congress that estimate the economic benefits and costs of various provisions of the Act. Finally, the benefit and cost estimates developed in national rule making can help in setting research or legislative priorities. In summary, a complete, accurate, and credible analysis of the benefits and costs of a given rule can have broad impacts, even if the analysis does not determine whether a currently proposed rule should be promulgated.

In conducting RIAs, EPA is subject to requirements specified by OMB guidance, and all EPA benefit assessments are subject to OMB oversight and approval. As noted in chapter 2, OMB's Circular A-4 (OMB, 2003) makes it clear that Executive Order 12866 requires an economic analysis of the benefits and costs of proposed rules conducted in accordance with the methods and procedures of standard welfare economics. In the context of national rule making, the terms benefit and cost thus have specific meanings. To the extent possible, EPA must assess the benefits associated with changes in goods and services as the result of a rule, judged by the sum of the individuals' willingness to pay for these changes. Similarly, the costs associated with regulatory action are to be evaluated as the losses experienced by people, and measured as the sum of their willingness to accept compensation for those losses. EPA must begin the analysis by specifically describing environmental conditions in affected areas, both with and without the rule. EPA must then value these changes based on individual willingness to pay and to accept compensation, aggregated over the people (or households) experiencing them. Although other valuation methods described in chapter 4 may yield monetary estimates of value, monetizing values using multiple methods and then aggregating the resulting estimates would mean combining estimates that are based on quite different theoretical constructs, as well as diverse underlying assumptions. Thus, for both theoretical and empirical consistency – as well as compliance with OMB guidance – monetization of benefits in the context of an RIA should be based on economic valuation.

Circular A-4 recognizes that it may not be possible to express all benefits and costs in monetary terms. In these cases, it calls for measurement of these effects in biophysical terms. If that is not possible, there should be a qualitative description of the benefits and costs (OMB, 2003, p. 10). Circular A-4 is clear about what should be included in regulatory analyses, but it does not preclude the inclusion of information drawn from non-economic valuation methods. Nonetheless, it implies that when conducting ecological valuation in the context of national rule making, EPA must seek to monetize benefits and costs using economic valuation methods as much as possible.

Although economic valuation methods are well- developed and there is a large literature demonstrating their application, applying these methods to the eco logical benefits of a national-level rule raises significant challenges. A key challenge is the difficulty of deriving a national estimate of the effect of an EPA rule on ecosystems and the services derived from these ecosystems. Such a national estimate requires information about changes in stressors resulting from the action, as well as information about how the changes in stressors will affect ecosystems and the flow of services nationally. In many rule-making contexts, predicting the changes in stressors is difficult. Often, the rule prescribes adoption of a particular technology or a particular behavior (e.g., adoption of best management practices) rather than a specific change in stressors (e.g., discharge limits). The aquaculture rule associated with the Clean Water Act, described in text box 1, provides an example. In those cases, to estimate associated benefits, EPA must predict the changes in stressors that would likely result from the required behavioral change.

A rule will often involve many stressors with complex interactions, which greatly complicates the development of quantitative estimates of changes in stressors. The CAFO rule, described in chapter 2 and below, is an illustration.

Changes must also be defined relative to a baseline, and few national-level databases useful for this purpose exist. For example, in the RIA for the aquaculture rule, it was difficult to quantify the changes in stressors because, in some cases, baseline data on stressor levels were not available.

Even if changes in stressor levels can be predicted at the national level, mapping these into national- level changes in ecosystem characteristics or services using ecological production functions is generally very difficult. There may be a long chain of ecological interactions between the stressors and the ecosystem services of interest – and often many of links in that chain are not fully understood by scientists, particularly at the level required for comprehensive national analysis. Scientific knowledge is especially lacking on the ecological impacts of substances such as heavy metals, hormones, antibiotics, and pesticides. However, these substances can have important and far-ranging impacts at the national level. In addition, the nature and magnitude of impacts can be very site-specific because they vary substantially both within and across regions of the country. As a result, predictions of biophysical impacts in one region generally cannot readily be transferred to other regions where the characteristics of the relevant ecosystems, as well as the affected population, are different.

TEXT BOX 1: VALUATION AND THE AQUACULTURE EFFLUENT GUIDELINES

Title III of the Clean Water Act gives EPA authority to issue effluent guidelines that govern the setting of national standards for wastewater discharges to surface waters and publicly owned treatment works (municipal sewage treatment plants). The standards are technology-based, i.e., they are based on the performance of available treatment and control technologies. The proposed effluent guidelines for the concentrated aquatic animal production industry (aquaculture) would require that all applicable facilities prevent discharge of drugs and pesticides that have been spilled. In addition, facilities must minimize discharges of excess feed and develop a set of systems and procedures to minimize or eliminate discharges of various potential environmental stressors. The rule also includes additional qualitative requirements for flow- through and recirculating discharge facilities and for open water system facilities (EPA, 2004a).

The Agency identified the following potential ecological stressors that might be affected by the rule: solids; nutrients; biochemical oxygen demand from feces and uneaten food; metals (from feed additives, sanitation products, and machinery and equipment); food additives for coloration; feed contaminants (mostly organochlorides); drugs; pesticides; pathogens; and introduction of non-native species. Some of these (e.g., drugs and pathogens) were thought by the Agency to be very small in magnitude and not require further analysis. To this list, C-VPESS would add habitat alteration from changes in water flows.

For most of these stressors, it is not possible to specify the change that would result from the rule for two reasons. First, the rule called for adoption of "best management practices" rather than imposing specific quantitative maximum discharge levels. Second, for most of these stressors, baseline data on discharges in the absence of the rule were not available.

The Agency analyzed the effects of the rule on dissolved oxygen, biochemical oxygen demand, total suspended solids, and nutrients (nitrogen and phosphorus). There appear to have been three reasons why the remaining endpoints were not quantified:

- The Agency lacked data on baseline stressor levels.
- The rule called for adoption of "best management practices" rather than imposing specific quantitative maximum discharge levels, and the Agency lacked information on how these requirements would change the levels of stressors.
- The Agency did not use a model capable of characterizing a wide range of ecological effects. The Agency used QUAL2E rather than the available AQUATOX model. The choice of QUAL2E appears to have been driven largely by the ability to link its outputs with the Carson and Mitchell valuation model (1993).

The Agency estimated benefits for recreational use of the waters and non-use values. To estimate these values, the Agency estimated changes in six water quality parameters for 30-mile stretches downstream from a set of representative facilities and calculated changes in a water quality index for each facility. The Agency then used an estimated willingness-to-pay function for changes in this index taken from Carson and Mitchell. Carson and Mitchell had asked a national sample of respondents to state their willingness to pay for changes in a water quality index that would move the majority of water bodies in the United States from one level on a water quality ladder to another, resulting in improvements that would allow for boating, fishing, and swimming in successive steps. The aggregate willingness-to-pay for the change in the water quality index for each representative facility was then used to extrapolate to the population of facilities of each type affected by the rule.

Even if the national impact of the rule can be estimated, the Agency must then seek to monetize the value of that impact using economic valuation techniques if possible. Because EPA generally does not have the time or resources required to conduct significant original economic valuation research for specific national assessments of benefits and costs, the

Agency typically must rely heavily on benefits transfer, i.e., using results from previous studies and adapting those results for the specific valuation context of interest. However, most of the previous ecological valuation studies that might serve as study sites for benefits transfer are not national in scope and generally have focused on only a limited number of ecosystem characteristics or services. Because they were designed for different purposes, previous studies have not selected either the study sites or the assessed services to facilitate national assessments of ecological benefits that might be important in a rule making context. Rather, they usually have involved specialized case studies selected because data were available or a specific change was readily observable. In addition, the studies generally measure tradeoffs for small, localized changes affecting a limited regional population.[47]

TEXT BOX 2: VALUATION AND THE CAFO EFFLUENT GUIDELINES

In December 2000, in response to structural changes in the industry, EPA proposed a new rule to govern discharges from CAFO facilities. The new rule, which was finalized in December 2003, requires facilities to implement comprehensive nutrient management plans designed to reduce the runoff of pollutants from feedlots and from the land application of manure. The rule focuses on the largest operations that represent the greatest environmental threats.

Manure from livestock operations produces a variety of potential pollutants that can migrate to ground water, streams, rivers, and lakes. These pollutants include nitrogen, phosphorus, sediments and organic matter, heavy metals, salts, hormones, antibiotics, pesticides, and pathogens (over 150 pathogens found in manure are human health risks). CAFO facilities also release a variety of gases and material into the atmosphere including particulates, methane, ammonia, hydrogen sulfide, odor-causing compounds, and nitrogen oxides.

Of the water-polluting materials covered in the CAFO rule, excess nutrients can directly affect human water supply through excess nitrates, adversely affect agriculture through excess salts in irrigation waters, and cause eutrophication of water bodies, anoxia, and toxic algal blooms. These latter effects can result in fundamental changes in the structure and functioning of aquatic ecosystems, including cascading effects that reduce water quality and species diversity. Uncontrolled releases of animal wastes have resulted in massive fish mortality.

Pathogens in polluted waters are a health hazard, both directly and through the food chain. The potential human health impacts of antibiotics and hormones in wastes have not been well identified but are of concern.

Of all the potential environmental impacts, the CAFO economic benefits analysis focused to a large extent on the nutrient runoff from land where manure has been applied and the economic benefits that would accrue from the manure management requirements of the CAFO rule. To estimate the benefits, the analysis utilized the GLEAMS model (Groundwater Loading Effects of Agricultural Management Systems). The outputs include nutrients, metals, pathogens, and sediments in surface runoff and ground-water leachate. This model was applied to model farms of different sizes, animal types, and geographic regions. From this model the reductions in pollutant loading of nutrients, metals, pathogens, and sediments were estimated for large- and medium-sized CAFO.

Seven categories of economic benefits were estimated: water-based recreational use (by far the largest category), reduced numbers of fish kills, increased shellfish harvest, reduced ground water contamination, reduced contamination of animal water supplies, reduced eutrophication of estuaries, and reduced water treatment costs. Reductions in fish kills and animal water supply contamination were valued using replacement cost. Increased shellfish harvests were valued using estimated changes in consumer surplus. Water-based recreation was valued using the Carson and Mitchell (1993) study. Ground water contamination was valued using economic benefits transfer based on a set of stated-preference studies. There was no national estimate of the economic benefits of reduced eutrophication of estuaries, but there was a case study on one estuary focusing on recreational fishing and using economic benefits transfer based on revealed- preference random utility models.

A number of potential impacts were not included in the economic benefits analysis relating to the water quality improvements of the rule including human health and ecological impacts of metals, antibiotics, hormones, salts, and other pollutants; eutrophication of coastal and estuarine waters due to nitrogen deposition from runoff; nutrients and ammonia in the air; reduced exposure to pathogens due to recreational activities; and reduced pathogen contamination of drinking water supplies. These impacts were not monetized mainly because of a lack of models and data to quantify the impacts and, in some cases, the lack of methods to perform the monetization.

Perhaps the most relevant area for which considerable economic valuation has been conducted is recreation demand. Many economic valuation studies have estimated the recreation benefits stemming from hypothetical or predicted changes in environmental characteristics of recreation sites. For example, several studies have used random utility models (a revealed-preference approach) to link physical descriptors of water quality to recreation behavior and estimate the willingness-to-pay or willingness-to-accept per recreational trip for a given change in water quality.[48] However, these studies value only localized changes and cannot be directly used to provide national-level benefit estimates.

Previous studies have also estimated the benefits associated with changes in ecological services that affect the well-being of homeowners living near the ecological systems. Examples include water regulation, flood control, and the amenities associated with healthy populations of plants and animals. The willingness of residents to pay for these services is capitalized into housing prices and can be estimated using hedonic property value methods. Examples illustrating this approach to valuing ecosystem services include Leggett and Bockstael (2000), Mahan et al. (2000), Netusil (2005), and Poulos et al. (2002). Estimates from such studies could also be candidates for use in an economic benefits transfer. However, as with the recreation studies, these studies are almost exclusively local rather than national in scope, which makes extrapolation to national- level benefit assessments difficult. Some exceptions that do provide national-level benefits assessments are Chay and Greenstone, 2005, and Deschenes and Greenstone, 2007. If sufficient high-quality original valuation studies are available, it might be possible to combine estimates of economic benefits from local studies in meta-analyses for use in benefits transfer (e.g., Smith and Pattanayak, 2002; Bergstrom and Taylor, 2006; and chapter 4). However, using meta-analysis to estimate benefits at a specific policy site can raise a number of issues. These include issues of consistency and those related to the scope of the resource changes valued in the original studies (e.g., whether they valued localized changes or changes at the national level). A meta-analysis of studies that valued localized changes can, at best, generate values for similar localized changes. It cannot generate values for changes that would occur at the national level unless individuals care only about localized effects. Therefore, even structurally based meta functions from local studies generally do not provide a functional relationship that can be used to estimate benefits at the national level, based on characteristics of the affected population. For example, using a meta function of unit values for a localized ecosystem change to predict average willingness-to-pay per person (e.g., evaluating the meta function using mean population characteristics) and then multiplying the resulting value by the relevant national population would generally not provide a valid measure of national-level benefits.

Despite the challenges described, the Agency has, in some cases, generated defensible estimates of economic benefits at the national level for a limited set of ecosystem services. For example, in the prospective benefit assessment of the Clean Air Act Amendments, described in text box 3, EPA used the best available economic and ecological models to estimate commercial forestry and agricultural benefits. However, in other cases, the Agency's efforts to provide monetized ecological benefit estimates using benefits transfer have generated benefit estimates that are much less defensible or have led the Agency to focus on a very limited set of ecosystem services.

Chapter 2 of this chapter addresses the benefit assessment for the CAFO rule and highlights the committee's concern about EPA's approach. As discussed in that chapter, EPA estimated recreational benefits using a water quality survey conducted in the early 1980s (Carson and Mitchell, 1993). The principal advantage of this approach was that it utilized a national survey and presented a simple willingness-to-pay relationship for improvements in water quality that allowed national- level benefits to be estimated relatively quickly without new research. This study was also used in the assessment of EPA's aquaculture rule (in text box 1). However, in addition to being more than 20 years old, the survey was not designed for those uses. The water quality index employed in the study was highly simplified and only reflected ecological services related to fishing, swimming, and boating. Thus, the benefits transfers were considerably outside the domain of what was envisioned in the design of the original survey and what could have been known by the people who responded to it in the early 1980s.

The desire to use value estimates from the Carson and Mitchell study apparently also influenced the choice of ecological models used to predict water quality impacts. In both the CAFO and aquaculture assessments, EPA chose to use the QUAL2E water quality model (Barnwell and Brown, 1987) apparently because it could readily be linked to this valuation study. Although this model can estimate the interactions among nutrients, algal growth, and dissolved oxygen, it is not capable of ascertaining the impacts of total suspended solids, metals, or organics on the benthos and the resulting cascading effects on aquatic communities that might have important water quality impacts.

TEXT BOX 3 – ECOLOGICAL BENEFIT ASSESSMENT AS PART OF THE PROSPECTIVE STUDY OF THE ECONOMIC BENEFITS OF THE CLEAN AIR ACT AMENDMENTS

The first prospective benefit-cost analysis mandated by the 1990 Clean Air Act Amendments included estimates of the ecological benefits resulting from the expected reductions in air pollutants (EPA, 1999).

The Agency included qualitative discussions of several potential ecological effects of atmospheric pollutants based on a review of the peer-reviewed literature (chapter 7, and pp. E-2 to E-9), including acid deposition, nitrogen deposition, mercury and dioxins, and ozone.

The Agency used two criteria to narrow the scope of work for quantification of impacts: First, the endpoint must be an identifiable service flow. Second, a defensible link must exist between changes in air pollution emissions and the quality or quantity of the ecological service flow, and quantitative economic models must be available to monetize these damages.

The Agency provided estimates of three categories of economic benefits related to ecosystems based on standard economic models and methods: a) benefits to commercial agriculture associated with reductions in ozone; b) benefits to commercial forestry associated with reductions in ozone; and c) benefits to recreational anglers in the Adirondacks lake region due to reductions in acidic deposition.

For agriculture, the Agency used crop-yield loss functions to estimate changes in yields, which were then fed into a model of national markets for agricultural crops (AGSIM) to estimate changes in consumers' and producers' surplus.

For commercial forestry, the PnET-II model was used to estimate the effects of elevated ambient ozone on timber growth. The PnET-II model relates ozone-induced reductions in net photosynthesis to cumulative ozone uptake. Analysis of welfare effects used the U.S. Department of Agriculture Forest Service Timber Assessment Market Model (EPA, 1999, pp. 92-93) to translate the increased tree growth from a reduction in ozone to an increase in the supply of harvested timber and computed the changes in consumers plus producer surplus based on the associated price changes. Because of the lack of data and relevant ecological models, the Agency did not quantify or monetize aesthetic effects, energy flows, nutrient cycles, or species composition in either commercial or non-commercial forests.

To estimate the recreational economic benefits of reducing acid deposition in Adirondacks lakes, the Agency used a published study of recreational angling choices of households in New York, New Hampshire, Maine, and Vermont (Montgomery and Needelman, 1997). Measured pH of lakes was used as an indicator of the level of ecological services from each lake. The literature on the economics of recreational angling shows that likelihood of success as measured by numbers of fish caught is a major determinant of demand for recreational angling (Phaneuf and Smith, 2005; Freeman, 1995). To the extent that populations of target species are correlated with pH levels, pH is a satisfactory proxy for fish populations and angling success rates. There was no attempt to quantify other ecosystem services of water bodies likely to be affected by acid deposition.

The Agency also presented an estimate of the economic benefits of reducing nitrogen deposition in coastal estuaries along the east coast of the United States. Although the Agency was able to estimate changes in nitrogen deposition for the three estuaries covered in the prospective analysis, it was not able to establish the necessary ecological linkages to quantify the effects on recreational and commercial fishing. The assumed avoided costs were the costs of achieving equivalent reductions in nitrogen reaching these water bodies through control of water discharges of nitrogen from point sources in these watersheds. As noted in chapter 4 of this chapter, avoided cost is a valid measure of economic benefits only under certain conditions, including a showing that the alternative whose costs are the basis of the estimate would actually be undertaken in the absence of the environmental policy being evaluated. Because it was not possible to make this showing in the case of controlling nitrogen deposition, the Agency chose not to include the avoided cost benefits in its primary estimate of economic benefits, but only to show them as an illustrative calculation.

The committee also notes that EPA has concentrated on a limited set of ecosystem services because of its focus on monetization. Estimating some of the ecological benefits of a given rule certainly provides better information than not estimating any of them, and the committee commends the Agency for its efforts to provide some information about ecological benefits. In addition, if the benefit estimates derived from a limited set of services are sufficiently large to "justify" the costs (as required by Executive Order 12866) and the only objective of the analysis is to make this determination, omitting detailed consideration of other impacts can save scarce resources without affecting the conclusion. However in some cases, the benefits from a limited set of services might not justify the costs, but a more complete assessment of benefits very well might. In these cases, focusing on only a subset of services could lead to incorrect conclusions or inferences about relative benefits and costs. Perhaps more importantly, even if estimated benefits based on a limited set of services are sufficient to justify costs, a more complete assessment of benefits could provide useful information about whether a more stringent rule is warranted. In addition, representing the benefits from a limited set of ecosystem services as the total economic benefits associated with a rule can be misleading and confusing to policy makers and the public if they have a broader conception of the rule's possible benefits.

6.1.3. Implementing The Proposed Approach

While recognizing the many challenges posed by ecological valuation in the context of national rule making, the committee believes that the valuation approach proposed in this chapter can be usefully applied in this context and can improve on the Agency's current approach to these challenges. Implementing the proposed approach would entail some short-term steps that could be incorporated into EPA's valuation processes using the existing knowledge base, as well as some longer-term strategies for research and data/method development that would improve ecological valuation for national rule makings in the future.

6.1.3.1. Implementation In The Short Run

A key premise of the committee's approach is that valuation should include early identification of the socially important impacts of an EPA rule. This requires information about both the potential biophysical effects of the Agency's actions and the ecological services that matter to people. As discussed in chapter 3, the Agency should develop a conceptual model early in the valuation process and then use that model to guide the valuation process. Conceptual models can allow the Agency to take a broad view of the complexities involved in ecological changes and ensure that impacts that are potentially important to people are included in the analysis. It should be standard practice for the Agency to develop such a conceptual model before other analytical work begins on an ecological benefit assessment.

Development of a conceptual model requires both an interdisciplinary team of experts and input about what matters to the public. To determine the relevant ecological effects to include in the conceptual model, EPA can draw on technical studies of impacts and their magnitudes. It can also solicit expert opinion regarding the physical and biological effects of a regulatory change. Figure 4, developed by the committee, illustrates that CAFOs can affect ecosystems in multiple ways and at multiple scales. The environmental effects of CAFOs extend beyond water quality impacts. For example, CAFOs generate interactive pollutants that affect air as well as water. The figure, however, does not provide a full conceptual model that maps EPA actions or decisions to potential ecological responses and ecosystem services. Instead, it provides a starting point for constructing a comprehensive overview of the potential ecological services that might be affected by an EPA rule.

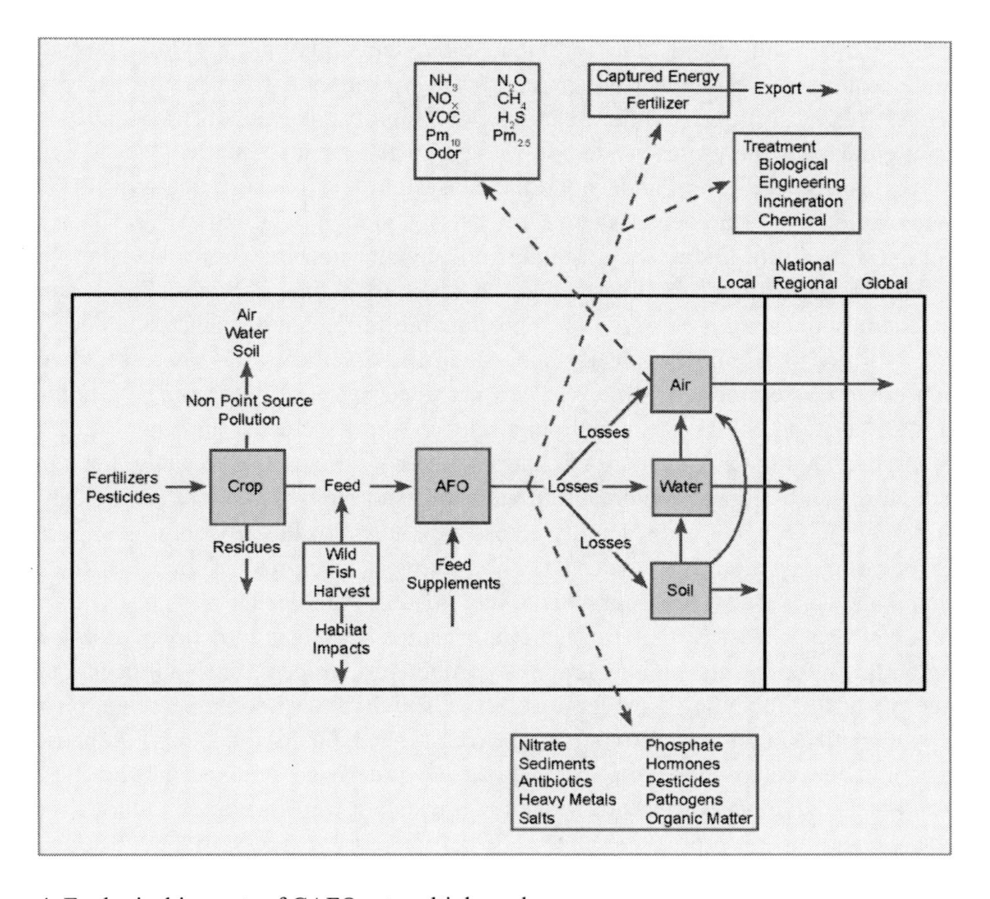

Figure 4. Ecological impacts of CAFOs at multiple scales

The conceptual model should reflect not only ecological science but also information about the changes that are likely to be of greatest importance to people. This information cannot be derived deductively. Rather, it requires input about public concerns and preferences. Although Circular A-4 requires use of economic valuation methods to estimate benefits and costs, at this early stage, EPA can use a variety of methods to identify the public concerns associated with a given rule. For example, EPA can glean this information from the existing knowledge base or actively solicit it through an interactive process. Approaches using the existing knowledge base might include:

- Inventorying the reasons invoked in similar rule- making processes in other jurisdictions (e.g., state and local)
- Inventorying the concerns expressed in public hearings at various governmental levels or in previous participatory processes through, for example, content analysis of transcripts (perhaps with weightings based on the frequency of concerns raised)
- Studying previously conducted surveys providing information about related public concerns
- Analyzing relevant initiatives, referenda, or community decisions revealing preferences for various types of ecosystem services or the avoidance of various risks

An important consideration in identifying socially important impacts is the extent to which the public understands the role that ecosystems play in providing services that contribute to human well-being. When relying on information from public expressions of preferences (e.g., surveys, public hearings, community decisions) to identify socially important impacts, the Agency should assess whether the public, when expressing preferences, was aware of and understood the ecosystem services sufficiently well to provide informed responses. Many ecosystem services – although well known to the scientific community – are little known or misunderstood by the general public (Weslawski et al., 2004). This is more likely for intermediate services than final services. For example, the public generally does not understand or appreciate the full chain of connections described in figure 4. Nor does the public typically understand the organisms and processes involved in breaking down waste products or the services provided by those processes. While the public need not understand all of the underlying science and associated linkages, they need to understand the magnitude and nature of changes they are being asked to value. Lack of public understanding can be more problematic in national-level analyses, where ecological impacts and vulnerabilities can vary substantially across locations. For this reason, it is important that queries regarding preferences and values be framed in terms of impacts that people understand and can value (see discussions in sections 2.1.4 and 3.3.2).

EPA can also at least partially mitigate information problems in national assessments by seeking public input through an interactive or participatory process. Such a process could take a number of forms, including focus groups, active solicitation of comments on a preliminary list of potentially important ecosystem services, or mediated modeling. A participatory process could also educate the participants about the underlying science and thus increase the likelihood that individuals expressing value judgments are well-informed. Although time and resource constraints may preclude use of a participatory process in many contexts, the committee suggests that EPA pilot the use of such processes (e.g., by holding open meetings with the public and Agency staff) to aid in identifying ecological changes that are important both biophysically and socially.

When properly conducted, the development of the conceptual model should identify a list of ecosystem effects or changes in ecosystem services that are potentially important to people, as well as the associated complexities, interactions, variability, and sources of uncertainty, including gaps in information. The Agency should ensure that the call for monetization, coupled with the need to generate national-level benefit estimates, does not unduly restrict the types of ecological changes considered in the benefit assessment or lead to inappropriate application of economic valuation methods or benefits transfer.

Toward this end, once EPA has identified a list of potentially important changes in ecosystems and their services, the Agency should assess the extent to which each of these can be monetized, quantified, or characterized. More specifically, the Agency should categorize potentially important effects identified in the conceptual model into five categories:

- Category 1: Effects for which benefits can be assessed and monetized using available ecological models and appropriate economic valuation methods, including benefits transfer.
- Category 2: Effects for which benefits cannot be monetized, but that can be quantified in biophysical terms using available ecological models and for which some indicator(s) of economic benefits exist.
- Category 3: Effects that can be quantified in biophysical terms but for which no indicators of economic benefits exist.
- Category 4: Effects that can be qualitatively described and generally related to benefits based on available ecological and social science, even if they cannot be quantified.
- Category 5: Effects that are likely to generate important non-economic values.

Categories 1 through 4 are designed to provide as much information as possible about economic benefits, as required by Circular A-4 (p.18). They thus fit conceptually within a benefit-cost framework. Category 5 corresponds to supplemental information about other values that could be of interest to policy makers and the public but are not based on the principles that underlie benefit-cost analysis. Thus, category 5 is conceptually distinct from categories 1-4. Note that some effects might fall into multiple categories. For example, a rule that affects a given fish population might have benefits not only for commercial fishing that can be monetized (category 1) but also cultural value to native populations that could be included in category 5.

In compliance with the OMB circular, EPA should try to include benefits in category 1 to the extent possible, and it is important for EPA to support the research needed to include more benefits in that category in the future. Nonetheless, explicit identification of benefits in categories 2 through 4 can help ensure that these effects receive sufficient attention in benefit assessments, even though they cannot be monetized with currently available data and models.

The analysis of economic benefits and other values under the committee's proposed approach differs across these different categories. With regard to the first category, estimation of monetized benefits requires three elements: a prediction of the change in relevant stressors resulting from the rule, a prediction of how that change will affect the ecosystem and ultimately the provision of ecosystem services, and an estimation of the benefits associated with the effect. To do this, the conceptual model must be linked with one or more ecological models that capture the essential linkages embodied in the conceptual model and are parameterized to reflect the range of relevant scales and regions. These ecological models must generate outputs that can be used as inputs in a benefits transfer or other economic valuation method. Because many existing ecological models do not satisfy this requirement, in the short run this requirement represents a significant constraint on the ecosystem effects for which benefits can be assessed and monetized and highlights the need for research to develop new ecological models.

Although in principle economic valuation methods can fully capture the benefits associated with changes in ecological systems and services, in practice there are significant limitations that can make this very difficult, particularly at the national level. However, even when benefits cannot be monetized using available ecological models and reliable information about economic values, the associated ecological changes may still be quantifiable. Here again, EPA should focus on quantifying all of the ecological changes that are potentially important to people. If EPA is not limited to effects for which the benefits can be monetized, the Agency can choose from among a broader set of ecological models, because the ecological models used need not directly link to existing information about economic benefits. As with monetized benefits, EPA can address the site-specificity of ecological impacts by using a bottom-up approach, and – if the relevant information about the joint distribution of the characteristics of ecosystems and human populations exists – aggregate the resulting estimates to the national level.

When monetization of benefits is not possible, the Agency should also seek to identify scientifically-based indicators of those benefits to the extent possible, i.e., it should seek to identify indicators in category 2.

Some of the valuation methods discussed in chapter 4 might be useful for this purpose. To the extent these methods generate biophysical and other measures that economic theory suggests are likely to be correlated with economic benefits, such measures can provide useful information about benefits when direct monetary estimates of those benefits are not available. For example, economic theory suggests that total economic benefits associated with an increase in wetlands in a specific area will depend, among other things, on the number of people who visit the area for recreational purposes. Other things being equal, the more people who visit the area, the higher the benefit associated with an increase in wetlands acreage. Likewise, the more people who live in the vicinity of an affected ecosystem, the greater the benefit associated with protecting that ecosystem. Similarly, if all other factors are equal, the more people who judge the protection of a given ecosystem service to be "somewhat important" or "very important" in a survey of attitudes and judgments, the higher the aggregate willingness-to-pay to protect that service is likely to be. Although these indicators do not typically reflect tradeoffs people are willing to make and hence would not provide monetary estimates of benefits that can be compared to cost, they can provide important information or signals about public preferences linked to possible benefits.

Care must be taken to avoid misinterpreting these indicators. For example, just because a large population lives in the vicinity of an affected ecosystem does not necessarily mean that a change in that ecosystem has a large value. If the change relates to a service that is not important to people, the value of that change (e.g., the willingness to pay for it) would be low regardless of the number of people living in the vicinity. To draw correct inferences, EPA needs information not only about the number of people affected but also about the importance that individuals attach to the service, as revealed through surveys or other methods.

If ecological effects can be quantified but indicators of the associated benefits are not available (category 3), EPA should report the effects in the most relevant biophysical units and discuss the basis for and expected direction of their link to possible benefits . For potentially important benefits for which quantification of the associated ecological changes is not possible (category 4 above), the Agency should characterize the changes as carefully as possible. It should discuss in detail why the changes are potentially important but not quantifiable, citing relevant literature. A carefully developed and scientifically based conceptual model can serve as the basis for a qualitative but detailed description of the ecological impacts of a given change. A simple summary of possible impacts is not sufficient. EPA should also provide justification based on the conceptual model and associated theoretical and empirical scientific literature. To the extent possible, the Agency should use the existing literature to draw inferences about the likely magnitude or importance of different effects, even if only qualitatively (e.g., high, medium, low).

Although benefit-cost analysis requires the use of economic valuation to estimate benefits, regulatory impact assessments need not be limited to information generated for use in comparing benefits and costs. Information about other sources of value that are not fully captured by the theoretical framework underlying benefit-cost analysis (category 5 above) can still be of interest to policy makers when making decisions on ecological protection. For example, the spiritual or cultural value of some ecosystems and services may be an important consideration not adequately captured by direct measures or indicators of economic benefits. Several of the valuation methods described in chapter 4 can provide information about these other sources of value.

An additional complexity, beyond the five categories for characterizing effects, arises from the national scale analysis required for most rule makings. Even when ecological models directly link to valuation, using these models to generate national-level estimates of the biophysical impacts of an EPA rule is very challenging, given the variability of ecosystem impacts within and across regions. The SAB has noted and discussed this point in other benefit assessment contexts, including the impact of Superfund sites (EPA Science Advisory Board, 2006c). To address variability across sites within a national assessment, the Agency should explore the use of a bottom-up approach to valuation. Under this approach, a number of case studies that reflect different types of ecosystems could be conducted. If information is available in a given rule making about the distribution of the ecosystem types and populations affected, EPA could aggregate these case studies to provide national-level estimates of changes in ecosystem services resulting from the rule. Even without full information about the distribution of ecosystem types and populations, the case studies could still provide information about the range of impacts and their dependence on ecosystem characteristics. This information could be useful not only for the specific policy decision at hand, but also in guiding future research. For example, it could suggest key ecosystem characteristics that would be useful in categorizing ecosystems for future valuation analyses and for which additional distribution information is needed.

Once changes in ecosystem services are estimated, those changes must still be valued to generate national benefit estimates. The appropriate valuation approach will depend on the nature of the ecosystem services. For services that generate only local benefits, benefits transfer based on comparable previous studies of localized impacts can be used, provided the benefits transfer is conducted appropriately (see discussion in section 4.3). The local or regional benefit estimates can then be appropriately aggregated to the national level. However, for ecosystem services for which local impacts generate broader national benefits, use of localized studies for benefits transfer can be problematic, as noted above. For these services, benefits transfer should instead be based on studies that have generated value information at the national level, such as national surveys of willingness to pay for national-level changes in ecosystem services. However, few surveys of this type exist. To date, EPA has relied primarily on the national water quality survey of Carson and Mitchell (1993), which was conducted over 20 years ago.[49] Additional research is needed to generate estimates of economic benefits and other values that could appropriately be used for transfers of ecological values in national assessments.

In addition to its implication for how ecological valuations are conducted, the committee's valuation approach also has implications for reporting value estimates in national benefit assessments. To increase transparency EPA should document in economic benefit assessments and RIAs the conceptual model used to guide the analysis and how decisions underlying the model were made. The assessments should describe how the ecosystem services were identified and the rationale for key choices regarding the focus of the assessment.

Consistent with the guidance in Circular A-4, benefit assessments should also clearly identify the five categories of values outlined above. If methods other than economic valuation are used to provide quantitative or qualitative information about benefits, the RIA should include a discussion of the extent to which the methods provide indicators of willingness to pay or to accept. If non-economic methods are used to capture sources of value other than those typically reflected in willingness to pay, the RIA should describe the methods used.

When monetized economic benefits are aggregated, the resulting sum should always be described as "total economic benefits that could be monetized" rather than "total benefits." In the past, EPA has sometimes reflected non-monetized benefits in aggregate measures of benefits by including an entry such as +X or +B in the summary table of benefits and costs to indicate the unknown monetary value that should be added to the benefits if the value could be determined. Although this approach indicates that the measure of monetary benefits is incomplete, the +X or +B designation provides insufficient information and can be easily overlooked in using the results of the benefit assessment. Designating the sum as "total economic benefits that could be monetized" provides a continual reminder of what is, and is not, included in this measure. EPA can provide a more accurate and complete indication of total benefits as called for by Circular A-4 by including key quantified but non-monetized impacts that are measured in biophysical units or in terms of expressed social importance or attitudes, if economic theory suggests those measurements are likely to be correlated with benefits, along with indicators of economic benefits and a detailed description of the non-quantifiable impacts.

Because of the difficulties of estimating the biophysical impacts of an EPA rule and the associated values, the Agency must also characterize the uncertainty associated with its assessment. EPA should include a separate chapter on uncertainty characterization in each assessment. This chapter should discuss the scope of the assessment, the different sources of uncertainty (e.g., biophysical changes and their impacts; social information relevant to values; valuation methods, including transfer of willingness-to-pay or willingness-toaccept information), and the methods used to evaluate uncertainty. At a minimum, the chapter should report ranges of values and statistical information about the nature of uncertainty for which data exist. For each type of uncertainty, EPA should report information similar to that reported in the Agency's prospective analysis of the benefits and costs of the Clean Air Act Amendments (EPA, 1999) and should provide a summary of this information in the executive summary of the assessment. Specifically, EPA should report potential sources of errors, the direction of potential bias for the overall monetary benefits and other value estimates and the likely significance relative to key uncertainties in the overall assessment.

6.1.3.2. Research Needs For Improvements In Future Valuation

EPA can take the steps suggested above in the short run to improve ecological valuation, but additional improvements will require longer-term investments in research in at least three areas: national-level databases to support prediction of ecological impacts; means of mapping changes in stressors to changes in ecosystem services; and transfers of value-related information.

Research is needed to develop national-level databases to predict ecological impacts (including baseline data on ecological conditions) and to value those impacts (including data on affected human populations). The current availability of national-level databases with this information is limited. In addition, research is needed on the joint distribution of relevant ecosystem and human population characteristics across local or regional sites that can be used to aggregate case studies in a bottom-up approach to national-level assessments. As noted above, case studies provide a means of incorporating heterogeneity regarding both ecological impacts and values. However, to generate national-level estimates for use in national rule making, results from case studies must be aggregated using weights that reflect the distributions of the relevant combinations of biophysical and population characteristics. Research to identify both the key relevant characteristics and their joint distributions is needed.

As discussed in chapter 3, research is needed to develop ecological production functions and associated models that can map changes in stressors to changes in ecosystem services. In the past, EPA has often been unable to estimate certain values because the Agency was not able to predict how a given rule would change stressors and how those changes would in turn affect ecosystem services. Both baseline data and the development of ecological models that focus on ecosystem services, as well as other ecosystem characteristics of importance to people, are needed. The datasets and models should support aggregate, national-level assessments.

Finally, additional research related to benefits transfer and transfers of other value-relevant information is needed, including both research on methodological issues that arise in transferring values across different contexts, as well as additional original valuation studies that the Agency can use for this purpose. These new studies should focus on value estimates that can be applied in multiple contexts (e.g., recurring rule makings) and across a broader geographical scale. Loomis and Rosenberger (2006) suggest features of study design that facilitate the use of a study's results in benefits transfer. These include use of objective quantitative measures of quality: measured in policy- relevant physical units; the evaluation of realistically small changes; the provision of information about relevant baselines; and the full and consistent reporting of results. New studies should also expand the range of ecosystem services valued so that transfers can be applied to a wider range of services and/or ecological impacts.

In addition to localized studies that could be used as study sites, national-level studies are also needed for ecological valuation in national rule making. National economic valuation surveys (such as the one conducted by Carson and Mitchell [1993]) that have recent data and a specific focus on ecosystem services have the potential to contribute significantly to the Agency's ability to conduct ecological benefit assessments to support national rule making, provided they are conducted in accordance with state-of-the-art survey procedures (see SAB Web site at http://yosemite.epa.gov/Sab/Sabproduct.nsf/WebFiles/SurveyMethods/$File/Survey_methods.pdf for detailed information about the use of survey methods for ecological valuation.). Because conducting surveys for individual rule makings is prohibitively costly in both time and resources, the Agency should focus on conducting a limited number of surveys designed to provide value information usable in multiple rule makings.

Toward this end, the Agency should develop a research program (both internally and through extramural grants) focused on developing methods and value estimates specifically for use in recurring rule makings (e.g., for rule making associated with National Ambient Air Quality Standards or Effluent Guidelines). In past years, EPA has targeted some of its Science To Achieve Results (STAR) grant resources toward benefits transfer, but a larger and more concerted effort focused on ecological valuation and the use of transfers in national rule making is clearly needed.

6.1.4. Summary Of Recommendations

To develop more comprehensive estimates of the value of ecological changes associated with national rules and regulations, the Agency needs a broader approach to ecological valuation than it has typically used in the past. The expanded approach to valuation proposed in this chapter can and should be applied to national rule making. This would entail challenges, but important opportunities for improvement as well. EPA can implement some of the committee's recommendations using the existing knowledge base. Other recommendations call for research to enhance the Agency's future capacity to conduct high-quality, scientifically-based ecological valuation for national rule making.

The Agency can improve ecological valuation for national rule making in the short run by incorporating the following recommendations:

- The Agency should begin each valuation exercise with the development of a conceptual model of the ecological system being analyzed and the ecosystem services that it generates. This model should serve as a guide or road map for the assessment.
- EPA should develop the model using input about both the relevant science and public preferences and concerns to ensure that it incorporates important ecological functions and processes as well as related ecosystem characteristics and services that are potentially important to people. Public concerns can be identified through a variety of methods, drawing on either existing knowledge or an interactive process to elicit public input.
- Once the Agency has identified a list of potentially important ecological effects and associated services, it should categorize those effects according to the extent to

which they can be quantified and monetized at the national level using economic valuation techniques including benefits transfer.

- To address site-specific variability in the impact of a rule, the Agency's assessments should include case studies for important ecosystem types. Aggregation across these case studies can be carried out if information about the joint distribution of ecosystem types and characteristics of affected human populations is available. This bottom-up approach would establish separate estimates for each locality or region and then aggregate them to obtain a national estimate.
- For ecosystem services for which the benefits are primarily local, EPA can rely on scientifically-sound benefits transfer using prior valuations at the local level. However, for services valued more broadly, the Agency should draw from studies with broad geographical coverage (in terms of both the changes that are valued and the population whose values are assessed).
- EPA should report aggregated monetized economic benefits as "total economic benefits that could be monetized" rather than "total benefits") (category 1 on page 74)
- To assess the benefits associated with effects that cannot be monetized for national rule making using scientifically sound valuation methods (including benefits transfer), EPA should:
  - provide a scientific basis for the importance of any projected ecological changes, whether they are monetized or not
  - include characterization of potentially important effects related to categories 2, 3, and 4 discussed on page 74:
    - Category 2: Effects for which benefits cannot be monetized, but that can be quantified in biophysical terms using available ecological models and for which some indicator(s) of economic benefits exist.
    - Category 3: Effects that can be quantified in biophysicial terms but for which no indicators of economic benefits exist.
    - Category 4: Effects that can be qualitatively described and generally related to benefits based on available ecological and social science, even if they cannot be quantified.
- EPA should also consider assessing values associated with effects that are likely to generate important noneconomic values (based, for example, on moral or spiritual convictions) (category 5 on page 74). Even though these values do not properly fit within a formal economic benefit-cost analysis, they can provide important additional information to support decision making. When such value estimates are included in RIAs, the RIA should discuss both the valuation method and the results in a separate section.
- EPA should include a separate chapter on uncertainty characterization in each assessment.

To enhance the Agency's capacity to conduct future ecological valuations, EPA should support research specifically designed to facilitate ecological valuation for national rule making, particularly for recurring rule makings. The committee recommends that EPA focus on at least three areas of research:

- EPA should support the development of national level databases to support valuation, including data on the joint distribution of ecosystem and human population characteristics that are important determinants of the value of ecological changes.
- EPA should support the development of quantitative ecosystem models and baseline data on ecological stressors and ecosystem service flows that can support national-level predictions of the consequences of changes in ecological stressors on the production of ecosystem services.
- EPA should support the development of additional methodological and original valuation studies designed to enhance national-level ecological values transfer, including national surveys relating to ecosystem services with broad (rather than localized) effects that can generate value estimates for use in multiple rule making contexts.

6.2. Valuation in Regional Partnerships

6.2.1. EPA's Role In Regional-Scale Value Assessment

Significant opportunities exist to use regional-scale valuations of ecosystem services to guide decision making by EPA and sub-national governments to protect and restore the environment. Many important ecological processes take place at a landscape scale. For example, habitat connectivity on landscapes, water and nutrient flows through watersheds, and patterns of exposure and deposition from air pollution in an airshed pose issues larger than a particular site and thus require regional- scale analysis.

An increase in data and methods, supported by EPA research, has also opened new frontiers for regional- scale analysis of ecosystems and their services. Publicly available, spatially explicit data on environmental, economic, and social variables have increased dramatically in recent years. At the same time, the ability to display data visually in maps and to analyze spatially explicit data using a variety of analytical models and statistical methods has expanded. An active EPA program in ecological research is underway for regional-scale analysis of ecosystems and services. As part of that program, EPA has funded research relating to restoration of water infiltration in urbanizing watersheds in Madison, Wisconsin, restoration of multiple ecosystem functions for the Willamette River in Oregon, decision-support tools to meet human and ecological needs in New England rivers, and the provision of multiple services from agricultural landscapes in the upper Midwest. As discussed in section 6.2.3.2, EPA Region 4: Southeast has developed a tool for regional ecological assessment. Other regions have also undertaken assessments of ecosystem services.

There is great potential – largely untapped to date – to use this type of analysis to aid regional decision making. Municipal, county, regional, and state governments make many important decisions affecting ecosystems and the provision of ecosystem services. Examples include land- use planning and watershed management. Unfortunately, local and state governments rarely have the technical capacity or the necessary resources to undertake regional-scale analyses of the value of ecosystems or their services or to incorporate these values into their decision making processes.

Regional partnerships among EPA, other governmental agencies, and the private sector offer the potential for expanding national, state, and local capacity to value and protect ecosystems and their services. EPA regional offices have many opportunities to collaborate at a regional scale with local and state governments, regional offices of other federal agencies, non-governmental environmental organizations, and private industry. Through collaborating with such groups, EPA can enhance environmental protection by engaging the public, gaining access to regional expertise, and promoting effective decision making on important regional-scale environmental decisions. Local and state partners can gain from access to EPA technical expertise and resources. Such partnerships can expand the knowledge base for decision making and improve the analysis of the value of ecosystems and services.

Unlike national rule making, where specific statutes or regulatory mandates often constrain analysis, regions have more freedom to use novel approaches to valuing ecosystems and their services. Such use, even on a pilot basis, may lead to improved methods and practices of valuation with potential positive impacts well beyond the region that pioneers the innovations. For example, EPA can use regional-level partnerships as a mechanism for testing and improving various valuation methods that might ultimately be used at the national level.

Because of the absence of legal or statutory requirements that EPA value ecosystems or services at the regional scale, there have been few regional ecological valuation efforts to date. In addition, regional offices may have lacked the time, resources, and expertise to undertake some of the crucial steps recommended in this chapter for valuing ecosystems and their services. For example, few regional offices have economists or other social or behavioral scientists on staff who can work on valuations. Partly for these reasons, many of the potential advantages of regional partnerships for valuing ecosystems or their services have not been realized to date.

To analyze opportunities for regional partnerships for valuation, the committee, through the SAB Staff Office, surveyed regional offices for examples of where the Agency or other governmental agencies have engaged in regional valuation efforts (EPA Science Advisory Board Staff, 2004). This section explores three case studies from Chicago; Portland, Oregon; and the Southeast. The case studies illustrate several general lessons about regional-scale analysis of the value of ecosystems and services and the potential usefulness of regional partnerships.

6.2.2. Case Study: Chicago Wilderness

Chicago Wilderness is an alliance of more than 180 public and private organizations. The overall goal of Chicago Wilderness, as stated in its *Biodiversity Recovery Plan,* is "to protect the natural communities of the Chicago region and to restore them to longterm viability, in order to enrich the quality of life of its citizens and to contribute to the preservation of global biodiversity" (Chicago Wilderness, 1999, p. 7). Chicago Wilderness is a bottom-up organization. No single decision maker or agency controls or guides Chicago Wilderness. It pursues objectives, as defined by its members, through consensus. Chicago Wilderness pursues its goals and objectives by promoting a green infrastructure to support biodiversity and to maintain ecosystems and services linked to quality of life in the Chicago metropolitan area.

As a member of Chicago Wilderness, EPA Region 5 (serving Illinois, Indiana, Michigan, Minnesota, Ohio, Wisconsin, and 35 Tribes) has provided technical and financial assistance and facilitates the partnership. EPA expertise in Region 5, particularly in natural sciences, has contributed to quantifying ecosystem services and understanding how potential stressors affect ecosystems and the provision of services. Chicago Wilderness has produced several reports, as well as a *Biodiversity Recovery Plan* and a green infrastructure map for the region.[50] The Chicago Wilderness Web site (http://www. chicagowilderness.org/) contains a chronology and links to many relevant documents, including the *Biodiversity Recovery Plan.*

Chicago Wilderness has been interested in valuing ecosystems and services, but has only begun to explore the opportunities. Although no specific legal authority mandates valuation of ecosystems or services as part of the work of Chicago Wilderness, quantifying values associated with the conservation of green space and biodiversity could help Chicago Wilderness meet its own stated objectives and communicate its analysis to other groups and the general public. The possible uses of valuation identified by Chicago Wilderness members include:

- Informing decisions on the establishment of green infrastructure, including priorities for acquisition of land by, for example, forest preserve districts or soil conservation districts
- Assessing the value of preserving ground water and ecosystem services related to clean water
- Assessing the relative value of conventional versus alternative development and demonstrating conditions in which development decisions that have positive impacts on the environment might be in the financial interest of the developer
- Communicating effectively with residents of the Chicago region regarding the value of green infrastructure and biodiversity and how these relate to residents' quality of life
- Assessing the relative value of investing in different research projects to establish priorities for funding decisions

Members of Chicago Wilderness, however, possess only limited technical expertise and practical experience in valuing the protection of ecological systems and services. EPA Region 5 also has limited capacity to value ecosystem services.

In sum, Chicago Wilderness, like many regional partnerships, would gain much from the ability to analyze the value of ecosystems and services, but it is constrained by lack of expertise and resources.

6.2.2.1. An Example Of How Valuation Could Support Regional Decision Making

Valuation of ecosystems and services is most useful when done in the context of specific decisions affecting the environment. The committee therefore chose a specific decision context – county open space referenda in the Chicago metropolitan area – to explore how this chapter's approach to valuation could support regional decisions.

Voters in four counties in northeastern Illinois have passed referenda authorizing bonds to purchase land for open space preservation or watershed protection. In November 1997, voters in DuPage County passed a $70 million open space bond. In November 1999, voters in Kane County and Will County passed bond issues totaling $70 million for open space acquisition or improvement. In 2001, the voters in McHenry County passed a $68.5 million bond for watershed protection. Although these multi-million dollar bond proposals have provided substantial funding to preserve open space and ecological processes in the region, the funds are insufficient to protect all worthwhile open space and watersheds. Given this shortfall, input about the most important lands to purchase or management actions to undertake to maintain or restore natural communities would help ensure that counties invest these funds wisely.

This section of the report looks at how valuation could help inform conservation investments under the local county bonds. The section examines three sources of values derived from protecting natural systems:

- Conservation of species and ecological systems
- Water quality and quantity
- Recreation and amenities

The discussion of water quality and quantity focuses on McHenry County because the bond issue there related directly to watershed protection. Following the process outlined in chapter 2 of this chapter, the section explores: the role of public involvement and input in determining ecosystem services of interest, predicting ecological impacts in terms of effects on these ecosystem services, and assessing and characterizing the values of these effects on the ecosystem services.

6.2.2.2. Public Involvement, Scientific And Technical Input, And Public Participation

The planning documents and activities of Chicago Wilderness illustrate several of the themes from chapter 2 of this chapter, including broad public involvement and interdisciplinary collaboration. Chicago Wilderness has made extensive efforts to engage the local community in determining the most important features of regional ecosystems and services. Two of the strengths of the organization are the broad range of groups involved and its commitment to open processes. Chicago Wilderness participants themselves define the objectives, goals, and priorities of the organization. As a result of the open, democratic process and the efforts to include multiple views and voices, the group's goals and objectives largely reflect what people in the region view as important to conserve. Engaging local communities is a vital first step in the process of valuing ecosystems and services. Engagement helps to focus scarce agency resources on issues of prime local importance, as well as to promote partnership and dialogue.

The inclusive planning process followed by Chicago Wilderness has included developing a common statement of purpose, setting up three working groups (steering, technical, and advisory committees), and working through nine planning steps (from visioning, development of inventories, assessment of alternative actions, to adopting a plan). In its early stages, Chicago Wilderness conducted workshops and meetings to define implementation strategies and to prioritize among its long- and short-term goals, which focus on the restoration and conservation of biodiversity. For priority setting, several of the workshops used valuation exercises to derive qualitative rankings of importance. Chicago Wilderness also referenced other valuation measures, such as polls and The Nature Conservancy's global rarity index.[51]

Chicago Wilderness conducted eight workshops to assess status and conservation needs for natural communities in the area: four workshops on species, addressing birds, mammals, reptiles, amphibians, and invertebrates; and four consensus-building workshops on natural communities, addressing forests, savanna, prairie, and wetlands. The natural-communities workshops developed overall relative rankings based on the amount of area remaining, the amount protected, and the quality of remaining areas (incorporating fragmentation and current management). The workshops assessed relative biological importance for community types, based on "species richness, numbers of endangered and threatened species, levels of species conservation, and presence of important ecological functions (such as the role of wetlands in improving water quality in adjacent open waters)" (*Biodiversity Recovery Plan,* chapter 4, p. 41). The workshops identified visions of what the areas should look like in 50 years.

Two different groups of scientists and land managers developed a classification scheme for aquatic communities based on physical characteristics. The groups assigned recovery goals (i.e., protection, restoration, rehabilitation, and enhancement) to streams and priority levels (i.e., exceptional, important, restorable, and other, based on Garrison, 1994-95) to lakes. The groups assessed streams using the index of biotic integrity, species or features of concern, the macroinvertebrate biotic index, and abiotic indicators. The groups also assessed threats and stressors to streams, lakes, and near-shore waters of Lake Michigan.

One disadvantage of Chicago Wilderness' broad engagement of local communities is the time-consuming nature of community involvement processes. The organization is not well placed to make rapid analyses or provide feedback on decisions that occur over a short time period.

6.2.2.3. Predicting Ecological Impacts In Terms Of Changes In Ecosystem Services

Because Chicago Wilderness is committed to the value of protecting biodiversity, it is interested in predicting impacts on the conservation of species and ecological systems at the landscape scale. It has collected spatially explicit information relevant to land use, open space, recreation, biodiversity conservation, water quality, and water quantity. It has also successfully applied a variant of the conservation value method to identify and prioritize conservation actions through spatial representation and analysis of unique and threatened species and ecosystems. Use of the method demonstrates how conservation science can be used for planning, and how a transparent approach to mapping conservation goals can be useful in a regional partnership.

However, for this spatially-explicit information to be relevant to decisions affecting ecosystems, Chicago Wilderness needs cause-and-effect relationships that can predict how policy choices will affect ecosystems and services. It does not have the information to estimate ecological production functions. Although it can be effective in providing descriptive information – particularly in the form of maps – it is limited in its ability to analyze alternative policies and make recommendations about which alternatives are preferable. For example, Chicago Wilderness would be able to provide only limited guidance to a decision maker in McHenry County concerning how to invest the $68.5 million approved by voters for watershed protection in a way that would maximize the value of ecosystems and services, because it would not be able to martial information about how particular actions affect systems and services.

Possible Ecological Impacts and Provision of Services from the Protection or Restoration of Watersheds

Watersheds figure prominently in Chicago Wilderness' work. The protection or restoration of watersheds can have a number of impacts on ecosystem services, including water quality, water quantity, and the support of ecological communities.

Surface water

- Availability – More water will be retained in the watershed because there is less runoff from impervious surfaces.
- Periodicity of flows – Changes in the hydrograph are mitigated because precipitation will be captured in the soil and vegetation, and subsequently released more slowly.
- Maintenance of minimum flows – There is a greater chance of maintaining adequate minimum flows because of the dampening effects of intact watersheds and continuation of subsurface flows.
- Flooding – Flooding is reduced because of the retention capabilities of the intact watershed.

Subsurface water

- Availability for domestic and industrial use – Availability will be increased because percolation and subsurface recharge will be enhanced by natural soil surface and vegetation.
- Maintenance of wetlands – Those habitats that depend on the water table or subsurface flow will be enhanced because natural percolation and recharge processes will be maintained.

Water quality

- Pollution dilution – Increased flows will dilute concentrations of organic and inorganic pollutants.
- Assimilation of biotic pollutants – Increased stream flows will permit greater opportunity for the assimilation of biological materials.
- Biological communities – Habitats that depend on increased quantities of water in the watershed and containing protected species will enjoy increased persistence.
- Specific habitats – Increased water quantity and more uniform stream flows will support regionally important ecological communities, e.g., in-stream communities, bottomland forests, wetlands, and wet prairies.

For illustrative purposes, suppose Chicago Wilderness wished to characterize impacts in McHenry County on three ecosystem services: minimizing flooding, maintaining or increasing groundwater recharge, and maintaining or increasing wetland communities. To predict impacts related to flooding, Chicago Wilderness could make use of a geographic information system (GIS) database it developed that includes layers depicting rivers, streams, wetlands, forest lands, and floodplains. As a first approximation, Chicago Wilderness could use historical records of flooding in McHenry County watersheds to identify watersheds with the greatest potential for flooding. It could then evaluate the potential for restoring floodplain forests and wetlands for mitigating flooding. To estimate whether a development option would adequately maintain or increase groundwater resources, it could use the maps of aquifers and soils in the GIS database that describe run-off and percolation rates for each soil type. Watersheds could be compared in terms of potential for aquifer recharge. Chicago

Wilderness could then consider the effects of alternative land use decisions on recharge (Arnold and Friedel, 2000). To address whether wetland communities would be maintained or increased, topographic maps and GIS data on rivers, streams, floodplains, forests, wetlands, and land cover could be used to rank watersheds within McHenry County in terms of potential wetlands minus current wetlands. The potential for expanding existing wetlands or restoring wetlands within watersheds could then be measured.

A number of GIS data files for McHenry County thus could assist in understanding how the protection of a given part of a watershed contributes to ecosystem processes and services. What is often lacking, however, is a cause and effect relationship that could be used to predict how alterations in management or policy would change the provision of ecosystem services. It might be possible to transfer results from studies of ecological services from other regions. For example, Guo et al. (2000) measured the water flow regulation provided by various forest habitats in a Chinese watershed. If these relationships are transferable, estimates of the effect of a policy of restoring forest habitat on water flow could be generated. Changes in water flow could then be used to predict impacts on aquatic organisms and their production functions such as waterfowl, fisheries, and wildlife viewing (Kremen, 2005).

In trying to predict how policy choices will affect ecosystems and the provision of services, experts must be careful not to substitute their own values for those of the public. Different judgments used in models may give rise to different recommendations.

6.2.2.4. Valuation Of Changes In Ecosystems And Services

Government decisions about what lands to conserve can involve tradeoffs among different ecosystem services of importance to the public. A study conducted in the Chicago metropolitan area, for example, found a tradeoff between desires to locate open space access close to people's homes and desires to locate open space to conserve species (Ruliffson et al., 2003). When there are such tradeoffs among different services, decision makers need information about the value of various aspects of ecosystems and services in order to determine what alternatives are more beneficial for the community. This information about relative values goes beyond understanding the ecological impacts of the management and policy alternatives and also reflects people's concerns and desires.

This section begins with a discussion of the potential contributions that valuation could make to Chicago Wilderness and briefly examines possible valuation methods that could be applied for different types of ecosystem services. This discussion goes well beyond what Chicago Wilderness has actually done in the valuation realm. The organization has conducted very few quantitative valuation studies and largely lacks the resources and the expertise to do so.

In one sense, however, Chicago Wilderness carried out an important valuation exercise at its very outset when it engaged its member organizations and gathered feedback on what the community felt was important. This process resulted in an important statement about the values held by the collection of organizations that constitute Chicago Wilderness. As noted earlier, its overall goal is "to protect the natural communities of the Chicago region and to restore them to long-term viability, in order to enrich the quality of life of its citizens and to contribute to the preservation of global biodiversity."

Given this clear goal statement, formal valuation studies that try to value the benefits of alternatives in monetary terms may be of secondary importance. Of primary importance is to understand how various potential strategies contribute to the protection and restoration of natural communities and the ecosystem services they provide. As noted earlier, Chicago Wilderness has used a variant of the conservation value method to identify and prioritize conservation actions that would contribute to this goal through spatial representation and analysis of biodiversity and conservation values. Not surprisingly, Chicago Wilderness has devoted most of its attention to biophysical measures of the status of natural communities. It has devoted much less attention to quantitative measures of value, monetary or otherwise.

With a clearly stated overall goal "to protect the natural communities of the Chicago region and to restore them to long-term viability," economic analysis may be largely restricted to estimating the cost of various potential strategies to achieve that objective. Cost-effectiveness analysis addresses how best to pursue a specific objective, given a budget constraint. Information about how potential strategies contribute to the protection and restoration of natural communities and about the cost of these strategies is the main information needed. There is no need to estimate the value of protecting natural communities or other ecosystem services.

Of course, things are rarely so clear. Even with a single overall goal, there are often multiple dimensions and tradeoffs among those dimensions that require an analyst to go beyond cost-effectiveness analysis. For example, in protecting natural communities, there may be tradeoffs between protecting one type of natural community versus another. When there are multiple natural communities or ecosystem services of interest, it becomes important to address questions of value – a practical matter when investment of bond monies is at stake. Is it more valuable to allocate resources to restoring upland forest or wetlands? Is it more valuable to mitigate flood risk or improve water quality? Such questions can be addressed only by comparing the relative value attached to different natural communities or services.

Economic valuation of the protection of natural communities may be important for Chicago Wilderness and the public at large for several reasons. First, when there are multiple sources of value generated by protecting natural communities (e.g., species conservation, water quality, flood control, recreational opportunities, aesthetics, etc.), monetary valuation provides a way to establish the relative importance of various sources of value. With prices or values attached to different ecosystem services, one can compare alternatives based on the overall economic value generated. Second, some biological concepts such as biodiversity are multi-faceted. How one makes tradeoffs among different facets of biodiversity conservation or among different natural community types is ultimately the same question as how one makes tradeoffs among multiple objectives. Establishing prices on different components of biodiversity or on different natural communities allows for analysis of tradeoffs among components and an assessment of the overall value of alternatives. Finally, monetary valuation may facilitate communication about the importance of protecting and restoring natural communities in terms readily understood by the public.

Non-monetary valuation can also be useful. If decisions involve tradeoffs among different natural communities or services, surveys containing attitude questions may be helpful. In some cases, people may find it easier to say whether they think it more important to provide additional protection of forests versus wetlands than to state the monetary value of protecting forests rather than wetlands.

People may value natural communities because of the ecosystem services they provide or because of their existence or intrinsic values. Of these two sources of value, the ecosystem services are generally the easier to value. Consider how Chicago Wilderness might value protecting wetlands and other watershed lands for flood control and water quality. To measure the value of flood control, it might measure avoided damages. Several studies of the value of preserving wetlands for flood control have been undertaken in Illinois, including studies of the Salt Creek Greenway (Illinois Department of Conservation, 1993; USACE, 1978) and the value of regional floodwater storage from forest preserves in Cook County (Forest Preserve District of Cook County, Illinois, 1988). The Cook County study found estimated flood control benefits of $52,340 per acre from forest preserves. The value of providing clean drinking water to the public is extremely high, far exceeding the costs of supplying it either by natural or human-engineered means. Because the question is how, not whether, to supply clean drinking water, replacement cost (e.g., the cost of building a filtration system to replace lost water purification services provided by wetlands) can be used to value the contribution of ecosystems to the provision of clean drinking water.

A large literature in environmental economics exists on estimating the values of various recreational opportunities and environmental amenities created by the natural environment. As discussed below, typical methods used to estimate the monetary value of recreation and environmental amenities include hedonic property price analysis, travel cost, and stated preference. A smaller literature uses referenda voting to infer values for open space and other environmental amenities.

Hedonic property price analysis is a common method for estimating the value of environmental amenities, especially in urban areas because of the availability of large data sets on the value of residential property values. Analysts have used the hedonic property price model to estimate the value of air quality improvements (e.g., Ridker and Henning, 1967; Smith and Huang, 1995), living close to urban parks (e.g., Kitchen and Hendon, 1967; Weicher and Zeibst, 1973; Hammer et al., 1974), urban wetlands (Doss and Taff, 1996; Mahan et al., 2000), water resources (e.g., Leggett and Bockstael, 2000), urban forests (e.g., Tyrvainen and Miettinen, 2000), and general environmental amenities (e.g., Smith, 1978; Palmquist, 1992). Although Chicago Wilderness has not used this method to date, the large number of residential property sales in the Chicago area and spatially explicit databases on many environmental attributes offers great potential to use hedonic property price analysis to estimate the values of environmental amenities.

A large literature has used the travel cost method to value recreation sites. With the large number of visitors to Lake Michigan beaches, forest preserves, and parks in the Chicago metropolitan area, Chicago Wilderness could also apply travel cost to estimate the value of recreational activities. Several studies have applied the travel cost method in urban areas (e.g., Binkley and Hannemann, 1978; Lockwood and Tracy, 1995; Fleischer and Tsur, 2003).

Stated-preference methods can also be used to estimate the value of recreational opportunities and environmental amenities. In one such study completed for Chicago Wilderness, Kosobud (1998) used a contingent valuation survey to estimate willingness to pay for the recovery or improvement of natural areas in the Chicago region. Kosobud found an average willingness to pay for expanded natural areas of approximately $20 per household per year. Extrapolating over the number of households in the region, expansion of natural areas in the region would generate about $50 million per year in benefits.

Finally, there is a small but growing literature that estimates values from voting behavior in referenda involving environmental issues. In particular, studies have analyzed the value of open space using results of voting on open-space referenda (Kline and Wichelns, 1994; Romero and Liserio, 2002; Vossler et al., 2003; Vossler and Kerkvliet, 2003; Schläpfer and Hanley, 2003; Schläpfer et al., 2004; Howell-Moroney, 2004a, 2004b; Solecki et al., 2004;

Kotchen and Powers, 2006; Nelson et al., 2007). As noted earlier, several counties in the Chicago metropolitan area have passed referenda authorizing bonds to purchase open space or protect watersheds. Although the number of referenda is relatively small, making it difficult to generalize or make comprehensive statements about values, analysis of these referenda could provide insights into the values different segments of the public place on various environmental amenities.

The only methods currently accepted by economists for estimating non-use values, such as the existence value of natural communities or biodiversity, are stated- preference methods such as contingent valuation and conjoint analysis. To estimate the existence value of protecting species and ecological systems, Chicago Wilderness could survey respondents in the Chicago area. Alternatively, it could attempt to use economic benefits transfer by applying the results of relevant surveys done in other locations. The advantage of obtaining a monetary value for the conservation of species and ecological systems through contingent valuation or conjoint analysis is that it would allow Chicago Wilderness to calculate a total economic value for alternative strategies. Without contingent valuation or conjoint analysis, non-use value could not be included, and only a partial economic value estimate for each strategy could be derived.

Any effort to estimate a monetary non-use value raises the question of whether monetary values fully reflect the values held by Chicago residents related to the protection of natural communities. In discussing the importance of protecting biodiversity, Chicago Wilderness emphasizes that a survey of Chicago focus groups found that "responsibility to future generations and a belief that nature is God's creation were the two most common reasons people cited for caring about conservation of biodiversity" (*Biodiversity Recovery Plan*, p. 14). Contingent valuation of the bequest value of biodiversity might be consistent with measuring responsibility to future generations, although the respondents in the focus group were presumably thinking in moral rather than monetary terms. Strong differences of opinion exist on whether it is appropriate to try to capture such notions as stewardship or moral values in monetary terms using stated preference methods (Sunstein et al., 2002; Sen, 1977).

Citizen juries or decision-science methods also provide a useful means of evaluating tradeoffs among potential strategies in the Chicago Wilderness context. With citizen juries, experts could work with a small group of selected individuals in the Chicago area to determine comparative values for parcels of land through a guided process of reasoned discourse. These methods might enable participants to develop more thoughtful and informed valuations, better analyze tradeoffs among multiple factors, and engage in a more public-based consideration of values. Decision science methods could provide either monetary comparisons of the values of alternative properties or weights that could be used to aggregate multiple layers of data.

Monetary values derived through citizen juries or decision science approaches may differ considerably from traditional economic measures based on individual welfare, for various reasons. Monetary values derived through citizen juries, for example, may differ both because of the consent-based choice rules employed and the explicitly public-regarded nature of the valuation exercise. Recent analysis suggests that deliberative valuations, in which a small, select group of individuals explores the values that should guide collective decisions through a process of reasoned discourse, may aggregate individual values in a manner that systematically departs from the additive aggregation procedures of standard benefit-cost analysis (Howarth and Wilson, 2006).

Although valuation information could be of great use to decision makers in evaluating alternative strategies and in communicating consequences of the alternatives to the public, Chicago Wilderness has undertaken very little valuation research or analysis. Despite some attempts to collect information about the value of protecting natural communities and ecosystem services (e.g., Kosobud, 1998), Chicago Wilderness' efforts have not been comprehensive or systematic. This contrasts with its major efforts to garner broad public involvement and input in setting the goals for the organization and its large-scale effort to collect technical and scientific knowledge to characterize the status of ecosystems and species. In part, the lack of valuation activity has been the result of the mix of expertise of the individuals involved in Chicago Wilderness. In part, the lack of valuation activity is the result of the organization's choice regarding the set of activities most important to it (which is a different sort of revealed preference). Chicago Wilderness is interested in using economic and other social-science approaches to study the value of protecting natural communities but has not yet enjoyed the right mix of expertise and circumstances to make this a reality.

6.2.3. Other Case Studies

6.2.3.1. Portland, Oregon's Assessment Of The Value Of Improved Watershed Management

In the early 2000s, Portland, Oregon, decided to analyze the ecosystem benefits and ecosystem-service values that would result from improved watershed management. Portland officials hoped to find more effective approaches to watershed management that could both save the city money and improve the well-being of its citizens. The city was particularly interested in impacts on flood abatement, water quality, aquatic species (salmon in particular), human health, air quality, and recreation. The city's Watershed Management Program requested David Evans and Associates and ECONorthwest to undertake the study, completed in June 2004 (David Evans and Associates and ECONorthwest, 2004). Although not an example of a regional partnership with EPA, the project provides one of the best examples of the kind of landscape-scale analysis of the value of ecosystems and services recommended by this chapter.

City officials realized that they understood only a portion of the contributions to well-being from improved watershed management. To be able to make more intelligent decisions about watershed management, these officials wished to have a more complete accounting. The project aimed to expand the range of ecological changes that were valued, focusing on those changes in ecosystems and services likely to be of greatest concern to the population. The study monetized the economic benefits from a variety of ecosystem services, including flood abatement, biodiversity maintenance (represented by improvement of avian and salmon habitat), air quality improvement, water quality improvement (measured by reduction of water temperature), and cultural services (which the study defined as including the creation of recreational opportunities and increase of property values).

The project commissioned both biophysical and economics analyses. The biophysical analyses included studies of hydrology and flooding potential, water quality, water temperature, habitat for salmon and other aquatic species, habitat for birds and other terrestrial species along riparian buffers, and air quality impacts (ozone, sulfur dioxide, carbon monoxide, carbon, and particulates). The economic analyses included studies of the impact of ecosystem changes on property values (including public infrastructure and residential and commercial property), flood risks, recreation, and human health.

The project used an approach that closely resembles the ecological production function approach advocated in this chapter The approach linked management changes, such as flood project alternatives, to a range of ecological changes. These ecological changes were then analyzed for their effect on various ecosystem services. Finally, the analysis attempted to economically value the changes in ecosystem services. Although conducted by separate teams, the project closely linked the ecological analyses and economic valuation.

Of particular note was the emphasis on estimating the change in values that would occur under various management alternatives. Rather than provide a static description of current conditions, which is the predominant form of information collected by Chicago Wilderness, Portland's approach tried to estimate causeand-effect relationships that would allow the systematic appraisal of alternative policy or management decisions. This focus, along with a systems approach capable of incorporating multiple economic benefits, made this an effective vehicle to study the net economic benefits of alternative management options.

The Portland study illustrates a number of good practices in conducting an integrated, regional-level analysis. The project solicited input from the public and important stakeholder groups in the design of the project so that it captured the impacts of greatest interest to the public. The project presented its results with a graphical interface that allowed stakeholders to run scenarios and see the resulting impacts based on underlying biophysical and economic models. The analysis effectively deployed existing methods and estimates, although it did not attempt to develop or test new approaches or methods.

The project also illustrates some of the potential problems and limitations in undertaking detailed quantitative landscape-scale analysis. Inevitably, there are gaps in data and understanding in this type of analysis. Gaps in understanding include how changes in management actions will affect ecological systems, and how this will affect the provision of ecosystem services and consequent value. For example: How will songbird populations change in response to changes in the amount and degree of habitat fragmentation? What is the value to residents of Portland of changes in songbird populations? Because of a lack of local information, the study often had to use economic benefits transfer, drawing on cases quite different from the Portland context to generate estimates of values.

The project was commissioned by the City of Portland and although it had minimal EPA involvement, the project is a good example of the type of systematic and integrated approach to valuing the protection of ecosystems and services recommended by this chapter. The project aptly illustrates the sequence of steps, from public input, to characterizing change in ecosystem functions under various policy and management options, to valuation of services under these alternatives. The project shows the great potential that this type of analysis offers in providing important and useful information to decision makers.

6.2.3.2. Southeast Ecological Framework Project

The Southeast ecological framework (SEF) project represents a regional geographic information system (GIS) approach for identifying important ecological resources to conserve. The Southeast region, which encompasses Alabama, Florida, Georgia, Kentucky, Mississippi, North Carolina, South Carolina, and Tennessee, is one of the fastest growing regions in the country, yet it still harbors a significant amount of globally important biodiversity and other natural resources. The SEF seeks to enhance regional planning across political jurisdictions and help focus federal resources to support state and local protection of ecologically important lands. The Planning and Analysis Branch of EPA Region 4 and the University of Florida completed the work in December 2001.

The SEF created a new regional map of priority natural areas and connecting corridors, along with GIS tools and spatial datasets. The project also identified 43 percent of this land that should be protected and managed for its specific contributions to human well-being. The project developed additional applications for conservation planning at the sub-regional and local scales.

The SEF offers a good tool to carry out regional analysis of ecological components, particularly habitat conservation. The SEF focused narrowly on conservation value, defined as the ability to sustain species and ecological processes. Because of its focus, the level of scientific knowledge underpinning the SEF is, in general, far higher than in the other case studies examined here.

The SEF, however, does not reflect the broad, integrated approach to valuation recommended by this chapter. The SEF focuses almost exclusively on habitat conservation rather than on a broad suite of ecosystem services. The SEF did not undertake extensive public involvement to determine its objective; it started with a focus on habitat conservation. It also did not attempt to combine its ecological analysis with an effort to value the protection of ecosystems or services in monetary or other terms. An important challenge facing regional analysis, particularly at a broad scale like the eight-state Southeast region, is how to incorporate all of these essential elements – a rigorous ecological approach capable of showing the range of ecological impacts from alternative policy and management decisions, public involvement and input on what consequences are of greatest importance to them, and rigorous evaluation of changes in value under alternative decisions.

6.2.4. Summary And Recommendations

Regional-scale analysis holds great potential to inform decision makers and the public about the value of protecting ecosystems and services. Recent increases in publicly available, spatially-explicit data and a parallel improvement in the ability to display and analyze such data make it feasible to undertake comprehensive regional-scale studies of the value of protecting ecosystems and services. Municipal, county, regional, and state governments make many important decisions affecting ecosystems and the provision of ecosystem services at a regional scale, but local and state governments rarely have the technical capacity or the necessary resources to undertake regional-scale analyses of the value of ecosystems or services. Regional-scale partnerships between EPA regional offices, local and state governments, regional offices of other federal agencies, environmental non-governmental organizations, and private industry could aid both EPA and regional partners. Such partnerships offer great potential for improving the science and management for protecting ecosystems and enhancing the provision of ecosystem services.

At present, however, this potential is largely unrealized. Valuation of ecosystems and services has not been a high priority for EPA regional offices largely because of tight agency budgets and the lack of specific legal mandates and authority. To date, regional offices have not undertaken the valuation of ecosystems and services at a regional scale in a comprehensive or systematic fashion. As the case studies have shown, however, various regional EPA offices and local governments have pursued some innovative and promising directions despite limited budgets and lack of specific mandates.

The committee sees great value in undertaking a comprehensive and systematic approach to valuing ecosystems and services at a regional scale. Realizing the great potential of regional-scale analyses, however, will require a significant increase in resources for regional offices and, in some cases, a somewhat different mode of operation. To reach the potential for regional-scale analysis of the value of ecosystems and services, the committee recommends that:

- EPA should encourage its regions to engage in valuation efforts to support environmental decision making, following the recommendations of this chapter.
- EPA regional staff should be given adequate resources to develop expertise necessary to undertake comprehensive and systematic studies of the value of protecting ecosystems and services. Increased expertise is needed in several areas:
  - Economics and social science: Expertise is very limited at the regional level to undertake economic or other social assessments of value. A pressing need exists to increase expertise in this area among regional offices.
  - Public involvement processes.
  - Ecology: Regional staffs have greater expertise in ecology than in public involvement processes economics, or other social sciences, but doing systematic valuations of ecosystem services will require additional ecological staff. Of greatest utility would be ecologists with expertise in assessing impacts on ecosystem services through ecological production functions to evaluate alternative management options.
- A systematic and comprehensive approach to valuing the protection of ecosystems and services requires that ecologists and other natural scientists work together with economists and other social scientists as an integrated team. Regional-scale analysis teams should be formed to undertake valuation studies. Teams composed of social scientists and natural scientists should participate from the beginning of the project to design and implement plans for public involvement, ecological production functions, and valuation.
- Gathering public input is of great importance in establishing the set of ecological consequences of greatest importance to the community. Where feasible, all regional-scale analyses of the value of ecosystems and services should involve the public at an early stage to ensure that subsequent ecological, economic, and social analyses are directed toward those ecosystem components and services deemed of greatest importance by affected communities. Generally, the process should proceed bottom-up, as opposed to top-down. Rather than asserting what is valuable, EPA must seek to understand what an informed public views as being valuable. An important question that should be addressed by EPA regional offices is how to develop effective public involvement at broader regional scales.
- Some EPA staff have expressed a desire to be provided a value for an ecosystem component or service that they can then apply to their region (e.g., a constant value per acre of wetland or wildlife habitat). Such short cuts to the valuation process are uninformed by local social, economic, and ecological conditions and can generate results that are not meaningful. This approach to valuation should be avoided.

- Regional staffs need to be able to learn effectively from valuation efforts being undertaken by other regional offices and by work within EPA's Office of Research and Development. EPA regional offices should document valuation efforts and share them with other regional offices, The National Center for Environmental Economics, and EPA's Office of Research and Development (which should in turn collaborate with the regional offices). Each regional office should also publish its studies.
- Future calls by the Agency for extramural research should incorporate the research needs of regional offices for systematic valuation studies. Doing so will maximize the probability that future grant funding will be useful for EPA's regional offices.
- Regional staff should form partnerships with local and state agencies or local groups where doing so advances the mission of EPA directly or indirectly by promoting the ability of partner organizations to value the effect of their actions on ecosystems and services and to protect environmental quality.

6.3. Valuation for Site-Specific Decisions

6.3.1. Introduction

The Environmental Protection Agency makes many decisions at the local level, including the issuance of permits (air, water, and waste), policies that influence the boundaries for establishing permits (e.g., impaired water bodies designations), and administrative orders related to environmental contamination. The social and ecological implications of such decisions, like the decisions themselves, generally are local in nature, affecting towns, townships, and counties rather than entire states or regions. Therefore, the decision processes need to rely on valuation approaches that also are local in nature and are robust enough to adapt to a range of local ecological conditions and public interests.

In this section, the committee focuses on the regulatory processes associated with one set of local decisions: the remediation and redevelopment of historically contaminated sites. That focus includes the Superfund program and its efforts to assess the contributions to human well-being from ecosystem services related to site remediation and redevelopment efforts (Davis, 2001; Wilson, 2004). As part of this committee's study, the SAB staff, with assistance from the Agency's National Regional Science Council, surveyed the regional offices to assess their needs for valuation information. Seven of the eight responding regions indicated that they need information to help value the protection of ecosystems in the management and remediation of contaminated sites (EPA Science Advisory Board Staff, 2004). The discussion that follows is applicable to any remediation and redevelopment processes for contaminated properties that contain the following elements:

- Identification, selection, and prioritization of sites
- Site characterization – establishment of site condition
- Site assessment – evaluation of risks and impacts
- Selection of remedial and redevelopment approaches
- Performance assessment of clean up and redevelopment
- Public communication of assessment results as well as proposed actions and outcomes

This section explores how valuation methods can positively influence individual steps in a remediation and redevelopment process and lead to a better outcome. As appropriate, the section identifies and discusses individual valuation approaches or methods relevant to specific steps. The section builds on a white paper funded by EPA's Superfund Program to evaluate the potential of valuation for redevelopment of contaminated sites (Wilson, 2004). The white paper assessed the improvement in ecosystem services and implied ecological value from the remediation and redevelopment of Superfund sites. Although the Wilson paper did not perform a formal valuation for any redeveloped property, it provides a useful starting point for exploring the utility of valuation methods in the remediation and redevelopment process. For his analysis, Wilson reviewed approximately 40 Superfund cases before selecting three case studies that represent urban (Charles-George landfill), suburban (Avtex Fibers), and exurban (Leviathan Mine) environments. This section analyzes and relies on these same three cases, as well as an additional urban example, the DuPage landfill, which provides a useful counterpoint to the Charles-George landfill example. The DuPage example shows how an early focus on ecosystem services can better identify potential ecosystem services that can be targeted during the remediation and restoration phases. A brief overview of each of these cases appears in section 6.3.2.

6.3.2. Opportunities For Using Valuation To Inform Remediation And Redevelopment Decisions

The Superfund process and its individual steps or stages are well defined (EPA CERCLA Education Center, 2005). Superfund and related remediation processes are focused on first defining a problem, then characterizing and assessing its potential and actual human health and environmental impacts, and finally developing and executing a technical strategy to alleviate or avoid those impacts. Since 1985, EPA's Brownfield Program has integrated consideration of upstream redevelopment into the remediation process (EPA, 2004b). The Agency developed the reuse assessment tool to integrate land use into the Superfund process (Davis, 2001). Integrating remediation and redevelopment demonstrates the need to consider ecological valuation into all steps and stage from the very beginning.

Figure 5 illustrates how valuation information can be integrated into the traditional process for remediation and redevelopment. In the committee's view, EPA and the community should define at the outset what the potential site should be after remediation and redevelopment and what ecological services are to be preserved, restored, or enhanced for use by the local community. This differs from the more traditional practice. This practice initially focuses on the type, degree, and extent of chemical contamination, and then on the human and ecological receptors currently exposed and therefore at risk under current chemical conditions.

In the traditional approach, the data collection for site characterization captures the degree and pattern of chemical contamination but does not collect information about the ecological condition of the site or the value of any services associated with the site in its current or proposed future conditions. In the traditional approach, moreover, the conceptual model that defines the exposure pathways to key receptors and therefore guides the design of the risk analysis is based on current rather than future conditions. This can lead to a risk assessment that selects receptors that are sensitive under current conditions but may not be sensitive or important under alternative future uses. This logic focuses remedy evaluation and selection on controlling the risks under current use. In the end, the traditional approach assumes that risk reduction and management, rather than the optimized reuse value for the community, are the ultimate performance goals. Such an approach may leave the community feeling that the risk is gone but still dissatisfied with the values gained by the cleanup.

Integrating future use considerations into the remediation process and focusing on value generation will lead to outcomes that better satisfy the public. To accomplish this metamorphosis, it is essential to find ways to introduce estimates of ecosystem services and values into management strategies and associated analytical processes. Early recognition of future uses and the ecosystem services that matter to people can inform site assessment and the ultimate selection of remedial actions and redevelopment options. Identifying expected or actual contributions to human well-being can also lead to more effective communication with the affected public. The rest of this section discusses the opportunities and utility of adapting valuation methods to this more integrated and forward-looking assessment and redevelopment process.

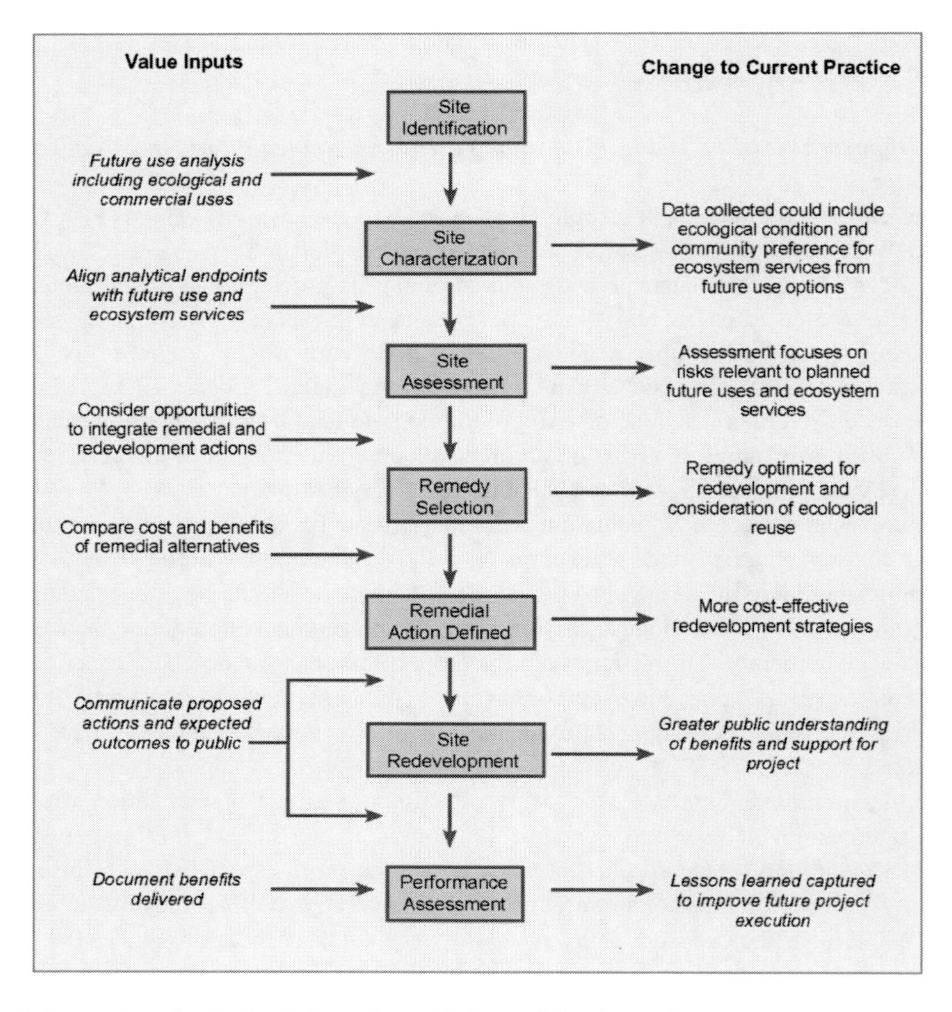

Figure 5. Integration of valuation information with the traditional remediation and redevelopment process

Valuation methodologies are important first in identifying how a site and its current or potential ecosystem services matter to the surrounding community. EPA should use valuation methods to determine how the site has contributed and can contribute to human well-being and how potential effects on ecological components may diminish those contributions. When the ecosystem services that matter to people are well-defined and when ecological risk assessments are coupled with these services, the remediation and redevelopment plan can target what matters to the local community. A key recommendation, therefore, is that EPA consider ecosystem services and their contributions to human well-being and other values from the earliest stages of addressing contaminated properties.

Even as early in the management process as site selection or prioritization, tools that can compare the potential of sites to provide ecosystem services could be informative. The contribution of ecological protection to human well-being should be considered in the design of any site characterization plan. A typical site characterization focuses on the aerial extent of chemicals and their range of concentration in site media (e.g., ground and surface water, soil, and biological tissue). A plan that also collects information to define and assess ecosystem services would better align ecological-risk assessments with economic benefits and other contributions to human well-being. Aligning risk assessments and assessments of contributions to human well-being should be a critical objective for the Agency. Alignment will help ensure that remedial actions address the restoration of contributions to human well-being derived from important ecosystem service flows that have been diminished or disrupted. Aligning risk assessment endpoints with ecosystem services should also result in multiple benefits, including:

- Improved alignment with community goals
- Improved ability to perform meaningful assessments of economic benefits and other contributions to human well-being
- Improved ability to communicate proposed actions
- Improved ability to monitor and demonstrate performance

Successfully remediating and redeveloping contaminated sites depends in great part on the degree to which efforts either protect or restore ecosystem services that contribute to human well-being. If values have been broadly explored and effectively integrated into site assessment and remedy-selection processes, appropriate measures of performance will be apparent. Ecological measures of productivity or the aerial extent of conditions directly linked to valued ecosystem services will be useful in tracking the performance of remediation and redevelopment processes. Advancing the Agency's capability to evaluate performance both in real time and retrospectively will help the Agency better justify its overall performance record in the remediation and redevelopment of contaminated sites.

Finally, the remediation and redevelopment of a property encompasses more than just the science and engineering that historically have underpinned the remediation process. Effective communication with members of the public actively participating in the remedial and redevelopment process and with the general public is a critical element in the success of the management process. Both of these audiences bring values to the table when they evaluate proposed actions or the results of any action taken. A strong alignment between the ecosystem services valued by these audiences and expected or actual outcomes will facilitate effective remediation and redevelopment.

TEXT BOX 4: CHARLES-GEORGE LANDFILL: AN URBAN EXAMPLE

From the late 1950s until 1967, the Charles-George Reclamation Trust Landfill, located one mile southwest of Tyngsborough and four miles south of Nashua, New Hampshire, was a small municipal dump. A new owner expanded it to its present size of approximately 55 acres and accepted both household and industrial wastes from 1967 to 1976. The facility had a license to accept hazardous waste from 1973 to 1976 and primarily accepted drummed and bulk chemicals containing volatile organic compounds (VOCs) and toxic metal sludges. Records show that over 1,000 pounds of mercury and approximately 2,500 cubic yards of chemical wastes were landfilled. The state ordered closure of the site in 1983. That same year, EPA listed the site on the National Priorities List (NPL) and the owner filed for bankruptcy. Samples from wells serving nearby Cannongate Condominiums and some nearby private homes revealed VOCs and heavy metals in the groundwater. Approximately 500 people lived within a mile of the site in this residential/rural area; 2,100 people lived within three miles of the site. The nearest residents were located 100 feet away. Benzene, tetrahydrofuran, arsenic, 1 ,4-dioxane, and 2-butanone, among others, had been detected in the groundwater. Sediments had been shown to contain low levels of benzo(a)pyrene. People faced a potential health threat by ingesting contaminated groundwater. Flint Pond Marsh, Flint Pond, Dunstable Brook, and nearby wetlands were threatened by contamination migrating from the site.

EPA's involvement at the site began with groundwater testing conducted by an EPA contractor during 1981 and 1982. The site was proposed for the NPL on October 23, 1981, and finalized on the NPL in September 1983. In September 1983, EPA also allocated funds for a removal action at the site to replace the state's Department of Environmental Quality Engineering temporary water line with another temporary but insulated water line. Other removal work included construction of a security fence along the northwestern entrance to the landfill, regrading and placement of soil cover over exposed refuse, and installation of twelve gas vents.

A remedial investigation and feasibility study was also begun in September 1983. The basis for the removal action was documented in the first record of decision issued on December 29, 1983.

EPA *Web site history*

http://yosemite.epa.gov/r1/npl_pad.nsf/f52fa5c31fa8f5c885256adc0050b631/ABD28 6D719D254878525690 D00449682?OpenDocument

TEXT BOX 5: DUPAGE COUNTY LANDFILL: AN URBAN EXAMPLE

The 40-acre tract of land that is the Blackwell Landfill was originally purchased by the DuPage County Forest Preserve District (FPD) in 1960 and is centrally located within the approximately 1,200-acre Blackwell Forest Preserve, about 30 miles outside Chicago, Illinois. The landfill was constructed as a honeycomb of one-acre cells lined with clay. Approximately 2.2 million cubic yards of wastes were deposited in the landfill between 1965 and 1973. The principal contaminants of concern for this site were the volatile organic compounds (VOCs) 1,2-dichloroethene, trichloroethene and tetrachloroethene, detected in onsite groundwater at or slightly above the maximum contaminant level (MCL). Landfill leachate contained all kinds of VOCs and semivolatiles including benzene, ethylbenzene toluene, and dichlorobenzene, as well as metals such as lead, chromium, manganese, magnesium, and mercury. VOCs and agricultural pesticides had also been detected in private wells down gradient of the site but at low levels. Some metals (manganese and iron) had been detected above the MCLs in down-gradient private wells. Post-remediation, the site now consists mainly of open space, containing woodlands, grasslands, wetlands, and lakes, used by the public for recreational purposes such as hiking, camping, boating, fishing, and horseback riding. There are no residences on the FPD property, and the nearby population is less than 1,000 people. The landfill created Mt. Hoy, which is approximately 150 feet above the original ground surface.

EPA *Web site history*
http://cfpub.epa.gov/supercpad/cursites/csitinfo.cfm?id=0500606

6.3.3. Illustrative Site-Specific Examples

The following analysis applies the general recommendations of chapter 2 to the site-specific level. The committee illustrates these site-specific recommendations with lessons gleaned from a series of Superfund examples in urban (Charles-George and DuPage landfills), suburban (Avtex Fibers) and ex-urban (Leviathan Mine) contexts. The backgrounds on each of these cases appear in text boxes 4-7.

6.3.3.1. Determining The Ecosystem Services Important To The Community.

The urban examples of the Charles-George and DuPage County landfills show the value of engaging with the community at an early stage to determine the ecosystem services of importance to them. Although neither landfill apparently used formal valuation methods at the outset, DuPage County's focus on ecosystem services and the inclusion of additional experts (i.e., forestry experts) led to a more positive outcome.

At the Charles-George landfill, EPA did not consider ecological values or future uses at the start. The human health risks at this site were so salient at the time that they were the focus of subsequent decisions. EPA addressed the health and safety risks by capping the landfill site and extending the water system from the city of Lowell, Massachusetts, to the affected community. Although EPA published the record of decision more than 20 years ago, the site is still a fenced-off no-man's land, and the potential for ecosystem services remains untapped.

By contrast, the remediation and redevelopment of the DuPage County landfill site appears to have been motivated largely by the need to address existence values (e.g., the presence of hawks and other rare birds) and recreational values (e.g., hiking, bird watching, boating, camping, picnicking, and sledding). The remediation effort succeeded, and the site is now part of the Blackwell Forest Preserve. Listed as a Superfund site in 1990, "a once dangerous area is now a community treasure, where visitors picnic, hike, camp, and take boat rides on the lake" (EPA, 2004c).

The urban examples show that even the most rudimentary dialogue about future use can lead to an outcome with greater service to the community. At the DuPage landfill site, a qualitative focus on the utility of ecosystem services led to the recognition that in a very flat landscape, even a 150-foot hill, if properly capped and planted, would be a welcome refuge for people as well as wildlife. The DuPage Forestry District understood the ecological potential of the area, particularly for hawks, and recognized that, where hawks abound, birders will come to watch them. The difference was one not of methodology but of conception.

In working with the Avtex Fibers site (described in text box 6), a suburban location, EPA also engaged key members of the public. After the site was listed and a management process established, EPA undertook a clear effort to engage the public through a multi- stakeholder process in the development of the master

TEXT BOX 6: AVTEX FIBERS SITE: A SUBURBAN EXAMPLE

The Avtex Superfund site consists of 440 acres located on the bank of the Shenandoah River within the municipal boundaries of Front Royal, Virginia. The site is bordered on the east by a military prep school, on the south by a residential neighborhood, and on the west by the Shenandoah River. From 1940 to its closure in 1989, industrial plants on the site manufactured rayon and other synthetics. Tons of manufacturing wastes and byproducts accumulated on the site, infiltrated into groundwater under the site, and escaped into the Shenandoah River. The Avtex Fibers site was proposed for inclusion on the National Priorities List on October 15, 1984, and the site was formally added to the list on June 10, 1986. EPA began removal activities at the site in 1989 to address various threats to human health and the environment. The cleanup and restoration plan called for most remaining wastes to be consolidated on site, secured with a protective material where needed, and covered by a thick cap of soil and vegetation.

Front Royal is close to the Appalachian Trail, Shenandoah National Park, and George Washington National Forest, and a number of significant Civil War sites, making it a major tourist center for the Blue Ridge Mountains. Biologically, the Avtex site contains some residual forested areas, open meadows, small wetland areas, and more than a mile and a half of frontage along the Shenandoah River. The proposed master plan for redevelopment, created through a formal multi-stakeholder process, divides the site into three areas: a 240- acre river conservancy park along the Shenandoah River combining ecological restoration and conservation of native habitats; a 25-acre active recreation park with boat landings, picnic shelters, and a developed recreation area including a visitor center and soccer fields; and a 165-acre eco-business park, featuring the refurbished historic former Avtex administration building. Cleanup of the Axtex site is ongoing, and the redevelopment plan is being actively pursued by local government agencies and private industry groups.

EPA Web site history

http://www.epa.gov/superfund/accomp/success /avtex.htm

Stakeholders' Avtex Fibers Conservancy Park Master Plan

http://www.avtexfibers.com/Redevelopment/avtexWEB/avtex-Mp.html

TEXT BOX 7: LEVIATHAN MINE SUPERFUND SITE: AN EX-URBAN EXAMPLE

In May 2000, the EPA added the Leviathan Mine site in California to the National Priority List of Superfund sites. The site is currently owned by the state, but from 1951 until 1962 the mine was owned and operated by the Anaconda Copper Mining Company (a subsidiary of ARCO) as an open pit sulfur mine. The mine property is 656 acres in a rural setting near the Nevada border, 24 miles southeast of Lake Tahoe. The mine itself physically disturbed about 253 acres of the property plus an additional 21 acres of National Forest Service land. The site is surrounded by national forest. In addition, it lies within the aboriginal territory of the Washoe Tribe and is close to several different tribal areas.

The mine has been releasing hazardous substances since the time that open pit mining began in the 1950s. Releases occur through a number of pathways, including surface water runoff, groundwater leaching, and overflow of evaporation ponds. In particular, precipitation flowing through the open pit and overburden and waste rock piles creates acid mine drainage (AMD) in the form of sulfuric acid, which leaches heavy metals (such as arsenic, cadmium, copper, nickel, and zinc) from the ore. These releases are discharged into nearby Leviathan Creek and Aspen Creek, which flow into the East Fork of the Carson River. Pollution abatement projects have been underway at the site since 1983. Despite these efforts, releases continue today.

The releases of hazardous substances from the mine have significantly injured the area's ecosystem and the services it provides. In the 1950s, structural failures at the mine that released high concentrations of AMD into streams resulted in two large fish kills, and the trout fishery downstream of the mine was decimated during this time. More recently, data have documented elevated concentrations of heavy metals in surface water, sediments, groundwater, aquatic invertebrates, and fish in the ecosystem near the site. This suggests that hazardous substances have been transmitted from abiotic to biotic resources through the food chain, thereby affecting many trophic levels. A recent assessment identifies seven categories of resources potentially impacted by the site: surface water resources, sediments, groundwater resources, aquatic biota, (including the threatened Lahontan cutthroat trout), floodplain soils, riparian vegetation, and terrestrial wildlife (including the threatened bald eagle). These uses, in turn, help support recreational uses (including fishing, hiking, and camping); and tribal uses (including social, cultural, medicinal, recreational, and subsistence).

The process of determining compensatory damages and developing a response plan involves a number of different stages for which information about the value of these lost services would be a useful input.

For example, in accordance with the Natural Resource Damage Assessment (NRDA) regulations under the Comprehensive Environmental Response, Compensation and Liability Act, the trustees for the site conducted a pre-assessment screening to determine the damages or injuries that may have occurred at the site and whether a natural resource damage assessment should be undertaken. This required a preliminary assessment of the likelihood of significant ecological or other impacts from the contamination (corresponding to step 2 in figure 1 of this chapter).

The decision was made in July 1998 to move forward with a Type B NRDA and thus to assess the value of the ecosystem services that have been lost as a result of the site contamination. A Type B assessment involves three phases: an injury determination to document whether ecological damages have occurred; a quantification phase to quantify the injury and reduction in services (corresponding to step 4 of figure 1); and a damage determination phase to calculate the monetary compensation that would be required (corresponding to step 5 of figure 1).

In the Leviathan Mine case, the trustees proposed using resource equivalency analysis based on a replacement cost estimate of the lost years of natural resource services to determine damages for all affected services other than non-tribal recreational fishing. For this latter ecosystem service, they proposed using economic benefit transfer to estimate the value of lost fishing days. In the decision by EPA whether to list the site on the NPL and in the subsequent record of decision selecting a final remedy for the site, information about the value of the ecological improvements from cleanup could play an important role, although these decisions have often been based primarily on human health considerations.

EPA *Web site history* http://www.epa.gov/superfund/sites/npl/nar1580.htm

Leviathan Mine National Resource Damage Assessment Plan,
http://www.fws.gov/sacramento/ec/Leviathan%20NRDA%20Plan%20Final.pdf.

For sites like Avtex Fibers, deliberative group processes involving the public and relevant experts, including historians, could help identify and document ecosystem services of most concern to the public. In framing the dialogue with members of the public, methods such as ecosystem benefits indicators or the conservation value method might help EPA's site managers understand the ecosystem-service potential of future uses. Those methods could also provide inputs for further valuation using other methods described in chapter 4 (e.g., economic methods or decision science approaches).

The Leviathan Mine case, described in text box 7, illustrates how EPA often must consider a complex array of competing interests. The Agency in this case faces a clear dichotomy between the ecosystem services valued by the full-time resident population of American Indians and by occasional recreational users. Recreational users would gain from services associated with hiking, fishing, and camping. The Washoe Tribe, however, values the ecosystem as a provisioning service for food as well as for its spiritual and cultural services.

The Leviathan Mine case also highlights the need to consider the existence or intrinsic values of an ecosystem. The ecosystem near the Leviathan Mine provides a habitat for threatened species such as the Lahontan cutthroat trout and bald eagle. In considering site restoration or remediation, or in measuring damages from contamination at the mine, the Agency could miss the primary sources of value if it limited consideration to use value and did not consider existence or intrinsic value.

For the Leviathan Mine example, EPA could obtain information about the impacts of greatest concern to affected individuals in at least three ways. The first would be to gather information about the relative importance of the various services directly from affected individuals through focus groups, mental models, mediated modeling, deliberative processes, or anthropological or ethnographic studies based on detailed interviews. The second approach would be to gather basic information that could indicate the importance of different services. This information might be of the type used to construct ecosystem benefit indicators: water use data for the Washoe tribe and others in the vicinity of the site (e.g., sources, quantities, and purposes), harvesting information for the Washoe (e.g., what percent of their harvesting of nuts, fish, etc., comes from the area affected by the site), recreational use data (e.g., the number of people visiting the local national forest for hiking, camping, fishing, and wildlife viewing), data on flooding potential and what is at risk in the vicinity of the site, and data on spiritual/cultural land-use practices by the Washoe. The third approach would be to review related literature and previous studies to learn about impacts of concern in similar contexts. For example, previous social/psychological surveys not specific to this site or other expressions of environmental preferences (e.g., outcomes of referenda or civil court jury awards) might provide insight into what people are likely to care about in this context. Similarly, previous contingent valuation studies of existence value might provide some, at least partial, indication of the likely importance of impacts on species such as bald eagles. Likewise, previous studies of the value of recreational fishing (e.g., from travel cost models) could be coupled with use data to provide an initial indication of the importance of the impact on recreational fishing.

6.3.3.2. Involving Interdisciplinary Experts Appropriate For Valuation.

Interactions among experts and the affected public form a key component of any program of hazardous site assessment, planning, and implementation. Ideally, collaboration among all relevant experts, including physical, chemical, and biological scientists (e.g., ecologists and toxicologists) and social scientists (e.g., economists, social psychologists, and anthropologists), as well as communication with affected publics, must begin very early in the planning stages of remediation and redevelopment and continue throughout implementation and post-project monitoring and evaluation. Key areas for collaboration among experts are the development of alternative management scenarios and the translation of physical and biological conditions and changes into value-relevant outcomes that can be communicated to the public.

The Leviathan Mine illustrates the need for collaboration among multiple disciplines to understand how the population's values are affected. Because of the unique cultural and spiritual values associated with the site, anthropologists could play an important role in characterizing the value of the ecosystem services to the Washoe Tribe. Economists or others seeking to estimate existence value for an affected species would need to work closely with ecologists to determine the likely impact of any change or proposed project on that species so that the change could be readily valued.

6.3.3.3. Constructing Conceptual Models That Include Ecosystem Services

Ecological assessments associated with the remediation and redevelopment of contaminated property will better aid decision making if they incorporate ecological production functions that link remediation and redevelopment actions to ecosystem services. None of the four sites chosen by the committee conducted such assessments. Both the DuPage County landfill and the Aztex Fibers cases appear to have qualitatively considered ecosystem services, with commendable results, illustrating how more formal assessments using ecological models and production functions could further improve site-specific remediation and redevelopment efforts.

Although it is now standard practice to develop a conceptual model in performing ecological risk assessments for contaminated sites, EPA's analyses of adverse impact have generally not linked to ecosystem services. The primary focus of the Agency's remediation efforts has been to control anthropogenic sources of chemical, biological, and physical stress that could lead to adverse impacts to human health or the environment. Developing conceptual models that incorporate the linkage between ecological endpoints and community-identified services would better guide both for the valuation of ecological protection and site remediation and redevelopment.

The Avtex Fiber case highlights what EPA could gain from developing the capacity to use conceptual models that integrate ecological effects and ecosystem services. A noteworthy feature of the Avtex Fiber process was the development of a master plan, which included some consideration of ecosystem services. For example, early concerns about contamination of groundwater and the discharge of toxic substances into the Shenandoah River focused attention on water quality. Aquatic basins constructed to contain contaminants on site were designed to restore important ecosystem services, including safe habitat for waterfowl, runoff control, and water purification services. In this regard, the plan implied but failed to quantify or document a rudimentary ecological production function.

The development of a conceptual model that incorporated ecosystem services would have systematically facilitated greater integration of ecosystem services into remedial design and future uses. Recreational and aesthetic services were clearly important considerations for many features of the plan. However, because no comprehensive ecological model identifying ecosystem services apparently guided redevelopment at the site, it is unclear whether the particular pattern of restored forests and wetlands, recreation areas, and industrial parks produced the most valuable protection for ecosystem services. Different siting and design of soccer fields, for example, might have provided the same recreational value while achieving greater wildlife habitat, water quality, or aesthetic values for visitors, nearby residents, or both. The master plan's declared green focus for the industrial park implied that ecological concerns were important in the selection of industrial tenants and in the siting and design of facilities, but no ecological model for achieving this goal, or monitoring progress toward it, was presented. This omission leaves open the prospect that future industrial, recreational, and tourist developments and uses at the Avtex site might simply substitute one set of damages to ecosystems and ecosystem services for another.

6.3.3.4. Predicting Effects On Relevant Ecosystem Services

As discussed in chapter 3, development of a conceptual model should be followed with predictive analyses of the effects of EPA's actions on ecosystem services. Expanding ecological risk assessments to include assessments of the services that matter to people may present technical challenges, given the current focus of ecological risk assessments on toxicological data for a limited range of species and for toxic responses from individuals in those species. Such data will rarely link well to the ecosystem services that matter to a particular site-specific decision.

The Agency will need to develop its capacity to adapt and apply models that incorporate ecological production functions. These models are the real bridge between risk estimates and subsequent injury or damage projections and provide a major piece of the puzzle to quantify and value the impacts of chemical exposures under different remedial and restoration alternatives.

Incorporating ecological production functions into EPA's risk assessments will be important not only for EPA decisions on site remediation and redevelopment but also for natural resource damage assessments (NRDAs). Although trustee agencies, such as the National Oceanic and Atmospheric Administration and the U.S. Fish and Wildlife Service, are the regulatory leads for NRDAs, the ecological risk assessments and conceptual models produced by EPA in the remediation process are often the basis for damage assessment. If EPA could effectively conduct assessments that use ecological production functions to predict impacts on ecosystem services, those assessments would enhance the ability of resource trustees to assess injury, define restoration goals, and calculate damages. Predictive ecological production functions can play a critical role in such assessments.

The Leviathan Mine example illustrates how ecological impacts and damages are currently assessed. The Leviathan Mine natural resource damage assessment plan gives detailed information on concentrations of key pollutants (particularly heavy metals such as cadmium, zinc, copper, nickel, and arsenic) in surface water samples, groundwater samples, sediment samples, samples of fish tissues, and insect samples at various distances from the mine site. These concentration levels can be compared to concentration levels at reference sites (because historical information for the site itself is not available), toxicity data from the literature, and existing regulatory standards (e.g., water quality criteria or drinking water standards) to evaluate the potential for impact. Importantly, none of these comparisons is a direct demonstration of injury, which can only be measured through field observation and tests. EPA must rely on surrogates to estimate impact.

Once the impacts on water quality, sediments, etc., have been determined, ecological production functions could translate these impacts into predicted changes in ecosystem services. If recreational fishing is important, for example, EPA must estimate the site's impact on the fish population in the nearby water body. Such an analysis would require estimating the impacts of changes in water quality, streambed characteristics, bank sediments, and riparian vegetation on fish population, both directly and through impacts on the insects on which fish feed. If elevated levels of arsenic, copper, zinc, or cadmium exist in insects and fish tissue, EPA must also be able to use this information to predict an overall impact on the fish population.

EPA has already developed complex ecological risk assessment modeling tools (e.g., TRIM, EXAMS, and AQUATOX) to estimate the fate and effects of chemical stresses on the environment. In some cases, EPA has even coupled such exposure-effects models with ecological production models to estimate population level effects (Nacci and Hoffman, 2006; Nacci et al., 2002).

In many cases, an ecological model that links ecological processes at a site to ecosystem services of interest to that site do not currently exist, although it might be possible to adapt models from the literature to fit local conditions with site-specific field data if the scale and ecological components of the site are similar (using the criteria for selecting among existing models described in section 3.3.1). In the absence of such a site-specific model, EPA might look to the scientific literature for guidance on how sensitive the insects and fish species are to these types of stressors. It could then ask expert ecologists to judge the likely magnitude of the impacts in the specific case. As for transfer of ecological benefits, however, scientists must take into account the differences between the reference site and the contaminated site and define and communicate the assumptions and limitations of transferring the information.

TEXT BOX 8: NET ENVIRONMENTAL BENEFIT ANALYSIS

As described by Efoymson et al. (2003), "Net environmental benefits are the gains in environmental services or other ecological properties attained by remediation or ecological restoration, minus the environmental injuries caused by those actions. Net environmental benefit analysis (NEBA) is a methodology for comparing and ranking the net environmental benefit associated with multiple management alternatives. A NEBA for chemically contaminated sites typically involves the comparison of the following management alternatives: (1) leaving contamination in place; (2) physically, chemically, or biologically remediating the site through traditional means; (3) improving ecological value through onsite and offsite restoration alternatives that do not directly focus on removal of chemical contamination; or (4) a combination of those alternatives.

NEBA involves activities that are common to remedial alternatives analysis for state regulations and the Comprehensive Environmental Response, Compensation and Liability Act, response actions under the Oil Pollution Act, compensatory restoration actions under Natural Resource Damage Assessment, and proactive land management actions that do not occur in response to regulations, i.e., valuing ecological services or other ecological properties, assessing adverse impacts, and evaluating restoration options."

Figure 6, taken from Efroymson et al. (2003), "depicts the high-level framework for NEBA. It includes a planning phase, characterization of reference state, net environmental benefit analysis of alternatives (including characterizations of exposure of effects, including recovery), comparison of NEBA results, and possible characterization of additional alternatives." Dashed lines indicate optional processes; circles indicate processes outside the NEBA framework. Only ecological aspects of alternatives are included in this framework. "The figure also depicts the incorporation of cost considerations, the decision, and monitoring and efficacy assessment of the preferred alternative, although these processes are external to NEBA."

Because NEBA is a framework, the needed resources, data inputs, and limitations are associated with whatever ecological models and valuation tools are selected.

Currently, NEBA is being applied at a local scale, although the size of some contaminated properties and their impacts can extend to the regional scale (e.g., impact of releases from a contaminated site to a watershed). NEBA should be highly adaptable to different levels of data, detail, scope, and complexity.

The Leviathan Mine Natural Resource Damage Assessment Plan also suggests studying the fish population downstream from the mine and comparing it to the population in a reference location, assuming an appropriate reference site can be identified. More generally, it suggests comparing riparian vegetation, the composition of the benthic community, and wildlife populations near the mine and at an acceptable reference site. Such a comparison can help frame the types of damages resulting from the mining activity. Because reference sites and exposed sites may differ for a number of reasons not related to the contamination, such a comparison may not directly estimate the injury and will not take into consideration the impact of proposed remedial actions. Decisions about remediation and restoration require analysis of proposed actions, and it may not be reasonable to assume that remedial actions will be 100 percent effective in restoring relevant ecosystem services to their original level.

Comparative analyses of remedial actions using ecological production functions are needed and can be facilitated through comparative tools such as net environmental benefit analysis (Efroymson et. al., 2004). This analysis provides a framework for using valuation tools to compare alternative remedial strategies based on net impacts on ecological services.

6.3.3.5. Defining, Cataloging, And Accounting For Ecosystem Services

Accounting rules are needed to avoid double counting or undercounting the contributions to human well-being from ecosystem services. Ecosystems and their numerous components are linked in an intricate and complex network of biological, chemical, and energy flows. A focus on isolated impacts to individual organisms or components and their associated services can lead to double counting or undercounting contributions to human well-being generated by Agency actions.

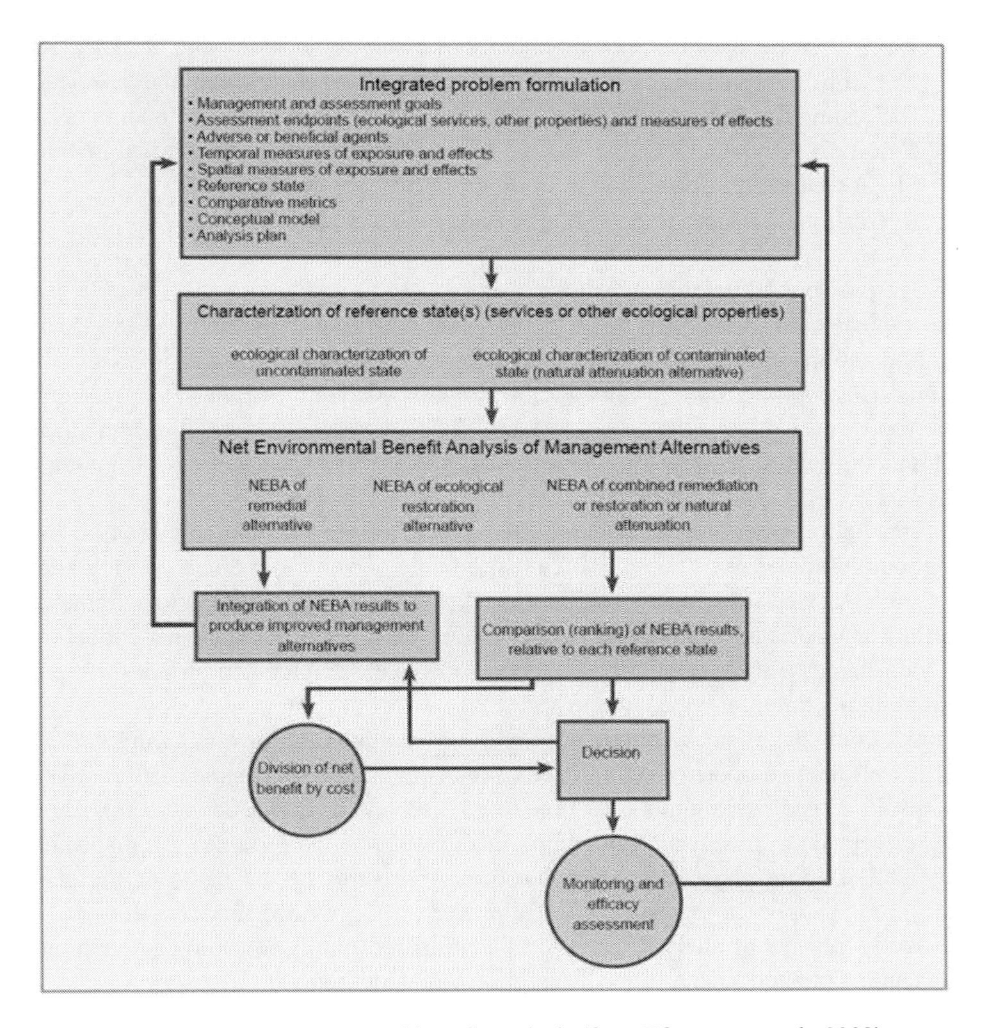

Figure 6. Framework for net environmental benefit analysis (from Efroymson et al., 2003)

For example, the listing of services (aquatic biota and habitat, riparian vegetation, terrestrial wildlife, recreational uses, and tribal uses) in the Leviathan Mine case does not seem to be useful for sorting out the different services to be valued. The listing fails to identify mutually exclusive services and presents a high likelihood of double counting. It also does not adequately distinguish between inputs and outputs. The significance of protecting habitat and riparian vegetation, for example, is not clearly addressed. Is it because society cares about the populations they support? Or is it because these populations are an input into something else of value, such as recreation? Consider insect populations. If society cares about the insects for their own sake, the insects generate unique existence value. If they are valued as a food source for fish and society cares about fish, there is value in the change in fish brought about by the change in insects. But in the latter case, insects should not be valued separately.

A better delineation of ecosystem services might involve identifying directly experienced, measurable, and spatially and temporally explicit services. For the Leviathan Mines example, such a list of ecosystem services might consist of the following:

- Water used by Washoe Tribe members and others for washing and drinking
- Non-consumptive uses of wildlife (e.g., viewing bald eagles and other species)
- Harvesting (hunting, fishing, and collecting fish) by Washoe tribal members
- Cultural, spiritual, and ceremonial values of land used by Washoe tribal members
- Flood control (e.g., reduction in flooding from snowmelt or runoff)
- Recreational services (e.g., fishing, hiking, and camping)

6.3.3.6. Expanding Valuation Methods

The typical comparison of remedial strategies currently includes two tests: whether a remediation action controls risk to an acceptable level, and if so, whether it is cost effective. Under this scheme, if a proposed remediation action is adequate with regard to risk reduction, the least costly alternative is the obvious choice. Such an approach decouples remediation and redevelopment, delays the development process, and may not maximize what matters to the public.

If remediation and redevelopment alternatives are to be compared based on their contributions to human well-being, EPA must be able to value the effect of each alternative on ecosystem services. As mentioned previously, NEBA offers a conceptual framework for comparing remedial and redevelopment alternatives on the basis of their net contributions to human well-being, whether monetized or non-monetized. Chapter 4 in turn describes a broad range of methods for valuing ecosystem services.

Habitat equivalency analysis (HEA) provides one approach for comparing contributions to human well-being associated with different remedial and redevelopment alternatives. HEA reports results in ecological units over time (e.g., discounted service acres years). The cost of creating or replacing those ecological units in monetary terms provides a replacement cost. Although these approaches do not provide direct measures of the value of the ecosystem services, they support a comparison of the services provided under different options. Alternatively, impacts of alternatives could be compared purely in ecological or biophysical terms through a method such as the conservation value method

EPA could also compare remediation and redevelopment alternatives using economic valuation. For example, EPA could use hedonic pricing studies to determine the economic impacts of the cleanup and redevelopment options on adjacent residential property values. New contingent valuation studies or studies using travel cost models could capture in monetary terms recreational or aesthetic values. Models might be used to compare expected gains to the local economy across the feasible set of redevelopment scenarios. Ecosystem benefit indicators, as discussed above, might also be used to evaluate the impacts of different remediation or redevelopment options.

If members of the public are involved in testing remediation and redevelopment alternatives, EPA could use decision-aiding processes to assess their preferences for or weighting of alternatives. Formal social-psychological surveys of potential recreational users, visitors, and tourists could measure the relative preferences of these groups among remediation and redevelopment plans. Parallel economic or monetary assessments, perhaps using contingent valuation or travel cost methods, could extend and cross-validate survey results. Decision science methods could provide weights to facilitate analyses of tradeoffs among recreation, tourism, and industrial development at a site.

6.3.3.7. Communicating Information About Ecosystems

EPA should explicitly address ecosystem services in communications about site remediation and redevelopment. Managers will be able to better communicate the reasoning behind their selection of preferred options if analyses effectively integrate ecosystem services and their contributions to human well-being. A focus on the ecosystem services that matter to the public should also lead to greater public understanding of the potential advantages of the options for remediation and redevelopment. Finally, performance measures defined in terms of contributions to well-being that the interested public understands and accepts as important should help facilitate communications about progress in the remediation and redevelopment process.

Scientific information can be complex and difficult to understand; visual communication approaches can help. For example, EPA might use perceptual representations (e.g., visualizations of revegetation options as viewed from adjacent homes and prominent tourist and recreation sites and passageways) to improve public understanding of the implications of the various restoration and redevelopment alternatives under consideration. Consider the restoration plan for the Avtex site, which included replanting and encouraging re-growth of three different forest types on appropriate locations within the site. Accurate visualizations of the reforestation projects, including their expected growth over time, would have been useful for communicating the implications of alternative plans. Effectively developing and using such visualizations would require collaboration between forest ecologists and visualization experts (such as landscape architects). These collaborations could lead to the creation of accurate and realistic representations of how the different forests would look from significant viewpoints at different stages of the restoration program for each management alternative. Psychologists, communications experts, and other relevant social or decision scientists might create appropriate vehicles and contexts for presenting the visualizations to relevant audiences. Computer graphics experts might also be helpful. Further interdisciplinary collaboration would be required if the visualizations were to be accompanied by information about expected wildlife or other ecological effects associated with each visualized forest condition. While this example may seem to be an intricate, exhaustive process, many contaminated properties are under redevelopment for years (or decades in the case of Superfund projects). With proportional resource allocations, this level of effort may be appropriate.

6.3.3.8. Fostering Information-Sharing About Ecological Valuations At Different Sites

The committee recommends that EPA pursue the broad and rapid transfer of experience within the Agency of integrating valuation concepts and techniques into the remediation and redevelopment of contaminated sites. The Agency can build its capacity to utilize valuation to inform local decisions through a systematic exchange of information about site-specific valuations. The lessons learned from trial efforts, whether successes or failures, need to be shared widely across the Agency with the regions, program offices, and tool-builders in research organizations. The Agency can catalog and share such experiences in a number of ways, such as reports, databases, or computer-based networks of users sharing best practices. The Agency is in the best position to know how to take advantage of the knowledge infrastructure provided by existing information exchange systems. Regardless of how it is done, information should be shared broadly.

6.3.4. Summary Of Recommendations For Valuation For Site-Specific Decisions

Incorporation of ecological valuation into decisions about site remediation and redevelopment can help maximize the ecosystem services provided in the long run by such sites and the sites' contributions to local well-being. To effectively value the protection of ecological systems and services in this context, the committee recommends that EPA:

- Provide regional offices with the staff and resources needed to incorporate ecological valuation into the remediation and redevelopment of contaminated sites.
- Determine the ecosystem services and values important to the community and affected parties at the beginning of the remediation and redevelopment process.
- Involve the mix of interdisciplinary experts appropriate for valuation at different sites.
- Construct conceptual models that include ecosystem services.
- Adapt current ecological risk assessment practices to include ecological production functions to predict effects on relevant ecosystem services.
- Define ecosystem services carefully and develop a standard approach for cataloging and accounting for ecosystem services for site remediation and redevelopment.
- Expand the variety of methods the Agency uses to assess the value of services lost or gained from current conditions or through proposed Agency action.
- Communicate information about ecosystem services in discussing options for remediation and redevelopment of sites with the public and affected parties.
- Create formal systems and processes to foster information-sharing about ecological valuations at different sites.

7. CONCLUSION

EPA's mission to protect human health and the environment requires that the Agency understand and protect ecosystems and the numerous and varied services they provide. Ecosystems play a vital role in our lives, providing such services as water purification, flood protection, disease regulation, pollination, recreation, aesthetic satisfaction, and the control of diseases, pests, and climate. EPA's regulations, programs, and other actions, as well as the decisions of other agencies with which EPA partners, can affect ecosystem conditions and the flow of ecosystem services at a local, regional, national, or global scale. To date, however, policy analyses have typically focused on only a limited set of ecological factors.

Just as policy makers at EPA and elsewhere need information about how their actions might affect human health in order to make good decisions, they also need information about how ecosystems contribute to society's well-being and how contemplated actions might affect those contributions. Such information can also help inform the public about the need for ecosystem protection, the extent to which specific policy alternatives address that need, and the value of the protection compared to the costs.

7.1. An Expanded, Integrated Valuation Approach

The committee advises EPA to use an "expanded and integrated approach" to ecological valuation. EPA's valuations should be "expanded" by seeking to assess and quantify a broader range of values than EPA has historically addressed and through a larger suite of valuation methods. The valuations should be "integrated" by encouraging greater collaboration among a wide range of disciplines, including ecologists, economists, and other social scientists, at each step of the valuation process.

The concept of value is complex. People may use many different concepts of value when assessing the protection of ecosystems and their services. Values, for example, can reflect people's preferences for alternative goods and services (as measured, for example, by economic methods, attitude surveys, and decision- science methods) or potential biophysical concerns (for example, biodiversity or energy flows).

To date, EPA has primarily sought to measure economic benefits, as required in many settings by statute or executive order. In addition, the Agency's valuation assessments have often focused on those ecosystem services or components for which EPA has concluded that it could relatively easily measure economic benefits, rather than on those services or components that may ultimately be most important to society. Such a focus can diminish the relevance and impact of a value assessment. The committee therefore advises the Agency to identify the services and components of likely importance to the public at an early stage of a valuation and then to focus on characterizing, measuring, and assessing the value of the responses of those systems and components to EPA's actions. The committee concludes that information based on some of the other concepts of value may also be a useful input into decisions affecting ecosystems, although members of the committee hold different views regarding the extent to which specific methods and concepts of values should be used in particular policy contexts. The methods discussed in this chapter are at different stages of development and validation and are of varying potential use depending on the policy context.

EPA should generally seek to measure the values that people hold and would express if they were well informed about the relevant ecological and human wellbeing factors involved. The committee therefore advises EPA to explicitly incorporate that information into the valuation process when changes to ecosystems and ecosystems services are involved. Valuation surveys, for example, should provide relevant ecological information to survey respondents, and valuation questions should be framed in terms of services or changes that people understand and can value. Likewise, deliberative processes should convey relevant information to participants. The committee also advises EPA to consider public education efforts where gaps exist between public knowledge and scientific understanding.

All steps in the valuation process, beginning with problem formulation and continuing through the characterization, representation, and measurement of values, also require information and input from a wide variety of disciplines. Instead of ecologists, economists, and other social scientists working independently, experts should collaborate throughout the process. Ecological models need to provide usable inputs for valuation, and valuation methods need to address important ecological and biophysical effects.

Of course, EPA conducts ecological valuations within a set of institutional, legal, and practical constraints. These constraints include substantive directives, procedural requirements relating to timing and oversight, and resource limitations (both monetary and personnel). For example, the preparation of regulatory impact analyses (RIAs) for proposed regulations is subject to OMB oversight and approval. OMB's Circular A-4 makes it clear that RIAs should include an economic analysis of the benefits and costs of proposed regulations conducted in accordance with the methods and procedures of standard welfare economics. At the same time, the circular recognizes that it might not be possible to monetize all potentially important benefits. In such cases, the circular instructs the Agency to quantify and report effects in value-relevant biophysical units, or, when quantification is not possible, describe the effect and associated value qualitatively. Regional and site-specific programs and decisions, which are not subject to the same legal requirements as national rule makings, can offer useful opportunities for testing and implementing a broader suite of valuation methods.

The remainder of this chapter summarizes the recommendations set out in the earlier chapters of this chapter. Some of these recommendations can be implemented in the short run, using the existing knowledge base, while others require investments in research and data or method development.

7.2. Early Identification of How Actions May Contribute to Human Welfare

As part of an expanded, integrated approach to ecological valuation, EPA should identify early in the valuation process the ecological responses that are likely to be of greatest importance to people, using information about ecological importance, human and social consequences, and public concerns. EPA should then focus its valuation efforts on those responses. This will help expand the range of ecological responses that EPA characterizes, quantifies, or values. To ensure early identification of the ecological responses of most public importance, EPA should:

- Begin each valuation by developing a conceptual model of the relevant ecosystem and the ecosystem services that it generates. This model should serve as a road map to guide the valuation.
- Involve staff throughout EPA, as well as outside experts in the biophysical and social sciences, in constructing the conceptual model. EPA should also seek information about relevant public concerns and needs.
- Incorporate new information into the model, in an iterative process, as the value assessment proceeds.

7.3. Prediction of Ecological Responses in Value-Relevant Terms

Another important aspect of an expanded, integrated approach to ecological valuation is that the Agency should predict ecological responses to governmental actions in terms that are relevant to valuation. Prediction of ecological responses is a key step in valuation efforts. To predict responses in value-relevant terms, EPA should focus on the effects of decisions on ecosystem services or other ecological features that are of most concern to people. This in turn will require the Agency to go beyond predicting only the biophysical effects of decisions and to map those effects to responses in ecosystem services or components that the public values.

Unfortunately, the science needed to do this has been limited, presenting a barrier to effective valuation of ecological systems and services. To better estimate ecological responses in value-relevant terms in the future, EPA should:

- Identify and develop measures of ecosystem services that are relevant to and directly useful for valuation. This will require increased interaction within EPA between natural and social scientists. In identifying and assessing the value of services, EPA should describe them in terms that are meaningful and understandable to the public.
- Where possible, use ecological production functions to estimate how effects on the structure and function of ecosystems, resulting from the actions of EPA or partnering agencies, will affect the provision of ecosystem services for which values can then be estimated.
- Where complete ecological production functions do not exist, Examine available ecological indicators that are correlated with changes in ecosystem services to provide information about the effects of governmental actions on those services.

- Use methods such as meta-analysis that can provide general information about key ecological relationships important in the valuation.
- Support all ecological valuations by ecological models and data sufficient to understand and estimate the likely ecological responses to the major alternatives being considered by decision makers.

7.4. Valuation

Central to an expanded, integrated valuation approach is the need to carefully characterize and, when possible, quantify and value the responses in ecosystem services or components. Three steps may be useful in this regard. First, EPA should consider the appropriate use of a broader suite of valuation methods than it has historically employed. As summarized in Table 3 at pages 42-43, the Committee looked at the possible use of not only economic methods, but also such alternative methods as measures of attitudes, preferences, and intentions; civic valuation; decision science approaches; ecosystem benefit indicators, biophysical ranking methods; and cost as a proxy for value. An expanded suite of valuation methods could allow EPA to better capture the full range of contributions stemming from ecosystem protection and the multiple sources of value derived from ecosystems – although it is important to recognize that different methods may measure different things and thus not be directly additive or comparable. Even when the Agency is required or chooses to base its assessment on economic benefits, other valuation methods may be useful in supporting, extending, and improving the basis and rationale for Agency decisions.

In considering what methods to use in specific contexts, EPA should keep in mind that many Agency actions affect not only ecosystems and ecosystem services but also other things that contribute to human well-being – e.g., human health. In these cases, valuation methods that focus solely on ecological effects will necessarily provide an incomplete picture of the consequences of EPA's actions, and the Agency should ensure that it uses valuation methods that capture information on the widest possible range of effects of Agency actions.

To move toward the possible use of a broader suite of valuation methods, EPA should:

- Pilot and evaluate the use of alternative methods where legally permissible and scientifically appropriate.
- Develop criteria to determine the suitability of alternative methods for use in specific decision contexts. Given differences in premises, goals, concerns, and external constraints, appropriate uses will vary among methods and contexts. As discussed, different methods are also at different stages of development and validation.

EPA also should more carefully evaluate the appropriate use of value transfers. EPA should identify relevant criteria for determining the appropriateness of value transfers. These criteria should consider similarities and differences in societal preferences and the nature of the biophysical systems between the study site and the policy site. Using these criteria, EPA analysts and those providing oversight should flag problematic transfers and clarify assumptions and limitations of the study-site results.

7.5. Other Cross-Cutting Issues

7.5.1. Deliberative Processes

Deliberative processes, in which analysts, decision makers, and/or members of the public meet in facilitated interaction, can be potentially useful in several steps of the valuation process. The committee particularly recommends that EPA consider using carefully-conducted deliberative processes to provide information about what people care about – especially where the public may not be fully informed about ecosystem services. Where EPA uses deliberative processes, it should provide the processes with the financial and staff resources needed to adequately address and incorporate relevant science and best practices.

7.5.2. Uncertainty

Because an understanding of the uncertainties underlying all aspects of ecological valuation will enable more informed policy making, the committee recommends that EPA more fully characterize and communicate uncertainty. In this regard, EPA should

- Go beyond simple sensitivity analysis in assessing uncertainty, and make greater use of approaches, such as Monte Carlo analysis, that provide more useful and appropriate characterizations of uncertainty in complex contexts such as ecological valuation.
- Provide information to decision makers and the public about the level of uncertainty involved in ecological valuation efforts. EPA should not relegate uncertainty analyses to appendices but should ensure that a summary of uncertainty is given as much prominence as the valuation estimate itself, with careful attention to how recipients are likely to understand the uncertainties. EPA should also explain qualitatively any limitations in the uncertainty analysis.

While EPA should improve its characterization and reporting of uncertainty, the mere existence of uncertainty should not be an excuse for delaying actions where the benefits of immediate action outweigh the value of attempting to further reduce the uncertainty. Some uncertainty will always exist.

7.5.3. Communication Of Valuation Information

The success of ecological valuations also depends on how EPA communicates ecological valuation information to decision makers and the public. To promote effective communications, the committee recommends that EPA design communications that are responsive to the needs of the users of the valuation information and follow basic guidelines for risk and technical communications. EPA's *Risk Characterization Handbook* provides one set of useful guidelines, including transparency, clarity, consistency, and reasonableness. To the extent feasible, EPA should communicate not only value information but also information about the nature, status, and changes to the ecological systems and services.

7.6. Context-Specific Recommendations

The use of an expanded, integrated approach can improve ecological valuation in multiple settings. Valuation of ecological systems and services, for example, is critical in national rule makings, where executive orders often require cost-benefit analyses and several statutes require weighing of economic benefits and costs. Regional EPA offices can find valuation important in setting program priorities and in assisting other governmental and non-governmental organizations in choosing among environmental options and communicating the importance of their actions to the public. Finally, ecological valuation can help EPA to enhance the cleanup of hazardous waste sites and make other site-specific decisions.

7.6.1. National Rule Making

Applying an expanded and integrated valuation approach to national rule making will entail some challenges but offers important opportunities for improvement as well. EPA can implement some, but not all, of the committee's recommendations using the existing knowledge base. The committee also recognizes that EPA must conduct valuations for national rule making in compliance with statutory and executive mandates. In the short run, EPA can take several actions to improve valuations for national rule making:

- EPA should develop a conceptual model at the beginning of each valuation, as discussed above, to serve as a guide or road map. To ensure that the model captures the ecological properties and services that are potentially important to people, EPA should incorporate input both from relevant science and about public preferences and concerns. EPA can identify public concerns through a variety of methods, drawing on either existing knowledge or interactive processes designed to elicit public input.
- The Agency should address site-specific variability in the impact of a rule by producing case studies for important ecosystem types and then aggregating across the studies where information is available about the joint distribution of ecosystem characteristics and human populations affected by them.
- EPA should not compromise the quality of its valuations by inappropriately applying value transfers. Where the values of ecosystem services are primarily local, the Agency can rely on scientifically-sound value transfers using prior valuations at the local level. However, for services valued more broadly, EPA should draw from studies with broad geographical coverage (in terms of both the changes that are valued and the population whose values are assessed).
- EPA should pilot and evaluate the use of a broader suite of valuation methods to support and improve RIAs. Although OMB Circular A-4 requires RIAs to monetize benefits to the extent possible using economic valuation methods, other methods could be useful in the following ways:
  - Helping to identify early in the process the ecosystem services that are likely to be of concern to the public and that should therefore be the focus of the benefit-cost analysis
  - Addressing the requirement in Circular A-4 to provide quantitative or qualitative information about the possible magnitude of benefits (and costs) when they cannot be monetized using economic valuation

  - Providing supplemental information outside the formal benefit-cost analysis about sources and concepts of value that might be of interest to EPA and the public but not fully reflected in economic benefits.
- To ensure that RIAs do not inappropriately focus only on impacts that have been monetized, EPA should also report on other ecological impacts in appropriate units where possible, as required by Circular A-4. The Agency should label aggregate monetized economic benefits as "total economic benefits that can be monetized," not as "total benefits."
- EPA should include a separate chapter on uncertainty characterization in each RIA or value assessment.

7.6.2. Regional Partnerships

The committee sees great potential in undertaking a comprehensive and systematic approach to estimating the value of protecting ecosystems and services at a regional scale, in part because of the effectiveness with which EPA regional offices can partner with other agencies and state and local governments. Regional-scale analyses hold great potential to inform decision makers and the public about the value of protecting ecosystems and services, but this potential is at present largely unrealized. The general recommendations of this chapter provide a guide for regional valuations. Regional valuations are a particularly appropriate setting in which to test alternative valuation methods because there are generally fewer legal or regulatory restrictions on what methods can be used.

In addition to recommending that regional offices adopt the general recommendations of this chapter in conducting ecological valuations, the committee advises EPA to:

- Encourage its regions to engage in valuation efforts to support decision making both by the regions and by partnering governmental agencies.
- Provide adequate resources to EPA regional staff to develop the expertise needed to undertake comprehensive and systematic studies of the value of protecting ecosystems and services.
- Ensure that regions can learn from valuation efforts by other regions. EPA regional offices should document valuation efforts and share them with other regional offices, EPA's National Center for Environmental Economics, and EPA's Office of Research and Development.

7.6.3. Site-Specific Decisions

Incorporation of ecological valuation into local decisions about the remediation and redevelopment of contaminated sites can help enhance the ecosystem services provided by such sites in the long run and thus the sites' contributions to local well-being. The general recommendations of the report again provide a useful guide for such site-specific valuations. The committee also advises the Agency to:

- Provide regional offices with the staff and resources needed to effectively incorporate ecological valuation into the remediation and redevelopment of contaminated sites.

- Determine the ecosystem services and values important to the community and affected parties at the beginning of the remediation and redevelopment process.
- Adapt current ecological risk assessment practices to incorporate ecological production functions and predict the effects of remediation and redevelopment options on ecosystem services.
- Communicate information about ecosystem services in discussing options for remediation and redevelopment with the public and affected parties.
- Create formal systems and processes to foster information-sharing about ecological valuations at different sites.

7.7. Recommendations for Research and Data Sharing

EPA should use its research programs to provide the ecological information needed for valuation, develop and test valuation methods, and share data. As an overarching recommendation, the report advises EPA to more closely coordinate its research programs on the valuation of ecosystem services and to develop links with other governmental agencies and organizations engaged in valuation and valuation research. It advises, at a more general level, fostering greater interaction between natural scientists and social scientists in identifying relevant ecosystem services and developing and implementing processes for measuring and estimating their value. Although the committee has identified those research areas that it believes are important in advancing EPA's ability to conduct valuations of ecological systems and services, the committee has not attempted to rank or prioritize among all of its research recommendations. EPA should develop a research strategy, building on the recommendations in this chapter, that identifies "low- hanging fruit" and prioritizes studies that are likely to have the largest payoff in both advancing valuation methods and providing valuation information of importance to EPA in its work.

To develop EPA's ability to determine and quantify ecological responses to governmental decisions, the Agency should:

- Support the development of quantitative ecosystem models and baseline data on ecological stressors and ecosystem service flows that can support valuation efforts at the local, regional, national, and global levels.
- Promote efforts to collect data that can be used to parameterize ecological models for site-specific analysis and case studies or that can be transferred or scaled to other contexts.
- Carefully plan and actively pursue research to develop and generate ecological production functions for valuation, including Office of Research and Development and STAR research on ecological services and support for modeling and methods development. The committee believes that this is a research area of high priority.
- Given the complexity of developing and using complete ecological production functions, continue and accelerate research to develop key indicators for use in ecological valuation. Such indicators should meet ecological and social science criteria for effectively simplifying and synthesizing underlying complexity and link to an effective monitoring and reporting program.

To develop EPA's capabilities for estimating the value of ecological responses to governmental decisions, EPA should:

- Support new studies and the development of new methodologies that will enhance the future use of ecological value transfers, particularly at the national level. Such research should include national surveys related to ecosystem services with broad (rather than localized) implications so that value estimates might be usable in multiple rule-making contexts.
- Invest in research designed to reduce uncertainties associated with ecological valuation through data collection, improvements in measurement, theory building, and theory validation.
- Incorporate the research needs of regional offices for systematic valuation studies in future calls by EPA for extramural ecological valuation research proposals.

To access and share information to enhance the Agency's capabilities for ecological valuation, EPA should:

- Work with other federal agencies and scientific organizations such as the National Science Foundation to encourage the sharing of ecological data and the development of more consistent ecological measures that are useful for valuation purposes. A number of governmental organizations, such as the United States Department of Agriculture and the Fish & Wildlife Service, are working on biophysical modeling and valuation, and EPA could usefully partner with them.
- Support efforts to develop Web-based databases of existing valuation studies that could be used in value transfers. The databases should include valuation studies across a range of ecosystems and ecosystem services. The databases should also carefully describe the characteristics and assumptions of each study, in order to increase the likelihood that those studies most comparable to new valuations can be identified for use.
- Support the development of national-level databases of information useful in the development of new valuation studies. Such information should include data on the joint distribution of ecosystem and human population characteristics that are important determinants of the value of ecosystem services.
- Develop processes and information resources so that EPA staff in one region or office of the Agency can learn effectively from valuation efforts being undertaken elsewhere within the Agency.

Appendix A: Web-Accessible Materials on Ecological Valuation Developed by or for the C-VPESS

The SAB Web site provides three sets of materials that supplement this chapter:

1. More detailed information on methods potentially useful for ecological valuation, as described in chapter 4 of this chapter.
2. A discussion of survey issues relevant to ecological valuation, including current best practices and recommendations for research
3. A summary of an SAB 2005 workshop on “Science for valuation of EPA’s ecological protection decisions and programs.”

These materials do not represent the consensus views of the committee, nor have they been reviewed and approved by the chartered Science Advisory Board. They are provided to extend the discussion of methods in chapter 4 of the main report and to encourage further deliberation within EPA and the broader scientific community about how to meet the need for an integrated and expanded approach for valuing the protection of ecological systems and services.

Methods Potentially Useful for Ecological Valuation

The SAB Web site provides descriptions of methods and approaches prepared by members of the C-VPESS as resources for the committee and others interested in ecological valuation (http://yosemite.epa.gov/sab/sabproduct.nsf/WebBOARD/C-VPESS_Web_ Meth ods_Draft?OpenDocument). Methods are described with specific reference to how they might be used by the EPA for valuing the protection of ecological systems and services within the valuation approach recommended by the committee. Some of the methods have already been used extensively in EPA policy and decision making. Some appear never or only rarely to have been used by the Agency, but are widely used by other agencies. Some are less proven in policy making contexts and should be considered experimental. All of the methods described have both conceptual and practical strengths and limitations.

The descriptions of these methods and approaches and of their utility for ecological valuation at EPA do not represent the consensus views of the committee, nor have they been reviewed and approved by the chartered Science Advisory Board. They are offered to extend and elaborate the very brief descriptions provided in chapter 4 of the main report and to encourage further deliberation within EPA and the broader scientific community about how to meet the need for an integrated and expanded approach for valuing the protection of ecological systems and services.

The descriptions provide suggestions for further reading, potential applications of the methods, and future research opportunities. The descriptions of specific methods and approaches are supplemented by a separate Web-accessible discussion (http://yosemite.epa. gov/Sab/Sabproduct.nsf/WebFiles/SurveyMethods/$File/Survey_ methods.pdf) of the use of survey techniques employed in some valuation methods.

Members of the C-VPESS do agree that EPA should carefully characterize and, when possible, quantify and value the responses in ecosystem services or components. They agree that a wider range of valuation methods can play a potential role throughout the expanded and integrated valuation process the committee envisions.

An expanded suite of valuation methods could allow EPA to better capture the full range of contributions stemming from ecosystem protection and the multiple types of value derived from ecosystems. At the same time, it is important to recognize that different methods may measure different values and thus not be additive or comparable. Even when the Agency is required or chooses to base its valuation assessment on economic values, however, use of additional methods may be useful in supporting, improving, or extending the valuation.

The descriptions of methods and approaches generally include the following kinds of information:

- Brief description of the method
- Status of the method
- Conceptual and practical strengths and limitations
- Treatment of uncertainty
- Research needs
- Key references

Methods and approaches described include:

- Measures of attitudes, preferences, and intentions
  - Surveys of attitudes, preferences, and intentions
  - Focus groups
  - Individual narratives
    - Mental model approaches
  - Emerging methods
- Economic methods
  - Market-based methods
  - Non-market methods – revealed preference
    - Travel cost
    - Hedonics
    - Averting behavior models
  - Non-market methods – stated preference
  - Combining revealed and stated preference methods
- Civic valuation
  - Referenda and initiatives
  - Citizen valuation juries
- Decision science methods
- Ecosystem benefit indicators
- Biophysical ranking methods
  - Conservation value method
  - Rankings based on energy and material flows
- Methods using cost as a proxy for value
  - Replacement costs
  - Tradable permits
  - Habitat equivalency analysis
- Deliberative processes
  - Mediated modeling
  - Constructed value approaches

Survey Issues For Ecological Valuation: Current Best Practices And Recommendations For Research

This document provides an introduction for EPA staff to questions posed to the C-VPESS pertaining to survey use for ecological valuation. It gives an overview of how recent research and evolving practice relating to those questions might assist the Agency. The document provides a definition of survey research, discusses survey design, identifies elements of a well-designed survey, addresses assessment of survey accuracy, and discusses challenges in using surveys for ecosystem protection valuation. The document can be found at the SAB Web site at: http://yosemite.epa.gov/ Sab/Sabproduct.nsf/WebFiles/SurveyMethods/$File/ Survey_methods.pdf.

Science Advisory Board Workshop Summary: Science For Valuation Of EPA's Ecological Protection Decisions And Programs. Summary Of Workshop Held December 13-14, 2005, Washington, DC

This document summarizes a public workshop held on December 13-14, 2005, in Washington, D.C., on "Science for valuation of EPA's ecological protection decisions and programs." The purpose of the workshop was to discuss the initial work of the SAB's C-VPESS; to provide an opportunity for members of the SAB, the Advisory Council on Clean Air Compliance Analysis, and Clean Air Scientific Advisory Committee to learn from each others' work relating to ecological valuation; and to feature feedback and insights from Agency clients and outside subject matter experts. The agenda included presentations and discussions with advisory committee members, Agency personnel, and invited speakers. . The workshop summary can be found on the SAB Web site at: http://yosemite.epa.gov/Sab/ Sabproduct.nsf/WebFiles/EcoWorkshop/$File/sab_ wksp_summary_12_13- 14_05.pdf.

APPENDIX B. TABLE OF ACRONYMS

AMD	Acid mine drainage
BTF	Benefit transfer
CAFO	Concentrated animal feeding operation
CERCLA	Comprehensive Environmental Response, Compensation, and Liability Act
C-VPESS	Committee on Valuing the Protection of Ecological Systems and Services
EVRI	Environmental Valuation Reference Inventory
FPD	Forest Preserve District
GEAE	Generic ecological assessment endpoints
GDP	Gross domestic product
GIS	Geographic Information System
GPRA	Government Performance and Results Act
HEA	Habitat equivalency analysis
LTER	Long-term Ecological Research
NCEE	National Center for Environmental Economics
NEBA	Net Environmental Benefit Assessment

NPL	National Priorities List
NRC	National Research Council
NRDA	Natural Resource Damage Assessments
NSF	National Science Foundation
OMB	Office of Management and Budget
RIA	Regulatory Impact Analysis
STAR	Science to Achieve Results
VOC	Volatile organic compound

REFERENCES

Adamowicz, W., Boxall, P., Wilhams, M. & Louviere. J. (1998). Stated preference approaches for measuring passive use values: Choice experiments and contingent valuation. *American Journal of Agricultural Economics*, *80*, 64-67.

Aldred, J. & Jacobs, M. (2000). Citizens and wetlands: Evaluating the Ely citizens' jury. *Ecological Economics*, *34*, 217-232.

Alvarez-Farizo, B. & Hanley, N. (2006). Improving the process of valuing non-market benefits: Combining citizens' juries with choice modelling. *Land Economics*, *82*, 465-478.

Antle, J., Capalbo, S., Mooney, S., Elliott, E., E. & Paustian, K. (2002). Sensitivity of carbon sequestration costs to soil carbon rates. *Environmental Pollution*, *116*, 413-422.

Arnold, T. L. & Friedel, M. J. (2000). *Effects of land use on recharge potential of surficial and shallow bedrock aquifers in the Upper Illinois River basin*. Water-Resources Investigations Report 00-4027. U.S. Department of the Interior, U.S. Geological Survey.

Arrow, K., et al. (1993). Report of the NOAA Panel on Contingent Valuation, *Federal Register*, *58*, 10 (Jan. 15, 1993), 4601-4614.

Arrow, K. J., et al. (1996). Is there a role for benefit-cost analysis in environmental, health, and safety regulation? *Science*, *New Series*, *5259*, 221-222.

Arvai, J. & Gregory, R. (2003). Testing alternative decision approaches for identifying cleanup priorities at contaminated sites. *Environmental Science & Technology*, *37*, 1469-1476.

Arvai, J. L., Gregory, R. & McDaniels, T. (2001). Testing a structured decision approach: Value-focused thinking for deliberative risk communication. *Risk Analysis*, *21*, 1065-1076.

Ascher, W. & Overholt, W. H. (1983). *Strategic planning and forecasting*. New York: Wiley.

Ayres, R. U. (1978). Application of physical principles to economics. In *Resources, environment, and economics: Applications of the materials/energy balance principle*, ed. R.U. Ayres, 37-71. New York: Wiley.

Balmford, A., Bruner, A., Cooper, P., Costanza, R., Farber, S., Green, R. E., et al., (2002). Economic reasons for saving wild nature. *Science*, *297*, 950-953.

Banzhaf, S., Burtraw, D., Evans, D. & Krupnick, A. (2004). *Valuation of natural resource improvements in the Adirondacks*. Washington, DC: Resources for the Future.

Banzhaf, S., Oates, W., Sanchirico, J. N., Simpson, D. & Walsh. R. (2006). Voting for conservation: What is the American electorate revealing? *Resources* 16. Washington, DC: Resources for the Future. http://ww.rff.org/rff/news/features/loader.cfm?url=/ commonspot/security/getfile. cfm&pageid=22017.

Barbier, E. B. & Strand, I. (1998). Valuing mangrove-fishery linkages. *Environmental and Resource Economics*, *12*, 151-166.

Barnwell, T. O. & Brown, L. C. (1987). *The Enhanced Stream Water Quality Models QUAL2E and QUAL2E-UNCAS: Documentation and User Manual*, EPA/600/3-87/007.

Bartik, T. J. (1988). Evaluating the benefits of non-marginal reductions in pollution using information on defensive expenditures. *Journal of Environmental Economics and Management*, *15*, 111-22.

Bateman, I. J. & K. G. Willis, eds. (1999). *Valuing environmental preferences: Theory and practice of the contingent valuation method in the US, EU, and developing countries.* Oxford: Oxford University Press.

Belden and Russonello Research and Communications. (1996). *Human values and nature's future: Americans' attitudes on biological diversity.* Washington, DC: Belden and Russonello Research and Communications.

Berelson, B. (1952). Democratic theory and public opinion. *Public Opinion Quarterly*, *16*, 313-330.

Bergstrom, J. C. & Taylor, L. O. (2006). Using meta-analysis for benefits transfer: Theory and practice. *Ecological Economics*, *60*, 351-360.

Binkley, C. Hanemann. & W. M. (1978). *The recreation benefits of water quality improvement: Analysis of day trips in an urban setting.* EPA-600/5-78-010.

Birdsey, R. A. (2006). Carbon accounting rules and guidelines for the United States forest sector. *Journal of Environmental Quality*, *35*, 1518-1524.

Bishop, I. D. & Rohrmann, B. (2003). Subjective responses to simulated and real environments: a comparison. *Landscape and Urban Planning*, *65*, 261-267.

Blamey, R. K., James, R. F., Smith, R. & Niemeyer, S. (2000). *Citizens' juries and environmental value assessment.* Canberra: Research School of Social Sciences, Australian National University. http://cjp.anu.edu.au/docs/CJ1.pdf.

Boardman, A. E., Greenberg, D. H., Vining, A. R. & Weimer, D. L. (2006). *Cost-benefit analysis: Concepts and practice.* 3rd ed. Upper Saddle River, NJ: Prentice-Hall.

Bockstael, N. E., Freeman, A. M., Kopp, R., Portney, P. R. & Smith, V. K. (2000). On measuring economic values for nature. *Environmental Science and Technology*, *34*, 1384- 1389.

Bockstael, N. B. & Mc Connell, K. E. (2007). *Environmental and Resource Valuation with Revealed Preferences: A Theoretical Guide to Empirical Models (The Economics of Non-Market Goods and Resources).* New York: Springer.

Bostrom, A., Fischhoff, B. & Morgan, M. G. (2002). Characterizing mental models of hazardous processes: A methodology and an application to radon. *Journal of Social Issues*, *48*, 85-100.

Boulding, K. E. (1966). The economics of the coming spaceship Earth. In *Environmental quality in a growing economy*, ed. H. Jarrett, 3-14. Baltimore, MD: Resources for the Future/ Johns Hopkins University Press.

Boyd, J. (2004). What's nature worth? Using indicators to open the black box of ecological valuation. *Resources.*

Boyd, J. & Banzhaf, S. (2006). What are ecosystem services? The need for standardized environmental accounting units. *RFF Discussion Papers*. Washington, DC: Resources for the Future.

Boyd, J., King, D. & Wainger. L. (2001). Compensation for lost ecosystem services: The need for benefit-based transfer ratios and restoration criteria. *Stanford Environmental Law Journal, 20.*

Boyd, J. & Wainger, L. (2002). Landscape indicators of ecosystem service benefits. *American Journal of Agricultural Economics, 84.*

Boyle, K. J. (2003). Contingent valuation in practice. In *A primer on non-market valuation*, ed. K. J. Boyle, & P. A. Champ. Boston: Kluwer Academic Publishers.

Brandenburg, A. M. & Carroll, M. S. (1995). Your place or mine? The effect of place creation on environmental values and landscape meanings. *Society & Natural Resources, 8*, 381-398.

Brenner, A. (1993). *Cartographic symbolization and design: ARC/INFO® methods*. U.S. EPA Office of Information Resources Management.

Brink, D. O. (1989). *Moral realism and the foundation of ethics*. Cambridge: Cambridge University Press.

Brouwer, R. (2000). Environmental value transfer: State of the art and future prospects. *Ecological Economics, 32*, 137-152.

Brouwer, R. & Bateman, I. J. (2005). Benefits transfer of willingness to pay estimates and functions for health- risk reductions: A cross-country study. *Journal of Health Economics, 24*, 591-611.

Brown, G. & Roughgarden, J. (1995). An ecological economy: Notes on harvest and growth. In *Biodiversity loss: Ecological and economic issues*, ed. C., Perrings, K. G., Maler, C., Folke, C. S. Holling, & B. O. Jansson, 150-189. Cambridge: Cambridge University Press. (Reprinted in: J. Barkley, Rosser, Jr., ed., 2004. *Complexity in economics*, vol. *III. The International Library of Critical Writings in Economics, 174*. Cheltenham, UK: Elgar.)

Brown, G. & Roughgarden, J. (1997). A metapopulation model with private property and a common pool. *Ecological Economics, 22*, 65-71.

Brown, N., Master, L., Faber-Langendoen, D., Comer, P., Maybury, K., Robles, M., Nichols, J. & Wigley, T. B. (2004). *Managing elements of biodiversity in sustainable forestry programs: Status and utility of NatureServe 's information resources to forest managers*. National Council for Air and Stream Improvement Technical Bulletin Number 0885.

Brown, T. C., et al. (1995). The values jury to aid natural resource decisions. *Land Economics, 71*, 250-260.

Budescu, D. F. & Wallsten, T. S. (1995). Processing linguistic probabilities: General principles and empirical evidence. In *Decision making from the perspective of cognitive psychology,* ed. J. R., Busemeyer, R. Hastie, & D. L. Medin, 275-318. New York: Academic Press.

Budnitz, R. J., Apostolakis, G., Boore, D. M., Cluff, L. S., Coppersmith, K. J., Cornell, C. A. & Morris, P. A. (1997). *Recommendations for probabilistic seismic hazard analysis: Guidance on uncertainty and the use of experts*. Washington, DC: U.S. Nuclear Regulatory Commission, Senior Seismic Hazard Analysis Committee.

D. Butler, & A. Ranney, eds. (1978). *Referendums*. Washington DC: American Enterprise Institute.

Carpenter, S., Brock, W. & Hanson, P. (1999). Ecological and social dynamics in simple models of ecosystem management. *Conservation Ecology*, *3*, 4. http://www. consecol.org/vol3/iss2/art4/.

Carpenter, S. R., DeFries, R., Dietz, T., Mooney, H. A., Polasky, S., Reid and W. V. & Scholes, R. J. (2006). Millennium ecosystem assessment: research needs. *Science*, *314*, 257-258.

Carson, R. T. & Hanemann, W. M. (2005). *Contingent Valuation. In Handbook of Environmental Economics*, Vol. *II*. Eds. in K. Göran-Mäler, & J. R. Vincent. Amsterdam: North Holland, 82 1-936.

Carson, R. T. & Mitchell, R. C. (1993). The value of clean water: The public's willingness to pay for boatable, fishable, and swimmable quality water. *Water Resources Research*, *29*, 2445-54.

P. A., Champ, K. J. Boyle, & T. C. Brown, eds. (2003). *A primer on non-market valuation*. Dordrecht: Klumer Academic.

Chay, K. & Greenstone, M. (2005). Does air quality matter? Evidence from the housing market. *Journal of Political Economy*, *113*.

Chicago Wilderness. (1999). *Biodiversity recovery plan*. http:// chicagowilderness.org /pubprod/brpindex.cfm.

Chichilnisky, G. & Heal, G. (1998). Economic returns from the biosphere. *Nature*, *391*, 629-630.

Chua, H. F., Yates, J. F. & Shah, P. (2006). Risk avoidance: Graphs versus numbers. *Memory & Cognition*, *34(2)*, 399- 410.

Cleaves, D. A. (1994). *Assessing uncertainty in expert judgments about natural resources*. New Orleans: U.S. Forest Service General Technical Report SO-1.

Clemen, R. T. (1996). *Making hard decisions: An introduction to decision analysis*. Boston: PWS-Kent Publishing Co.

Cleveland, C. J. (1987). Biophysical economics: Historical perspective and current research trends. *Ecological Modelling*, *38*, 47-74.

Cleveland, C. J., Costanza, R., Hall, C. A. S. & Kaufmann, R. (1984). Energy and the U.S. economy: A biophysical perspective. *Science*, *225*, 890-897.

Cliburn, D. C., Feddema, J. J., Miller, J. R. & Slocum, T. A. (2002). The design and evaluation of a collaborative decision support system in a water balance application. *Computers and Graphics*, *26*, 93 1-949.

Costanza, R. (1980). Embodied energy and economic valuation. *Science*, *210*, 1219-1224.

_____. (2004). Value theory and energy. In *Encyclopedia of energy*, vol. 6, ed. C. Cleveland. Amsterdam: Elsevier Science. Costanza, R., ed. 2000. Forum: The ecological footprint. *Ecological Economics*, *32*, 34 1-394.

Costanza, R., Farber, S. C. & Maxwell, J. (1989). The valuation and management of wetland ecosystems. *Ecological Economics*, *1*, 335-361.

Costanza, R. & Folke, C. (1997). Valuing ecosystem services with efficiency, fairness and sustainability as goals. In *Nature's services: Societal dependence on natural ecosystems,* ed. G.C. Daily. Washington, DC: Island Press.

Covich, A. P., Ewel, K. C., Hall, R. O., Giller, P. S., Goedkoop, W. & Merritt, D. M. (2004). Ecosystem services provided by freshwater benthos. In *Sustaining biodiversity and ecosystem services in soils and sediments*, ed. D. H. Wall, 45-72. *Vol. 64*. SCOPE. Washington, DC: Island Press.

Cowling, R. & Costanza, R. (1997). Valuation and management of fynbos ecosystems. *Ecological Economics*, *22*, 103-155.

Cronin, T. E. (1989). *Direct democracy: The politics of referendum, initiative and recall.* Cambridge: Harvard University Press.

Cropper, M. L., Deck, L. & McConnell, K. E. (1988). On the choice of functional forms for hedonic price functions. *Review of Economics and Statistics*, *70*, 668-75.

Daily, G. C., ed. (1997). *Nature's Services: Societal dependence on natural ecosystems.* Washington, DC: Island Press.

Dasgupta, P. & Maler, K. G. ed. (2004). *The Economics of Non-Convex Ecosystems.* Dordrecht:: Kluwer Academic Publishers.

David Evans and Associates Inc. & ECONorthwest. (2004). *Comparative valuation of ecosystem services: Lents project case study.*

Davis, E. F. (2001). Reuse assessment: A tool to implement the Superfund land use directive. Memorandum to Superfund national policy managers. Washington, DC: U.S. Environmental Protection Agency (OSWER 9355.7-06P).

Deacon, R. & Shapiro. P. (1975). Private preference for collective goods revealed through voting on referenda. *American Economic Review*, *65*, 793.

Deschenes, O. & Greenstone, M. (2007). The economic impacts of climate change: Evidence from agricultural output and random fluctuations in weather. *American Economic Review* (in press).

Desvousges, W. H., Johnson, F. R. & Banzhaf, H. S. (1998). *Environmental policy analysis with limited information: principles and the application of the transfer method.* Cheltenham, UK: Edward Elgar.

de Zwart, D., Dyer, S. D., Posthuma, L. & Hawkins. C. P. (2006). Predictive models attributer effects on fish assemblages to toxicity and habitat alteration. *Ecological Applications*, *16*, 1295-1310.

Dickie, M. (2003). Defensive behavior and damage cost methods. In *A primer on non-market valuation*, ed. P. A., Champ, K. J. Boyle, & T. C. Brown. Dordrecht: Kluwer Academic Press.

Dietz, T., Fitzgerald, A. & Shwom, R. (2005). Environmental Values. *Annual Review of Environment and Resources*, *30*, 335-372.

Dillman D. A. (1991). The design and administration of mail surveys. *Annual Review of Sociology*, *17*, 225-249.

Doss, C. R. & Taff, S. J. (1996). The influence of wetland type and wetland proximity on residential property values. *Journal of Agricultural and Resource Economics*, *21*, 120-129.

Dunford, R. W., Ginn, T. C. & Desvousges, W. H. (2004). The use of habitat equivalency analysis in natural resource damage assessment. *Ecological Economics*, *48*, 49-70.

Dunlap, R. E., Van Liere, K. D., Mertig, A. G. & Jones. R. E. (2000). Measuring endorsement of the New Ecological Paradigm: A revised NEP scale. *Journal of Social Issues*, *56*, 425-442.

Efroymson, R. A., Nicolette, J. P. & Suter, G. W. (2003). *A framework for net environmental benefit analysis for remediation or restoration of petroleum-contaminated sites.* ORNL/TM-2003/17. Oak Ridge National Laboratory, Oak Ridge, TN.

_____. (2004). A framework for net environmental benefits analysis for remediation or restoration of contaminated sites. *Environmental Management, 34.*

Encarnacao, J., Foley, J., Bryson, S., Feiner, S. K. & Gershon, N. (1994). Research issues in perception and user interfaces. *Computer Graphics and Applications, IEEE 14*, 67-69.

Failing, L. & Gregory, R. (2003). Ten common mistakes in designing biodiversity indicators for forest policy. *Journal of Environmental Management, 68*, 12 1-132.

Farber, S., Costanza, R., Childers, D. L., Erickson, J., Gross, K., Grove, M., et al. (2006). Linking ecology and economics for ecosystem management. *BioScience*, 117-129.

Fazio, R. H. (1986). How do attitudes guide behavior? In *Handbook of motivation and cognition,* ed., R. M. Sorrentino, & E. G. Giggins, 204-243. New York: Guildford Press.

Fischhoff, B. (1991). Value elicitation: Is there anything in there? *American Psychologist, 46*, 835-847.

(1997). What do psychologists want? Contingent valuation as a special case of asking questions. In *Determining the value of non-marketed goods,* ed. R. J., Kopp, W. W. Pommerehne, & N. Schwarz. Norwell, MA: Kluwer Academic Publishers.

(2005). Cognitive processes in stated preference methods. In *Handbook of environmental economics*, ed. K. G. Mäler, & J. Vincent. Amsterdam: Elsevier.

(2009). Risk Perception and Communication. In *Oxford Textbook of Public Health*, 5th ed., eds. R., Dete, R., Beaglehole, M. A. Lansang, & M. Gulliford. Oxford: Oxford University Press.

Fleischer, A. & Tsur, Y. (2003). Measuring the recreational value of open space. *Journal of Agricultural Economics*, *54*, 269-283.

Forest Preserve District of Cook County Illinois. (1988). *An evaluation of floodwater storage.*

Fox, D. (2007). Back to the no-analog future? *Science*, *316*, 823- 825.

Freeman, A. M., III (1995). The benefits of water quality improvements for marine recreation: A review of the empirical evidence. *Marine Resource Economics*, *10*, 385-406.

Freeman, A. M., III. (2003). *The measurement of environmental and resource values: Theory and methods.* 2nd ed. Washington, DC: Resources for the Future.

Freudenberg, W. R., Gramling, R. & Davidson, D. J. (2008). Scientific certainty argumentation methods (SCAMs): Science and the politics of doubt. *Sociological Inquiry*, *78*, 2-38.

Garrison, V. (1994-95). Vermont Lake protection classification system. *Proceedings of five regional citizens education workshops on lake management.* Madison, WI: North American Lake Management Society.

Gentner, D. & Whitley, E. W. (1997). Mental models of population growth: A preliminary investigation. In *Environment, ethics, and behavior: The psychology of environmental valuation and degradation, ed.* M. Bazerman, D. M., Messick, A. E. Tenbrunsel, & K. WadeBenzoni, 209-233. San Francisco, CA: New Lexington Press.

Gimblett, H. R., Daniel, T. C., Cherry, S. & Meitner, M. J. (2001). The simulation and visualization of complex human- environment interactions. *Landscape and Urban Planning*, *54*, 63-79.

Global Footprint Network. 2008. [cited 2008]. Available from http://www.foot printnetwork.org/

Greenleaf, S. S. & Kremen, C. (2006). Wild bees enhance honey bees' pollination of hybrid sunflower. *Proceedings of the National Academies of the United States of America*, 10. 1073/pnas.0600929103.

Gregory, R., Arvai, J. L. & McDaniels, T. (2001). Value-focused thinking for environmental risk consultations. *Research in Social Problems and Public Policy*, *9*, 249-275.

Gregory, R., Lichtenstein, S. & Slovic, P. (1993). Valuing environmental resources: A constructive approach. *Journal of Risk and Uncertainty*, *7*, 177-197.

Gregory, R., McDaniels, T. & Fields. D. (2001). Decision aiding, not dispute resolution: Creating insights through structured environmental decisions. *Journal of Policy Analysis and Management*, *20*, 415-432.

Gregory, R. & Slovic, P. (1997). A constructive approach to environmental valuation. *Ecological Economics*, *21*, 175-181.

Gregory, R. & Wellman, K. (2001). Bringing stakeholder values into environmental policy choices: A community- based estuary case study. *Ecological Economics*, *39*, 37-52.

Grossman, D. H. & Comer, P. J. (2004). *Setting priorities for biodiversity conservation in Puerto Rico*. Nature Serve Technical Report. Arlington, VA: NatureServ.

Guo, Z., Xiao. X. & Dianmo, L. (2000). An assessment of ecosystem services: water flow regulation and hydroelectric power production. *Ecological Applications*, *10*, 925-936.

Haab, T. C. & McConnell, K. E. (2002). *Valuing environmental and natural resources*. Cheltenham, UK: Edward Elgar.

Hall, C. A. S., Cleveland, C. J. & Kaufmann, K. (1992). *Energy and resource quality: The ecology of the economic process*. New York: Wiley.

Hammer, T. R., Coughlin, R. E. & Horn, E. T. (1974). The effect of a large urban park on real estate value. *Journal of the American Institute of Planners*, *40*, 274-277.

Hammond, J. S., Keeney, R. L. & Raiffa, H. (1998). Even swaps: A rational method for making trade-offs. *Harvard Business Review*, *76*, 137-138, 143-148, 150.

_____. (1999). *Smart choices: A practical guide to making better decisions*. Cambridge: Harvard Business School Press.

Hannon, B., Costanza, R. & Herendeen, R. A. (1986). Measures of energy cost and value in ecosystems. *The Journal of Environmental Economics and Management*, *13*, 391-401.

Harwell, M., Myers, V., Young, T., Bartuska, A., Gassman, N., Gentile, J. H., et al. (1999). A framework for an ecosystem integrity report card. *Bio Science*, *49*, 543-554.

Hays, S. P. (1989). *Beauty, health and permanence: Environmental politics in the United States*, 1955-1985. Cambridge: Cambridge University Press.

The H. John Heinz III Center for Science, Economics, and the Environment. (2008). *The state of the nation's ecosystems 2008*. Washington, DC: Island Press.

Herendeen, R. (2000). Ecological footprint is a vivid indicator of indirect effects. *Ecological Economics*, *32*, 357-35 8.

Higgins, S. I., Turpie, J. K., Costanza, R., Cowling, R. M., le Maitre, D. C., Marais, C. & Midgley, G. (1997.) An ecological economic simulation model of mountain fynbos ecosystems: dynamics, valuation, and management. *Ecological Economics*, *22*, 155-169.

Hitlin, S. & Piliavin, J. A. (2004). Values: Reviving a dormant concept. *Annual Review of Sociology*, *30*, 359-93.

Hoagland, P. & Jin, D. (2006). Science and economics in the management of invasive species. *BioScience*, *56*, 93 1-935.

Horowitz, J. A. & McConnell, K. E. (2002). A review of WTA / WTP studies. *Journal of Environmental Economics and Management*, *44*, 426-447.

Howarth, R. B. & Wilson, M. A. (2006). A theoretical approach to deliberative valuation: aggregation by mutual consent. *Land Economics*, *82*, 1-16.

Howell-Moroney, M. (2004a). Community characteristics, open space preservation and regionalism: Is there a connection? *Journal of Urban Affairs*, *26*, 109-118.

Hufbauer, G. & Elliott, K. A. (1994). *Measuring the costs of protection in the US.* Washington, DC: Institute for International Economics.

Hunsaker, C. T. (1993). New concepts in environmental monitoring: the question of indicators. *Science of the Total Environment Supplement part*, *1*, 77-96.

Hunsaker, C. T. & Carpenter, D. E. eds. (1990). *Ecological indicators for the Environmental Monitoring and Assessment Program.* EPA/600/3-90/060. Research Triangle Park, NC: Office of Research and Development, U.S. Environmental Protection Agency.

Illinois Department of Conservation, (1993). *The Salt Creek Greenway plan.* IL: Illinois Department of Conservation.

Jackson, L., Kurt, J. & Fisher, W. eds. (2000). *Evaluation guidelines for ecological indicators* (EPA/620/-99/005). Research Triangle Park, NC: National Health and Environmental Effects Research Laboratory, U.S. Environmental Protection Agency.

Jacobs, M. (1997). Environmental Valuation, Deliberative Democracy and Public Decision-making Institutions. In *Valuing nature? Ethics, economics, and the environment*, ed. J. Foster. London: Routledge.

Janis, I. (1982). *Groupthink: Psychological studies of policy decisions and fiascoes.* 2nd ed. New York: Houghton Mifflin.

Janssen, M. A. & Carpenter, S. R. (1999). Managing the resilience of lakes: A multi-agent modeling approach. *Conservation Ecology*, *3*, 15. http://www.consecol.org/vol3/iss2/art15/.

Johnson, B. B. (2003). Further notes on public response to uncertainty in risks and science. *Risk Analysis*, *23*, 78 1-789.

Johnson, B. B. & Slovic, P. (1995). Presenting uncertainty in health risk assessment: Initial studies of its effects on risk perception and trust. *Risk Analysis*, *15*, 485-494.

_____. (1998). Lay views on uncertainty in environmental health risk assessment. *Journal of Risk Research*, *1*, 26 1-279.

Johnston, R. J., Besedin, E. Y. & Wardwell, R. F. (2003). Modeling relationships between use and nonuse values for surface water quality: A meta-analysis. *Water Resources Research*, *39*, 1363-1372.

Jones, C. A. & Pease, K. A. (1997). Restoration-based compensation measures in natural resource liability statutes. *Contemporary Economic Policy*, *15*, 111-122.

Kahn, M. E. & Matsusaka, J. G. (1997). Demand for environmental goods: Evidence from voting patterns on California initiatives. *Journal of Law and Economics*, *40*, 137- 173.

Kanninen, B. J., ed. (2007). *Valuing environmental amenities using stated choice studies.* Dordrecht: Springer.

Karr, J. R. (1981). Assessment of biotic integrity using fish communities. *Fisheries*, *6*, 21-27.

_____. (1993). Defining and assessing ecological integrity: Beyond water quality. *Environmental Toxicology and Chemistry*, *12*, 1521-153 1.

Karr, J. R. & Chu, E. W. (1999). *Restoring Life in Running Waters.* Washington, DC: Island Press.

Kaufmann, R. K. (1992). A biophysical analysis of the energy/ real GDP ratio: Implications for substitution and technical change. *Ecological Economics*, *6*, 35-56.

Keeney, R. L. (1992). *Value-focused thinking. A path to creative decision making.* Cambridge: Harvard University Press.

Keeney, R. L. & Raiffa, H. (1993). *Decisions with multiple objectives: Preferences and value tradeoffs*. Cambridge: Cambridge University Press.

Kempton, W. (1991). Lay perspectives on global climate change. *Global Environmental Change*, *1*, 183-208.

Kenyon, W. & Hanley, N. (2001). *Economic and participatory approaches to environmental evaluation*. Glasgow, U.K.: Economics Department, University of Glasgow.

Kenyon, W. & Nevin, C. (2001). The use of economic and participatory approaches to assess forest development: A case study in the Ettrick Valley. *Forest Policy and Economics*, *3*, 69-80.

King, D. M. (1997). *Comparing ecosystem services and values: With illustrations for performing habitat equivalency analysis*. Service Paper Number 1. Washington, DC: National Oceanic and Atmospheric Administration.

Kitchen, J. W. & Hendon, W. S. (1967). Land values adjacent to an urban neighborhood park. *Land Economics*, *43*, 357-360.

Kline, J. & Wichelns, D. (1994). Using referendum data to characterize public support for purchasing development rights to farmland. *Land Economics*, *70*, 223-233.

Korsgaard, C. (1996). Two distinctions in goodness. In *Creating the kingdom of ends,* ed. C. Korsgaard, 249-274. Cambridge: Cambridge University Press.

Kosobud, R. F. (1998). *Urban Deconcentration and Biodiversity Valuation in the Chicago Region*: Report to the Chicago Wilderness Project Coalition.

Kotchen, M. & Powers, S. (2006). Explaining the appearance and success of voter referenda for open-space conservation. *Journal of Environmental Economics and Management*.

Kremen, C. (2005). Managing ecosystem services: What do we need to know about their ecology? *Ecology Letters*, *8*, 468-479.

Kremen C. & Chaplin, R. (2007). Insects as providers of ecosystem services: crop pollination and pest control. In *Insect Conservation Biology: Proceedings of the Royal Entomological Society's 23rd Symposium*, ed. A. J. A. Stewart, T. R. New, & O. T. Lewis, 349-382. Wallingford, U.K.: CABI Publishing.

Kremen, C., et al., (2007). Pollination and other ecosystem services produced by mobile organisms: A conceptual framework for the effects of land-use change. *Ecology Letters*, *10*, 299-3 14.

Krosnick, J. A. (1999). Survey research. *Annual Review of Psychology*, *50*, 537-567.

Krupnick, A. J., Morgenstern, R. D., et al. (2006). *Not a sure thing: Making regulatory choices under uncertainty*. Washington, DC: Resources for the Future. http://www.rff.org/rff/News/Features/Not-a-Sure-Thing.cfm.

Lear, J. S. & Chapman, C. B. (1994). *Environmental Monitoring and Assessment Program (EMAP) cumulative bibliography*. EPA/620/R-94/024. Research Triangle Park, NC: U.S. Environmental Protection Agency.

Leggett, C. G. & Bockstael, N. E. (2000). Evidence of the effects of water quality on residential land prices. *Journal of Environmental Economics and Management*, *39*, 121-144.

Leung, B., Finnoff, D., Shogren, J. F. & Lodge, D. (2005). Managing invasive species: Rules of thumb for rapid assessment. *Ecological Economics*, *55*, 24-36.

Lichtensten, S. & Slovic. P. (2006). *The construction of preference*. Cambridge: Cambridge University Press.

List, J. & Shogren, J. (2002). Calibration of willingnessto-accept. *Journal of Environmental Economics and Management*, *43*, 2 19-233.

Lockwood, M. & Tracy, K. (1995). Nonmarket economic valuation of an urban recreational park. *Journal of Leisure Research*, *27*, 155-167.

Loomis, J. B. & Rosenberger, R. S. (2006). Reducing barriers in future benefit transfers: Needed improvements in primary study design and reporting. *Ecological Economics*, *60*, 343- 350.

Louviere, J. J. (1988). Analysing decision making: Metric conjoint analysis. *Quantitative Applications in the Social Sciences*. Sage University Papers Series, N8 67. Newbury Park, CA: Sage.

Lupia, A. (1992). Busy voters, agenda control, and the power of information. *American Political Science Review*, *86*, 390-399.

Mace, B. L., Bell, P. A. & Loomis, R. J. (1999). Aesthetic, affective, and cognitive effects of noise on natural landscape assessment. *Society & Natural Resources*, *12*, 225-243.

MacEachren, A. M. (1995). *How maps work*. New York and London: The Guilford Press.Macmillan, D. C., et al. (2002). Valuing the non-market benefits of wild goose conservation: A comparison of interview and group-based approaches. *Ecological Economics*, *43*, 49-59.

Magleby, D. (1984). *Direct legislation: Voting on ballot propositions in the United States*. Baltimore, MD: Johns Hopkins University Press.

Mahan, B. L., Polasky, S. & Adams, R. M. (2000). Valuing urban wetlands: A property price approach. *Land Economics*, *76*, 100-113.

Malm, W., Kelly, K., Molenar, J. & Daniel, T. C. (1981). Human perception of visual air quality: Uniform haze. *Atmospheric Environment*, *15*, 1874-1890.

McConnell, K. E. & Bockstael, N. E. (2005). Valuing the Environment as a Factor of Production. In *Handbook of environmental economics*, ed. K. G. Maler, & J. R. Vincent. Amsterdam: North-Holland.

McDaniels, T. L., et al. (2003). Decision structuring to alleviate embedding in environmental valuation. *Ecological Economics*, *46*, 33-46.

McDowell, J. (1985). Values and secondary qualities. In *Morality and objectivity: A tribute to J.L.Mackie*, ed. T. Honderich, 110-129. Boston: Routledge and Kegan Paul.

Merton, R. K., Fiske, M. & Kendall, P. L. (1990). *The focused interview: A manual of problems and procedures*. 2nd ed. London: Collier MacMillan.

Millennium Ecosystem Assessment. (2005). *Ecosystems and human well-being: Synthesis*. Washington, DC: Island Press.

Millennium Ecosystem Assessment Board. (2003). *Ecosystems and human well-being: A report of the conceptual framework working group of the Millennium Ecosystem Assessment*. Washington, DC: Island Press.

Mitchell, R. & Carson, R. (1989). *Using surveys to value public goods: The contingent valuation method*. Washington, DC: Resources for the Future.

Montgomery, M. & Needelman, M. (1997). The welfare effects of toxic contamination in freshwater fish. *Land Economics*, *73*, 211-223.

Morgan, M. G., Fischhoff, B., Bostrom, A. & Atman, C. J. (2002). *Risk communication: A mental models approach*. Cambridge: Cambridge University Press.

Morgan, M. & Henrion, M. (1990). *Uncertainty: A guide to dealing with uncertainty in quantitative risk and policy analysis*. Cambridge: Cambridge University Press.

Moss, R. H. & Schneider, S. H. (2000). Uncertainties in the IPCC TAR: Recommendation to lead authors for more consistent assessment and reporting. In *Third assessment report: Cross cutting issues guidance papers,* ed. R., Pachauri, T. Taniguchi, & K. Tanaka, 33–51. Geneva, Switzerland: World Meteorological Organisation. From Global Industrial and Social Progress Institute, www.gispri. or.jp.

Murphy, J. J., Allen, P. G., Stevens, T. H. & Weatherhead, D. (2003). *A meta-analysis of hypothetical bias in stated preference valuation.*Working Paper No. 2003-8. Amherst: University of Massachusetts.

_____. (2005). A meta-analysis of hypothetical bias in stated preference valuation. *Environmental and Resource Economics*, *30*, 313-325.

Muthke, T. & Holm-Mueller, K. (2004). National and international benefit transfer testing with a rigorous test procedure. *Environmental and Resource Economics*, *29*, 323-336.

Nacci, D. E., Gleason, T. R., Gutjahr-Gobell, R., Huber, M. & Munns, W. R. Jr. (2002). Effects of chronic stress on wildlife populations: A modeling approach and case study. In *Coastal and estuarine risk assessment: risk on the edge*, ed. M. C., Newman, M. H., Roberts, Jr. & R.C. Hale, 247-272. New York: CRC/Lewis.

Nacci, D. E. & Hoffman, A. A. (2006). Genetic variation in population-level ecological risk assessment. In *Population- level ecological risk assessment*, ed. L. W., Barnthouse, W. R. Munns, Jr., & M. T. Sorensen. Pensacola, FL: SETAC.

National Research Council, 1989. (1996). *Understanding risk: Informing decisions in a democratic society*. National Academy Press.

_____. (2000). *Watershed management for potable water supply: Assessing the New York City strategy*. Washington, DC: National Academies Press.

_____. (2001). *The science of regional and global change: Putting knowledge to work*. Washington, DC: National Academies Press.

_____. (2004). *Valuing ecosystem services; toward better environmental decision-making*. Washington, DC: National Academies Press.

_____. (2007). *Models in environmental regulatory decision making*. Washington, DC. National Academies Press.

_____. (2008). *Public Participation in Environmental Assessment and Decision Making*. Washington, DC: National Academies Press.

Netusil, N. (2005). The effect of environmental zoning and amenities on property values: Portland, Oregon. *Land Economics*, *81*, 227.

W. D. Nordaus, & Kokkelenberg, E. C., eds. (1999). *Nature's numbers: expanding the U.S. national economic accounts to include the environment*. Washington, D.C. National Academies Press.

Pagiola, S., von Ritter, K. & Bishop, J. (2004). *Assessing the economic value of ecosystem conservation*. Environment Department Paper No. 101. Washington, DC: World Bank Environment Department Papers.

Palmquist, R. B. (1992). Valuing localized externalities. *Journal of Urban Economics*, *31*, 59-68.

_____. (2005). Hedonic models. In *Handbook of environmental economics, vol. 2,* ed. K. Mäler and J. Vincent. Amsterdam: North-Holland.

Payne, J. W., Bettman, J. R. & Johnson, E. J. (1992). Behavioral decision research: A constructive processing perspective. *Annual Review of Psychology*, *43*, 87-132.

Phaneuf, D. J. & Smith, V. K. (2005). Recreation demand models. In *Handbook of environmental economics , vol. 2,* ed. K. G. Maler, & J. R. Vincent, Amsterdam: North-Holland.

Poulos, C., Smith, V. K. & Kim, H. (2002). Treating open space as an urban amenity. *Resource and Energy Economics*, *24*, 107-129.

Randall, A. (1994). A difficulty with the travel cost method. *Land Economics*, *70*, 88-96.

Ready, R. S. Narvrud, B., Day, R., Dubourg, F., Machado, S., Mourato,. S., et al. (2004). Benefit transfer in Europe: How reliable are transfers between countries? *Environmental and Resource Economics*, *29*, 67-82.

Rees, W. E. Eco-footprint analysis: Merits and brickbats. *Ecological Economics*, *32*, 371-374.

Ribe, R. G., Armstrong, E. T. & Gobster. P. H. (2002). Scenic vistas and the changing policy landscape: Visualizing and testing the role of visual resources in ecosystem management. *Landscape Journal*, *21*, 42-66.

Ricketts, T. H., Daily, G. C., Ehrlich, P. R. & Michener, C. D. (2004). Economic value of tropical forest to coffee production. *PNAS*, *34*, 12579-12582.

Riordan, R. & Barker, K. (2003). *Cultivating biodiversity in Napa*. Geospatial Solutions November 2003.

Roach, B. & Wade, W. W. (2006). Policy evaluation of natural resource injuries using habitat equivalency analysis. *Ecological Economics*, *58*, 421-433.

Romero, F. S. & Liserio, A. (2002). Saving open spaces: Determinants of 1998 and 1999 "antisprawl" ballot measures. *Social Science Quarterly*, *83*, 341-352.

Rossi, P. H., Lipsey, M. V. & Freeman, H. E. (2003). *Evaluation: A systematic approach.* Thousand Oaks, CA: Sage Publications.

Roughgarden, J. (1995). Can economics protect biodiversity? In *The economics and ecology of biodiversity decline*, ed. T. Swanson, 149-156. Cambridge: Cambridge University Press.

_____. (1998a). *Primer of ecological theory.* Upper Saddle River, NJ: Prentice Hall.

_____. (1998b). Production functions from ecological populations: A survey with emphasis on spatially explicit models. In *Spatial ecology: The rule of space in population dynamics and interspecific interactions*, ed. D. Tilman and P. Kareiva. Princeton: Princeton University Press. 296-317.

_____. (1998c). How to manage fisheries. *Ecological Applications*, *8*, S160-164.

_____. (2001a). Guide to diplomatic relations with economists. *Bulletin of the Ecological Society of America*, *82*, 85-88.

_____. (2001b). Production functions from ecological populations: A survey with emphasis on spatially implicit models. In *Ecology: Achievement and challenge,* ed. M. C., Press, N. J. Huntly, & S. Levin. The 41st *Symposium of the British Ecological Society jointly sponsored by the Ecological Society of America*, Orlando, FL, April 10-13.

Roughgarden, J. & Armsworth, P. (2001). Managing ecosystem services. In *Ecology: Achievement and Challenge*, Blackwell Science, ed. M. N. Huntly, & S. Levin, 337-356.

Roughgarden, J. & Smith. F. (1996). Why fisheries collapse and what to do about it. *Proceedings of the National Academy of Sciences*, *93*, 5078-5083.

Ruliffson, J. A., Haight, R. G., Gobster, P. H. & Homans, F. R. (2003). Metropolitan natural area protection to maximize public access and species representation. *Environmental Science and Policy*, *6*, 291-299.

Russell, E. P., III. (1993). Lost among the parts per billion: Ecological protection at the Environmental Protection Agency, 1970-1993. *Environmental History*, *2*, 29-51.

Ruth, M. (1995). Information, order and knowledge in economic and ecological systems: Implications from material and energy use. *Ecological Economics*, *13*, 99-114.

Sagoff, M. (1998). Aggregation and deliberation in valuing environmental public goods: A look beyond contingent pricing. *Ecological Economics*, *24*, 213-230.

_____. (2004). *Price, principle, and the environment*. Cambridge: Cambridge University Press.

_____. (2005). *The Catskills parable*. PERC Report. Bozeman, MT: Political Economy Research Center.

Satterfield, T. & Slovic, S. eds. (2004). *What's nature worth: Narrative expressions of environmental values*. Salt Lake City: University of Utah Press.

Sayre-McCord, G. (1988). The many moral realisms. In *Essays on moral realism* ed. G. Sayre-McCord, 1-26. Ithaca: Cornell University Press.Schaeffer, N. C. & S. Presser. (2003). The science of asking questions. *Annual Review of Sociology*, *29*, 65-88.

Schiller, A., Hunsaker, C. T., Kane, M. A., Wolfe, A. K., Dale, V. H., Suter, G. W., Russell, C. S., Pion, G., Jensen, M. H. & Konar. V. C. (2001). Communicating ecological indicators to decision makers and the public. *Conservation Ecology*, *5*, 19. http://www.ecologyandsociety.org/vol5/iss1/art19/# FromValuesToValuedAspects.

Schläpfer, F. & Hanley, N. (2003). Do local landscape patterns affect the demand for landscape amenities protection? *Journal of Agricultural Economics*, *54*, 21-35.

Schläpfer, F., Roschewitz, A. & Hanley, N. (2004). Validation of stated preferences for public goods: a comparison of contingent valuation survey response and voting behaviour. *Ecological Economics*, *51*, 1-16.

Schläpfer, F., Schmitt, M. & Roschewitz, A. (2008). Competitive politics, simplified heuristics, and preferences for public goods. *Ecological Economics*, *65*, 574-589.

Schriver, K. A. (1989). Evaluating text quality: the continuum from text-focused toreader-focused methods, *Professional Communication. IEEE Transactions 32*, 238-255.

Scriven, M. (1967). The methodology of evaluation. In *Perspectives of curriculum evaluation*. American Educational Research Association Monograph Series on Curriculum Evaluation, No. 1. Chicago: Rand McNally.

Seidl, C. (2002). Preference reversal. *Journal of Economic Surveys*, *16*, 621-55.

Sen, A. K. (1977). Rational fools: A critique of the behavioral foundations of economic theory. *Philosophy and Public Affairs*, *6*, 317-344.

Settle, C., Crocker, T. D. & Shogren, J. F. (2002). On the joint determination of biological and economic systems. *Ecological Economics*, *42*, 301-311.

Shabman, L. A. & Batie, S. S. (1978). The economic value of coastal wetlands: A critique. *Coastal Zone Management Journal*, *4*, 231-237.

Shabman, L. & Stephenson. K. (1996). Searching for the correct benefit estimate: Empirical evidence for an alternative perspective. *Land Economics*, *72*, 433-49.

Shah, P. & Miyake, A. eds. (2005). *The Cambridge handbook of visuospatial thinking.* Cambridge: Cambridge University Press.

Shrestha, R. K. & Loomis, J. B. (2003). Testing a meta- analysis model for benefit transfer in international outdoor recreation. *Ecological Economics*, *39*, 67-83.

Silva, P. & Pagiola. S. (2003). A review of the valuation of environmental costs and benefits in World Bank projects. In *World Bank environmental economics series.* Washington, DC: The World Bank.

Simmons, C., Lewis, K. & Moore, J. (2000). Two feet – two approaches: A component-based model of ecological footprinting. *Ecological Economics*, *32*, 375-380.

Slovic, P. (1995). The construction of preference. *American Psychologist*, *50*, 364-371.

Smith, B. A. (1978). Measuring the value of urban amenities. *Journal of Urban Economics*, *5*, 370-387.

Smith, J. E., Heath, L. S., Skog, K. E. & Birdsey, R. A. (2006). *Methods for calculating forest ecosystem and harvested carbon with standard estimates for forest types of the United States.* Northeastern Research Station General Technical Report NE-343. U.S. Department of Agriculture Forest Service.

Smith, V. K. (1991). Household production functions and environmental benefit estimation. In *Measuring the demand for environmental quality*, ed. J. B. Braden and C. D. Kolstad. Amsterdam: North-Holland.___. 1997. Pricing What is Priceless: A Status Report on Non-Market Valuation of Environmental Resources. In *The International Yearbook of Environmental and Resource Economics*, eds. H. Folmer and T. Tietenberg. Cheltenham, U.K.: Edward Elgar, 156-204.

Smith, V. K. & Huang, J. C. (1995). Can markets value air quality? A meta-analysis of hedonic property value models. *Journal of Political Economy*, *103*, 209-27.

Smith, V. K. & Kaoru, Y. (1990a). Signals or noise? Explaining the variation in recreation benefit estimates. *American Journal of Agricultural Economics*, *72*, 419-433.

(1990b). What have we learned since Hotelling's letter? A meta-analysis. *Economics Letters*, *32*, 267-272.

Smith, V. K. & Pattanayak, S. K. (2002). Is meta-analysis a Noah's ark for non-market valuation? *Environmental and Resource Economics*, *22*, 271-296.

Smith, V. K., Poulos, C. & Kim, H. (2002). Treating open space as an urban amenity. *Resource and Energy Economics*, *24*, 107- 129.

Solecki, W. D., Mason, R. J. & Martin, S. (2004). The geography of support for open-space initiatives: A case study of New Jersey's 1998 ballot measure. *Social Science Quarterly*, *85*, 624-639.

Spyridakis, J. H. (2000). Guidelines for authoring comprehensible Web pages and evaluating their success. *Technical Communication*, *47*, 301-310. http://www.uwtc. washington.edu/people/faculty/jspyridakis.php.

Stoms, D. M., Comer, P. J., Crist, P. J. & Grossman, D. H. (2005). Choosing surrogates for biodiversity conservation in complex planning environments. *Journal of Conservation Planning*, *1*, 44-63.

Stone, E. R., Sieck, W. R., Bull, B. E., Yates, J. F., Parks, S. C. & Rush, C. J. (2003). Foreground: background salience: Explaining the effects of graphical displays on risk avoidance. *Organizational Behavior and Human Decision Processes*, *90*, 19-36.

Strecher, V. J., Greenwood, T., Wang, C. & Dumont, D. (1999). Interactive multimedia and risk communication. *Journal of the National Cancer Institute. Monographs*, *25*, 134-139.

Sturgeon, N. (1985). Moral explanations. In *Morality, reason, and truth*, ed. D. Cropp and D. Zimmerman, 49-7 8. Totowa, NJ: Rowman and Allenheld.

Sunstein, C. R., Kahneman, D., Schkade, D. & Ritov, I. (2002). Predictably incoherent judgments. *Stanford Law Review*, *54*, 1153-12 15.

Suter, G. W. II. (2006). *Ecological risk assessment*. 2nd ed. Boca Raton, FL: CRC Press.

Tourangeau, R. (2004). Survey research and societal change. *Annual Review of Psychology*, *55*, 775-80 1.

Turner, W. R., Brandon, K., Brooks, T. M., Costanza, R., de Fonseca, G. A. B. & Portela. R. (2007). Global conservation of biodiversity and ecosystem services. *BioScience*, *57*, 868-873.

Tyrvainen, L. & Miettinen, A. (2000). Property prices and urban forest amenities. *Journal of Environmental Economics and Management*, *39*, 205-223.

United Kingdom Department for Environment, Food and Rural Affairs. (2007). *An introductory guide to valuing ecosystem services*. London: Department for Environment, Food and Rural Affairs. http://www.defra.gov.uk/wildlife natres/pdf/eco_valuing.pdf.

U.S. Environmental Protection Agency. (1994). *Managing ecological risks at* EPA*: Issues and recommendations for progress*. EPA/600/R-94/183. Washington, DC: EPA.

_____. (1997). *Guiding principles for Monte Carlo analysis*. EPA/630/R-97/001. Washington, DC: EPA.

_____. (1999). *The benefits and costs of the Clean Air Act 1990 to 2010*. Washington, DC: EPA.

_____. (2000a). *Engaging the American people: A review of* EPA*'s public participation policy and regulations with recommendations for action*. Washington, DC: EPA.

_____. (2000b). *Guidelines for preparing economic analyses*. EPA 240-R-00-003. Washington, DC: EPA.

_____. (2000c). *Peer review handbook*. 2nd ed. EPA 100-B-00-001. Washington, DC: EPA.

_____. (2000d). *Risk characterization handbook*. EPA 100-B-00- 002. Washington, DC: EPA.

_____. (2002a). *Draft report on the environment*. EPA 260-R-02- 006. Washington, DC: EPA.

_____. (2002b). *Environmental and economic benefit analysis of final revisions to the National Pollutant Discharge Elimination System (NPDES) regulation and the effluent guidelines for concentrated animal feeding operations (CAFOs)*. EPA-821-R-03-003. Washington, DC: EPA.

_____. (2003). EPA *Strategic Plan: Direction for the Future*. EPA- 190-R-03-003. Washington, DC: EPA.

_____. (2004a). *Economic and environmental benefits analysis of final effluent limitations guidelines and new source per for mance standards for the concentrated aquatic animal production industry point source category*. Washington, DC: EPA.

_____. (2004b). *Innovating for better results: A report on* EPA *Progress from the Innovations Action Council*. EPA100-R-04-001. Washington, DC: EPA.

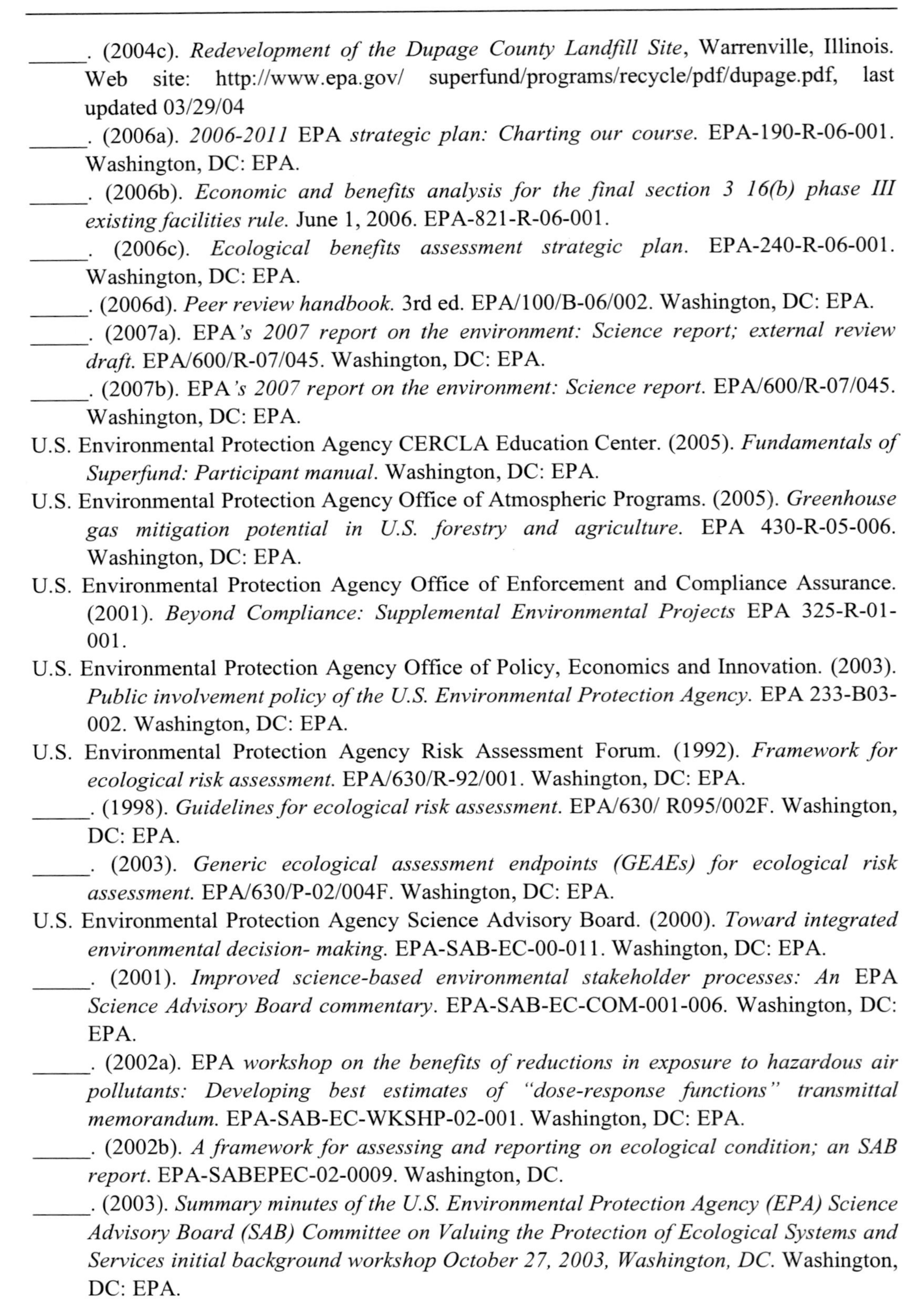

_____. (2004c). *Redevelopment of the Dupage County Landfill Site*, Warrenville, Illinois. Web site: http://www.epa.gov/ superfund/programs/recycle/pdf/dupage.pdf, last updated 03/29/04

_____. (2006a). *2006-2011* EPA *strategic plan: Charting our course.* EPA-190-R-06-001. Washington, DC: EPA.

_____. (2006b). *Economic and benefits analysis for the final section 3 16(b) phase III existing facilities rule.* June 1, 2006. EPA-821-R-06-001.

_____. (2006c). *Ecological benefits assessment strategic plan.* EPA-240-R-06-001. Washington, DC: EPA.

_____. (2006d). *Peer review handbook.* 3rd ed. EPA/100/B-06/002. Washington, DC: EPA.

_____. (2007a). EPA*'s 2007 report on the environment: Science report; external review draft.* EPA/600/R-07/045. Washington, DC: EPA.

_____. (2007b). EPA*'s 2007 report on the environment: Science report.* EPA/600/R-07/045. Washington, DC: EPA.

U.S. Environmental Protection Agency CERCLA Education Center. (2005). *Fundamentals of Superfund: Participant manual.* Washington, DC: EPA.

U.S. Environmental Protection Agency Office of Atmospheric Programs. (2005). *Greenhouse gas mitigation potential in U.S. forestry and agriculture.* EPA 430-R-05-006. Washington, DC: EPA.

U.S. Environmental Protection Agency Office of Enforcement and Compliance Assurance. (2001). *Beyond Compliance: Supplemental Environmental Projects* EPA 325-R-01-001.

U.S. Environmental Protection Agency Office of Policy, Economics and Innovation. (2003). *Public involvement policy of the U.S. Environmental Protection Agency.* EPA 233-B03-002. Washington, DC: EPA.

U.S. Environmental Protection Agency Risk Assessment Forum. (1992). *Framework for ecological risk assessment.* EPA/630/R-92/001. Washington, DC: EPA.

_____. (1998). *Guidelines for ecological risk assessment.* EPA/630/ R095/002F. Washington, DC: EPA.

_____. (2003). *Generic ecological assessment endpoints (GEAEs) for ecological risk assessment.* EPA/630/P-02/004F. Washington, DC: EPA.

U.S. Environmental Protection Agency Science Advisory Board. (2000). *Toward integrated environmental decision- making.* EPA-SAB-EC-00-011. Washington, DC: EPA.

_____. (2001). *Improved science-based environmental stakeholder processes: An* EPA *Science Advisory Board commentary.* EPA-SAB-EC-COM-001-006. Washington, DC: EPA.

_____. (2002a). EPA *workshop on the benefits of reductions in exposure to hazardous air pollutants: Developing best estimates of "dose-response functions" transmittal memorandum.* EPA-SAB-EC-WKSHP-02-001. Washington, DC: EPA.

_____. (2002b). *A framework for assessing and reporting on ecological condition; an SAB report.* EPA-SABEPEC-02-0009. Washington, DC.

_____. (2003). *Summary minutes of the U.S. Environmental Protection Agency (EPA) Science Advisory Board (SAB) Committee on Valuing the Protection of Ecological Systems and Services initial background workshop October 27, 2003, Washington, DC.* Washington, DC: EPA.

_____. (2005). *Advisory on* EPA*'s draft report on the environment.* EPA-SAB-05-004. Washington, DC: EPA.

_____. (2006a). *Review of agency draft guidance on the development, evaluation, and application of regulatory environmental models and models knowledge base by the regulatory environmental modeling guidance review panel of the* EPA *Science Advisory Board*. EPA-SAB-06-009. Washington, DC: EPA.

_____. (2006b). *Summary minutes of the U.S. Environmental Protection Agency (EPA) Science Advisory Board (SAB) Committee on Valuing the Protection of Ecological Systems and Services (C-VPESS) public meeting – October 5-6, 2006.* Washington, DC: EPA.

_____. (2006c). *Advisory on* EPA*'s Superfund benefits analysis.* EPA-SAB-ADV-06-002. Washington, DC: EPA.

_____. (2007a). *Comments on* EPA*'s strategic research directions and research budget for FY 2008: An advisory report of the U.S. Environmental Protection Agency Science Advisory Board*. EPA-SAB-07-004. Washington, DC: EPA.

_____. (2007b). *U.S .Environmental Protection Agency Science Advisory Board (SAB) Committee on Valuing the Protection of Ecological Systems and Services (C-VPESS) summary meeting minutes of a public teleconference meeting, June 12, 2007.* Washington, DC: EPA.

_____. (2008a). *SAB Advisory on the* EPA *ecological research program multi-year plan.* EPA-SAB-08-01 1. Washington, DC: EPA

_____. (2008b). *SAB advisory on* EPA*'s issues in valuing mortality risk reduction.* EPA-SAB-08-001. Washington, DC: EPA.

U.S. Environmental Protection Agency Science Advisory Board Staff. (2004). *Survey of needs for regions for science- based information on the value of protecting ecological systems and services*. Background document for C-VPESS Meeting Sept., 13-15, 2004. Washington, DC: EPA.

_____. (2008b). *Advisory on the* EPA *Ecological Research Program Multi-Year Plan.* EPA-SAB-08-011. Washington, DC: EPA.

U.S. National Oceanic and Atmospheric Administration. (1995). *Habitat equivalency analysis: An overview.* Policy and Technical Paper Series No. 95-1 (revised 2000). Washington, DC: National Oceanic and Atmospheric Administration.

U.S. National Oceanic and Atmospheric Administration. (1999). *Discounting and the treatment of uncertainty in natural resource damage assessment.* Technical Paper 99-1. Washington, DC: National Oceanic and Atmospheric Administration, Damage Assessment and Restoration Program.

U.S. National Oceanic and Atmospheric Administration. (2001). *Damage assessment and restoration plan and environmental assessment for the Point Comfort/Lavaca Bay NPL site recreational fishing service losses.* Washington, DC: National Oceanic and Atmospheric Administration.

U.S. National Oceanic and Atmospheric Administration. Coastal Service Center Web site. Habitat Equivalency Analysis. www.csa.noaa.gov/economics/habitatequ.htm.

U.S. Office of Management and Budget. (2003). *Circular A-4.* September 17. Washington, DC: OMB.

van den Belt, M. (2004). *Mediated modeling: A systems dynamics approach to environmental consensus building.* Washington, DC: Island Press.

Vossler, C. A. & Kerkvliet, J. (2003). A criterion validity test of the contingent valuation method: Comparing hypothetical and actual voting behavior for a public referendum. *Journal of Environmental Economics and Management*, *45*, 631-49.

Vossler, C., Kerkvliet, J., Polasky, S. & Gainutdinova, O. (2003). Externally validating contingent valuation: an open- space survey and referendum in Corvallis, Oregon. *Journal of Economic Behavior & Organization*, *51*, 26 1-277.

Wackernagel, M., Onisto, L., Bello, P., Linares, A. C., Falfan, I. S. L., Garcia, J. M., Guerrero, A. I. S. & Guerrero, M. G. S. (1999). National natural capital accounting with the ecological footprint concept. *Ecological Economics*, *29*, 375- 390.

Wackernagel, M., Schultz, N. B., Deumling, D., Linares, A. C., Jenkins, L., Kapos, V., Monfreda, C., Loh, J., Myers, N., Norgaard, R. B. & Randers, J. (2002). Tracking the ecological overshoot of the human economy. *Proceedings of the National Academy of Sciences*, USA, *99*, 9266-9271.

Wainger, L., King, D., Salzman, J. & Boyd, J. (2001). Wetland value indicators for scoring mitigation trades. *Stanford Environmental Law Journal:* 20.

Walsh, R. G., Johnson, D. M. & McKean, J. R. (1992). Benefit transfer of outdoor recreation demand studies, 1968-198 8. *Water Resources Research*, *28*, 707-713.

Wang, B. & Manning, R. E. (2001). Computer simulation modeling for recreation management: A study on carriage road use in Acadia National Park, Maine, USA. *Environmental Management*, *23*, 193-203.

Weicher, J. C. & Zeibst, R. H. (1973). The externalities of neighborhood parks: An empirical investigation. *Land Economics*, *49*, 99-105.

Weigel, R. H. & Newman, L. S. (1976). Increasing attitude- behavior correspondence by broadening the scope of the behavioral measure. *Journal of Personality and Social Psychology*, *33*, 793-802.

Weslawski, J. M., et al. (2004). Marine sedimentary biota as providers of ecosystem goods and services. In *Sustaining biodiversity and ecosystem services in soils and sediment*, ed. D. H. Wall, 73-98. Washington, DC: Island Press.

Willey, Z. & Chameides, B. ed. (2007). *Harnessing farms and forests in the low-carbon economy: How to create, measure, and verify greenhouse gas offsets*. Durham, NC: Duke University Press.

Wilson, M. A. (2004). *Ecosystem services at Superfund redevelopment sites: Revealing the value of revitalized landscapes through the integration of ecology and economics.* (Report prepared by Spatial Informatics Group, LLC under subcontract of Systems Research and Applications Corporation.) Washington, DC: EPA Office of Solid Waste and Emergency Response.

Wilson, M. & Hoehn, J. (2006). Valuing environmental goods and services using benefit transfer: The state-of-the-art and science. Editorial in *Ecological Economics*, *60*, 335-42.

Wilson, T. D., Lisle, D. J., Kraft, D. & Wetzel, C. G. (1989). Preferences as expectation-driven inferences: Effects of affective expectations on affective experience. *Journal of Personality and Social Psychology*, *56*, 45 19-530.

Winston, C. (1993). Economic deregulation: Days of reckoning for microeconomists. *Journal of Economic Literature.*

Worm, B., Barbier, E. B., Beaumont, N., Duffy, J. E., Folke, C., Halpern, B. S., Jackson, J. B. C., Lotze, H. K., Micheli, F., Palumbi, S. R., Sala, E., Selkoe, K. A., Stachowicz, J. J. &

Watson, R. (2006). Impacts of biodiversity loss on ocean ecosystem services. *Science*, *314*, 787-790.

Zacharias, J. (2006). Exploratory spatial behaviour in real and virtual environments. *Landscape and Urban Planning*, *78*, 1-13.

Zaksek, M. & Arvai, J. L. (2004). Toward improved communication about wildland fire: Mental models research to identify information needs for natural resource management. *Risk Analysis*, *24*, 1503-1514.

Endnotes

[1] Laws include: the Clean Air Act, Clean Water Act, Comprehensive Environmental Response, Compensation, and Liability Act, Federal Insecticide, Fungicide and Rodenticide Act, Toxic Substances Control Act, and Resource Conservation and Recovery Act.

[2] Although C-VPESS was initiated by the SAB, senior EPA managers supported the concept of this SAB project and participated in the initial background workshop that launched the work of the C-VPESS.

[3] The SAB Staff Office published a *Federal Register* notice on March 7, 2003, (68 FR 11082-11084) announcing the project and calling for the public to nominate experts in the following areas: decision science, ecology, economics, engineering, law, philosophy, political science, and psychology with emphasis in ecosystem protection. The SAB Staff Office published a memorandum on August 11, 2003, documenting the steps involved in forming the new committee and finalizing its membership.

[4] The committee developed the conclusions in this chapter after multiple public meetings, teleconferences and workshops including: (a) an Initial Background Workshop on October 27, 2003, to learn the range of EPA's needs for science-based information on valuing the protection of ecological systems and services from managers of EPA Headquarters and Regional Offices; (b) a Workshop on Different Approaches and Methods for Valuing the Protection of Ecological Systems and Services, held on April 13-14, 2004; (c) an advisory meeting focused on support documents for national rule makings held on June 14-15, 2004; (d) an advisory meeting focused on regional science needs, in EPA's Region 9 (San Francisco) Office on Sept. 13, 14, and 15, 2004; (e) advisory meetings held on January 25-26, 2005, and April 12-13, 2005, to review EPA's draft *Ecological Benefits Assessment Strategic Plan;* and *(*f) a Workshop on Science for Valuation of EPA's Ecological Protection Decisions and Programs, held on December 13-14, 2005, to discuss the integrated and expanded approach described in this paper. The committee also discussed text drafted for this chapter at public meetings on October 25, 2005; May 9, 2006; October 5-6, 2006, and May 1-2, 2007, and on ten subsequent public teleconferences.

[5] The committee also notes a report published shortly before this chapter was finalized (United Kingdom Department for Environment, Food and Rural Affairs, 2007).

[6] Likewise, this definition would not include goods or services such as recreation that are produced by combining ecological inputs or outputs with conventional inputs (such as labor, capital, or time). In addition, Boyd and Banzhaf (2006) advocate defining changes in ecosystem services in terms of standardized units or quantities, which requires that they be measurable in practice. Such an approach is consistent with the concept of "green accounting," which extends the principles embodied in measuring marketed products to the measurement and consideration of the production, or changes in the stock, of ecological or other environmental "products" (Nordaus and Kokkelenberg, 1999).

[7] Even the term "values" itself means very different things within different disciplines. For example, economists associate values with changes, while other disciplines associate values with beliefs or mental structures that influence behavior or provide a "moral compass." For discussions of the concepts of value used within different disciplines, see Dietz et al., 2005; Fischoff, 1991, 2005; and Hitlin and Piliavin, 2004.

[8] There is controversy over the meaning of intrinsic value (Korsgaard, 1996). Many people take intrinsic value to mean that the value of something is inherent in that thing. Some philosophers have argued that value or goodness is a simple non-natural property of things (see Moore, 1903 for the classical statement of this position), and others have argued that value or goodness is not a simple property of things but one that supervenes on the natural properties to which we appeal to explain a thing's goodness. This view is defended by, among others, contemporary moral realists; see McDowell (1985), Sturgeon (1985), Sayre-McCord (1988), and Brink (1989).

[9] Although table 1 lists concepts of value considered by the committee, these value concepts are not mutually exclusive. For example, values expressing attitudes or judgments can be based on the same utilitarian goals as those underlying economic values or on the same considerations that underlie civic values. Likewise,

constructed preferences can relate to self-interested attitudes or judgments (as economic values do) as well as expressed civic values.

[10] Some members of the committee argued that in a broader definition, "economic" should refer to all methods of assessing tradeoffs and contributions to human well-being, not just those based on willingness to pay or willingness to accept. However, in this chapter, "economic values" include only values reflecting preference-based tradeoffs.

[11] Monetized measures of economic output form the basis of national income accounts. While historically these have included only marketed outputs (such as agriculture and manufacturing), in principle the contributions of ecosystems and other natural assets to national income or aggregate output could be defined. A number of researchers have examined efforts to expand the national accounts to include these types of contributions (Nordhaus and Kokkelenberg, 1999). In the context of national income accounts, "value" is typically defined in terms of the dollar value of output, computed using prices and quantities. However, given the committee's focus on valuing changes resulting from EPA decisions or actions, this chapter defines economic values in terms of tradeoffs, consistent with standard welfare economics.

[12] See, for example, Seidl (2002) for a survey of the preference reversal literature.

[13] Environmental values are often defined in terms of a set of guiding principles, concepts or beliefs that guide decisions and evaluations. For a recent survey on environmental values, see Dietz et al. (2005).

[14] Under GPRA, the Office of Management and Budget requires EPA to periodically identify its strategic goals and describe both the social costs and budget costs associated with them. EPA's strategic plan for 2003-2008 described the current social costs and willingness-to-pay or willingnessto-accept analyses of EPA's programs and policies under each strategic goal area for the year 2002 (EPA, 2003).
This analysis repeatedly points out that EPA lacks data and methods to quantify willingness-to-pay or willingnessto-accept associated with the goals in its strategic plan. In addition, GPRA established requirements for assessing the effectiveness of federal programs, including the outcomes of programs intended to protect ecological resources. EPA must report annually on its progress in meeting program objectives linked to strategic plan goals and must engage periodically in an in-depth review [through the Program Assessment Rating Tool (PART)] of selected programs to identify their net contributions to human welfare and to evaluate their effectiveness in delivering meaningful, ambitious program outcomes. Characterizing ecological contributions to human welfare associated with EPA programs is a necessary part of the program assessment process.

[15] These interviews were conducted by one committee member, Dr. James Boyd, in conjunction with the Designated Federal Officer Dr. Angela Nugent, over the period September 22, 2004, through November 23, 2005. In seven sets of interviews, Dr. Boyd spoke with staff from the Office of Policy, Economics and Innovation, Office of Water, Office of Air and Radiation, and the Office of Solid Waste and Emergency Response.

[16] NCEE is typically brought in by the program offices to help both design and review RIAs. NCEE can be thought to provide a centralized "screening" function for rules and analysis before they go to OMB. NCEE is actively involved in discussions with OMB as rules and supporting analysis are developed and advanced.

[17] In addition, Circular A-4 states (p. 27) "If monetization is impossible, explain why and present all available quantitative information" and "If you are not able to quantify the effects, you should present any relevant quantitative information along with a description of the unquantified effects, such as ecological gains, improvements in quality of life, and aesthetic beauty" (p. 26).

[18] The committee reviewed and critically evaluated the CAFO Environmental and Economic Benefits Analysis at its June 15, 2004, meeting. As stated in the Background Document for SAB Committee on Valuing the Protection of Ecological Systems and Services for its session on June 15, 2004, the purpose of this exercise was "to provide a vehicle to help the Committee identify approaches, methods, and data for characterizing the full suite of ecological values' affected by key types of Agency actions and appropriate assumptions regarding those approaches, methods, and data for these types of decisions." The committee based its review on EPA's final benefits report (EPA, 2002b) and a briefing provided by the EPA Office of Water staff.

[19] In December 2000, EPA proposed a new CAFO rule under the federal Clean Water Act to replace 25-year-old technology requirements and permit regulations (66 FR 2959). EPA published its final rule in December 2003 (68 FR 7176). The new CAFO regulations; which cover over 15,000 large CAFO operations, reduce manure and wastewater pollutants from feedlots and land applications of manure and remove exemptions for stormwater-only discharges.

[20] The potential "use" benefits included in-stream uses (commercial fisheries, navigation, recreation, subsistence, and human health risk), near-stream uses (non-contact recreation, such as camping, and nonconsumptive, such as wildlife viewing), off-stream consumptive uses (drinking water, agricultural/irrigation uses, and industrial/commercial uses), aesthetic value (for people residing, working, or traveling near water), and the option value of future services. The potential "non-use" values included ecological values (reduced mortality/morbidity of certain species, improved reproductive success, increased diversity, and improved habitat/sustainability), bequest values, and existence values.

[21] These benefits were recreational use and non-use of affected waterways, protection of drinking water wells, protection of animal water supplies, avoidance of public water treatment, improved shellfish harvest, improved recreational fishing in estuaries, and reduced fish kills.

[22] These include reduced eutrophication of estuaries; reduced pathogen contamination of drinking water supplies; reduced human and ecological risks from hormones, antibiotics, metals, and salts; improved soil properties from reduced over-application of manure; and "other benefits".

[23] EPA apparently conducted no new economic valuation studies (although a limited amount of new ecological research was conducted) and did not consider the possible benefits of developing new information where important benefits could not be valued in monetary terms based on existing data.

[24] For example, while the report notes the potential effects of discharging hormones and other pharmaceuticals commonly used in CAFOs into drinking water sources and aquatic ecosystems, the nature and possible ecological significance of these effects is not adequately developed or presented. Similarly, the report does not adequately address the well-known consequences of discharging trihalomethane precursors into drinking-water sources.

[25] EPA used estimates based on a variety of public surveys in its benefit transfer efforts, including: a national survey (1983) that determined individuals' willingness to pay for changes in surface water quality relating to water-based recreational activities (section 4 of the CAFO Report); a series of surveys (1992, 1995, 1997) of willingness to pay for reduced/avoided nitrate (or unspecified) contamination of drinking water supplies (section 7); and several studies (1988, 1995) of recreational fishers' values (travel cost, random utility model) for improved/protected fishing success related to nitrate pollution levels in a North Carolina estuary (section 9).

[26] Although EPA later prepared more detailed conceptual models of the CAFO rule's impact on various ecological systems and services, EPA did not prepare these models until after the Agency finished its analysis.

[27] Contamination of estuaries, for example, might negatively affect fisheries in the estuary (a primary effect) but might have an even greater impact on offshore fisheries that have their nurseries in the estuary (a secondary effect).

[28] The goal of EPA's analysis was a national-level assessment of the effects of the CAFO rule. This involved the effects of approximately 15,000 individual facilities, each contributing pollutants across local watersheds into local and regional aquatic ecosystems. A few intensive case studies were mentioned in the report and used to calibrate the national scale models (e.g., NWPCAM, GLEAMS), but there was no indication that these more intensive data sets were strategically selected or used systematically for formal sensitivity tests or validations of the national-scale model results.

[29] This could include either a robust public involvement process following Administrative Procedures Act requirements (e.g., publication in the *Federal Register*), or some other public involvement process (see EPA's public involvement policy [EPA Office of Policy, Economics and Innovation, 2003] and the SAB report on science and stakeholder involvement [EPA Science Advisory Board, 2001]).

[30] In theory, one can value a final product *either* directly (output valuation) or indirectly as the sum of the derived value of the inputs (input valuation), but not both, because separately valuing both intermediate and final products leads to double counting. In some cases, it may be easier or more appropriate to value the intermediate service, while in other cases the change in the final product can be directly valued.

[31] Indicators therefore provide information on the direction and possible magnitude of the impact or response of an ecosystem to a stressor, rather than on merely the stressor itself.

[32] Note that these essential ecosystem characteristics are very similar to the seven ecological indicators in EPA's report on assessing ecological systems (EPA Science Advisory Board 2002b): landscape condition, biotic condition, chemical and physical characteristics, ecological processes, hydrology and geomorphology and natural disturbance regimes.

[33] The NSF has recently emphasized engaging social sciences in LTER, increasing the potential usefulness of LTER in building EPA's capacity for ecological valuations. Although most LTERs are in pristine areas, two urban sites and an agricultural site could be of special value to the Agency.

[34] These supplemental materials were compiled by the committee as background information, but they do not necessarily represent the consensus views of the committee, nor have they been reviewed and approved by the chartered Science Advisory Board (see appendix A for more detail).

[35] The U.S. federal government is one of the largest producers of survey data, which form the basis of many government policy making decisions (see the table on page 120.).

[36] This comparison of causal beliefs with formal decision models entails three steps. First is the construction of an expert decision model, generally through systematic, formal decision analysis involving scientists and other topical experts, individually or in groups. Following this is the analysis of semi-structured interviews with individuals from the population of interest, and comparison of these to the decision model. Third is the design and fielding of a survey to test the reliability of findings from the interviews in a representative sample of the population of interest or the public at large. The interviews and surveys employ mixed methods, and assess both how decision makers intuitively structure and conceptualize their environmental mitigation decisions, as well as how they react to structured stimuli and questions (Morgan et al., 2002).

[37] The discussion here deals with indicators that are based on a narrowly-defined objective. More generally, a number of "sustainability indicators" have been developed, which aggregate measures reflecting a variety of factors deemed relevant to this broad goal. See Parris and Kates (2003) for a discussion of sustainability indicators and the challenges that arise in their use.

[38] Both embodied energy analysis and ecological footprint analysis use a consistent set of accounting principles based on input-output analysis to compute these costs. An alternative biophysical method, emergy, on the other hand, also seeks to measure the energy cost of producing a good or service, but it does not follow these principles, and hence, does not generally satisfy basic adding-up properties. Rather, it focuses on converting inputs of varying quality to a common energy metric – usually solar energy equivalents – so that they can be combined into a cost estimate measured in those units.

[39] People using models may sometimes find that the implications of their models are surprising and unacceptable to them. For example, Slovic et al. (1982) found that people preferred a convex function (their general model) to express the value of varying numbers of lives lost, yet made choices in violation of this abstract model. They had not realized that the abstract model implied choices that were unacceptable to them. In the view of Slovic and others, modeling needs to be interactive and mixed with examples of the model's specific implications.

[40] While stakeholder processes are sometimes used as a decision mechanism *per se*, the C-VPESS considered them only as a way of providing informed input from the public into valuation processes. A 2001 SAB report assessed stakeholder processes involving environmental science and concluded that they are appropriate as a decision making mechanism *per se* in only a modest subset of environmental regulatory decisions under select conditions, if at all (SAB, 2001).

[41] Valuations also require a variety of other predictions, including predicting the anthropogenic response to EPA actions or decisions. Valuations sometimes ignore the need for such predictions. For example, many valuations assume that the regulated community will comply fully with regulations and not adjust other behavior in response to the regulation. In many cases, this assumption is incorrect. Where valuations do incorporate additional predictions, however, they again are subject to uncertainty.

[42] For a more detailed discussion of the sources and possible typologies of uncertainty, see Krupnick, Morgenstern et al. (2006).

[43] Depending on the context, explicit reporting of uncertainties may be perceived as indicating dishonesty or incompetence (Johnson, 2003; Johnson and Slovic, 1995, 1998) and are sometimes treated in public policy discussions as indicating junk science (e.g., Freudenberg et al., 2008). Despite these perceptions, it is important to convey that uncertainties are inherent in all science and that good science acknowledges the remaining uncertainties. Experts communicating uncertainty to policy makers or the public should beware of unintended effects and design and test their communications accordingly.

[44] The discussion of value in the National Research Council report (2001) and SAB review of the EPA's Draft *Report on the Environment* (EPA SAB, 2005) and related literature (e.g., Failing and Gregory, 2003) tends to focus more on qualitative rather than quantitative expressions. However, issues of scale and aggregation are important. Both the NRC report (2001) and the SAB review of the EPA's Draft Report on the Environment (EPA SAB, 2005) emphasize the importance of using regional and local indicators. Over-aggregating information can obscure critical ecological threats or problems. In general, allowing sensitivity analysis on disaggregated data is desirable if the data are aggregated at a regional or higher level. So while some authors recommend simple summary indicators (e.g., Schiller et al., 2001; Failing and Gregory, 2003), others emphasize disaggregating indicators (EPA SAB, 2003).

[45] For more information, see the analysis of survey techniques available on the SAB Web site at http://yosemite.epa.gov/sab/sabproduct.nsf/WebBOARD/CVPESS_Web_Methods_Draft?OpenDocument.

[46] This analysis evaluated the benefits and costs of amendments to the Clean Air Act passed by Congress in 1990. Its effort to evaluate the ecological benefits of these amendments raises many of the same issues that arise in evaluating the benefits of national rules. The prospective analyses compare the sequence of increasingly stringent rules called for under the 1990 Clean Air Act Amendments with a situation where the rules were held constant at their 1990 levels (e.g., with the regulatory regime prior to the amendments).

[47] The one exception is the national survey on water quality conducted in the 1980s by Carson and Mitchell (1993), but this survey is not appropriate for use by the Agency in valuing ecosystem services, for reasons discussed later in section 6.1.2 and in endnote 49.

[48] Random utility models are a form of discrete choice model in which each individual's choice of a recreation activity to take part in or recreation site to visit is assumed to depend on the characteristics of the available activities or sites as well as the individual's socio-economic characteristics and variables reflecting preferences. The estimated parameters of the model can be used to calculate the values revealed by the choices made. For more information, see the section on travel cost models available on the SAB Web site at http://yosemite.epa.gov/Sab/Sabproduct.nsf/WebFiles/ Non-MarketRevealedPref/$File/Nonmarket-revealed-pref-03-09-09.pdf.

[49] A more recent national survey regarding willingness to pay for water quality improvements was conducted in 2004 (see Viscusi et al., In press). This research was funded by EPA and used a nationally-representative web-based panel of over 4000 respondents from the Knowledge Network. Participants in the survey were asked

questions designed to reveal their willingness to pay to increase the percentage of water in their region that was rated "good" according to the EPA's National Water Quality Inventory ratings. EPA's rating scale is based primarily, albeit not exclusively, on end uses relating to swimming and fishing. Carson and Mitchell (1993) used the water quality ladder to describe with a quantitative scale changes in water quality. The scale was defined as an index with associated activities that related primarily to recreational uses. Viscusi et al. provide a definition of ratings of water quality as good based on standards used by EPA and the states. No scale is used to quantitatively link the rating to a measure of water quality conditions. Their description includes recreation and other services supported by "good" conditions, but there is not an explicit link to specific ecosystem services or to specific water bodies. In addition, the water bodies are described using geographical designations – national versus regional water bodies. Thus, the study does not clearly link values to specific changes in ecosystem services.

[50] See the table below that lists major Chicago Wilderness reports and a its chronology of valuation efforts. In one 1996 poll, only two out of ten Americans had heard of the term "biological diversity." Yet when the concept was explained, 87% indicated that "maintaining biodiversity was important to them" (Belden and Russonello, 1996, as cited in the Chicago Wilderness *Biodiversity Recovery Plan*, p. 117).

Examples of federal surveys (see endnote 35)		
Continuously Funded Surveys	**Agency Sponsor**	**Years**
Survey of Income and Program Participation	Census Bureau	1984-present
Consumer Expenditure Surveys	Census Bureau	1968-present
Survey of Consumer Attitudes and Behavior	National Science Foundation	1953-present
Health and Nutrition Examination Surveys	National Center for Health Statistics	1959-present
National Health Interview Survey	National Science Foundation	1970-present
American National Election Studies	National Science Foundation	1948-present
Panel Study of Income Dynamics	National Science Foundation	1968-present
General Social Survey	National Science Foundation	1972-present
National Longitudinal Survey	Bureau of Labor Statistics	1964-present
Behavioral Risk Factor Surveillance System	Centers for Disease Control and Prevention	1984-present
Monitoring the Future	National Institute of Drug Abuse	1975-present
Continuing Survey of Food Intake by Individuals	Department of Agriculture	1985-present
National Aviation Operations Monitoring System	National Aeronautics and Space Administration	2002-present
National Survey of Drinking and Driving	National Highway Traffic Safety Administration	1991-present
National Survey of Family Growth	National Center for Health Statistics	1973-present
National Survey of Fishing, Hunting, and Wildlife-Associated Recreation	Census Bureau	1991-present
National Survey of Child and Adolescent Well-Being	Department of Health and Human Services	1997-present
Survey of Earned Doctorates	National Science Foundation	1958-present

Examples of federal surveys (see endnote 35)		
Continuously Funded Surveys	**Agency Sponsor**	**Years**
National Survey on Drug Use and Health	Department of Health and Human Services	1971-present
Youth Risk Behavior Surveillance System	Department of Health and Human Services	1990-present
National Crime Victimization Survey	Bureau of Justice Statistics	1973-present
Schools and Staffing Survey	National Center for Educational Statistics	1987-present
Educational Longitudinal Survey	National Center for Educational Statistics	2002-present
Current Employment Statistics Survey	Bureau of Labor Statistics	1939-present

Other Major Federally-Funded Surveys	Agency Sponsor
National Survey of Distracted and Drowsy Driving	National Highway Traffic Safety Administration
National Survey of Veterans	Department of Veteran Affairs
National Survey of Children's Health	Health Resources and Services Administration's Maternal and Child Health Bureau
National Survey of Recent College Graduates	National Science Foundation
National Survey of Speeding and Other Unsafe Driving Actions	Department of Transportation

Major Chicago Wilderness reports and chronology of valuation effort (see endnote 50)	
Decision/document	**Date**
Biodiversity Recovery Plan	1999 (Award from APA in 2001 for best plan)
Chicago Wilderness Green Infrastructure Vision	Final report, March 2004
Green Infrastructure Mapping	2002
A Strategic Plan for the Chicago Wilderness Consortium	17 March 2005
Chicago Wilderness Regional Monitoring Workshop final report by Geoffrey Levin	February 2005
Center for Neighborhood Technology (CNT) – green infrastructure valuation calculator	Copyright 2004-2007

In: Environmental Stewardship and Ecological Protection
Editor: Daniel J. Moran
ISBN: 978-1-61209-341-3

Chapter 2

THE USE OF MARKETS TO INCREASE PRIVATE INVESTMENT IN ENVIRONMENTAL STEWARDSHIP[*]

Marc Ribaudo, LeRoy Hansen, Daniel Hellerstein and Catherine Greene

ABSTRACT

U.S. farmers and ranchers produce a wide variety of commodities for food, fuel, and fiber in response to market signals. Farms also contain significant amounts of natural resources that can provide a host of environmental services, including cleaner air and water, flood control, and improved wildlife habitat. Environmental services are often valued by society, but because they are a public good—that is, people can obtain them without paying for them—farmers and ranchers may not benefit financially from producing them. As a result, farmers and ranchers underprovide these services. This chapter explores the use of market mechanisms, such as emissions trading and eco-labels, to increase private investment in environmental stewardship. Such investments could complement or even replace public investments in traditional conservation programs. The report also defines roles for government in the creation and function of markets for environmental services.

Keywords: Eco-labeling, environmental service, emissions trading, market, public good, supply and demand, transaction cost

ACKNOWLEDGMENTS

This chapter benefited from the insightful comments and information provided by Marca Weinberg, Utpal Vasavada, Kitty Smith, Mary Bohman, Pat Sullivan, Scott Swinton, Cathy Kling, Virginia Kibler, Jan Lewandrowski, Lorraine Mitchell, and the Natural Resources

[*] This is an edited, reformatted and augmented edition of a United States Department of Agriculture publication, Report 64, dated September 2008.

Conservation Service staff. Thanks also to Cynthia Nickerson for her help on the maps for the water quality trading case study, to Linda Hatcher for the excellent editorial assistance, and to Wynnice Pointer-Napper and Curtia Taylor for the design and layout.

SUMMARY

U.S. farmers and ranchers produce a wide variety of commodities for food, fuel, and fiber in response to market signals. Farms also contain significant amounts of natural resources that can provide a host of environmental services, including cleaner air and water, flood control, and improved wildlife habitat. Environmental services are often valued by society, but because they are a public good—that is, people can obtain them without paying for them—farmers and ranchers may not benefit financially from producing them. As a result, farmers and ranchers underprovide these services.

What Is the Issue?

Farmers can provide environmental services by adopting conservation or production practices that improve the environment. Farmers often produce these services unintentionally, however, by maintaining grasslands, wetlands, or forests rather than converting them to cropland or by adopting practices that increase net returns but also improve environmental performance. Although society values these services, because of the services' public-goods nature, farmers usually cannot benefit financially by intentionally producing them. As a result, there are no naturally occurring markets for environmental services. If environmental services could be sold like other commodities, farmers would likely invest more to maintain wildlife habitat, woodlots, and wetlands. The U.S. Department of Agriculture (USDA) has expressed great interest in the creation of markets to provide environmental quality and other environmental services. Such markets would supplement existing conservation programs and provide an additional source of income for farmers.

What Did the Study Find?

Markets for environmental services may fail to form or function properly for several reasons.

- The public-goods nature of most environmental services is the primary reason that markets for them do not naturally develop. In addition, environmental services, such as improved water quality and wildlife preservation, are unintended consequences of the primary production activities on the farm. These characteristics can limit potential suppliers' ability to benefit financially from providing environmental services.
- Uncertainty about the quantity and quality of services a farmer can produce is a common problem that often hinders market function. Environmental services are

often difficult to observe, such as the nutrientfiltering capacity of wetlands or the sequestration (storing) of greenhouse gases from adopting conservation tillage. Farmers are reluctant to adopt management practices if potential returns are uncertain. Uncertain quality can also deter potential buyers from purchasing environmental services from farms.

- Environmental services are associated with the land and are not transportable to central markets. The costs of bringing buyers and sellers together may hinder the development of markets.
- Government conservation programs and markets for environmental services sometimes have common objectives and outcomes and may end up competing for the same land, the natural capital in the production of environmental services. Such competition could hinder the development of markets by driving up costs.

The consequence of these limitations is that markets for environmental services are rare. Even though public demand for environmental services is strong, farmers are unable to benefit financially by providing them.

Barriers to market development and function can be overcome in a number of ways.

- In some cases, regulation can be used to create a private good, and the demand for that good, that is closely related to an environmental service. For example, the Federal Government places caps on pollutant discharges from regulated firms and issues discharge allowances to each firm, specifying how much pollution the firm can legally discharge. A firm may be able to discharge more pollution than its original allocation by purchasing allowances from other firms that have cut their own pollution discharges below their own allowances or from unregulated sources of pollution, such as agriculture. This transaction is known as a trade. Discharge allowances, therefore, have characteristics of a private good. Farmers are often able to provide discharge reductions at a lower unit cost than industry can and to profit from the exchange.
- Uncertainty over the performance of agricultural management practices for the production of environmental services can be reduced through education and research. USDA and State efforts can play an important role in both areas. Research at the Agricultural Research Service, USDA, and the Conservation Effects Assessment Project at the Natural Resources Conservation Service, USDA, are quantifying the performance of management practices in different settings, and State extension services can convey this information to farmers. In addition, validation and certification services can bolster consumer confidence that, when they purchase environmental services, they are getting the service for which they paid. USDA has played an important certification role in the organic market.
- Improved market design can reduce the search and bargaining costs of bringing buyers and sellers together. Government or other entities can play the role of an aggregator or clearinghouse in a market, making it easier for geographically dispersed market participants to find each other, thereby reducing bargaining costs.
- Coordinating conservation programs and environmental service markets can enhance the performance of both. Targeting conservation programs to producers who need to meet minimum performance standards to enter a market would likely increase the

number of farmers willing to participate. Identifying program rules that prevent farmers from selling environmental services for which they have not received a government payment would also increase farmer interest in entering environmental service markets.

Creating markets for environmental services is not always possible or advis-able. Transactions costs associated with reducing uncertainty may be greater than the benefits of creating a market. The public-goods nature of environmental services may also prevent markets from developing, despite research and education. Even though people may be willing to pay for environmental services, the ability to acquire these services without paying for them reduces the incentive for farmers to provide them. In these cases, regulation or direct financial assistance through government programs may be the most cost- effective options.

How Was the Study Conducted?

The study used an extensive literature review and five case studies to explore important economic issues affecting the development of markets for environmental services. Because working markets for environmental services are rare, we used the literature to provide the reasons that markets are not developing and to provide insight into the role government might play in helping markets to form and to function.

We present case studies for environmental services for which attempts have been made to develop markets. These markets are as follows:

- Water quality trading—Firms with high pollution-control costs purchase pollution reductions from another source at lower cost.
- Carbon emissions trading—Same as water quality trading.
- Wetland mitigation—Loss in wetland services is offset by an improved wetland with similar services.
- Fee hunting—Hunters pay for access to land in order to hunt.
- Eco-labeling—Labels tout goods made in a way that avoids harming the environment.

These case studies provide a more detailed look at the issues surrounding markets for environmental services, as well as the steps that were taken to overcome market impediments. The findings of the case studies are used to identify some specific actions governments could take to support the creation and function of markets for environmental services. This chapter provides context for the actions USDA has recently taken to support markets for environmental services and for the Department's response to the Food, Conservation, and Energy Act of 2008.

1. INTRODUCTION

Farmers and ranchers produce a wide variety of agricultural commodities, which are sold in well-established markets. Farms and ranches can also produce a variety of environmental services that are often unintended consequences of production practices or land use decisions. Some examples are air and water, fl ood mitigation, drought mitigation, and wildlife. Even when unintended, these services provide benefits to people. Agricultural producers' actions can increase or decrease the provision of environmental services. Understanding how agricultural producers make their production and land management decisions is critical in designing strategies for enhancing those environmental services that people value.

Well-functioning commodity and input markets use prices to signal farmers and ranchers what to produce with their land and how to allocate resources most efficiently to maximize profits. In contrast, for a variety of reasons, markets for environmental services have generally not developed. As a result, producers' responses to market signals lead them to produce agricultural commodities rather than environmental services. Environmental services therefore may be underprovided from society's point of view.

Yet, with growing population and incomes, society increasingly values the environmental services agriculture can produce (Antle, 1999). Since markets typically undersupply environmental services, Federal, State, and local governments have developed a range of approaches for increasing their production (table 1.1). Most rely on policy tools, such as financial and technical assistance, regulation, and education. Although these approaches may be relatively simple to implement, basic economic principles suggest that they cannot allocate resources as efficiently as working markets, assuming such markets can exist.

The U.S. Department of Agriculture (USDA) and other groups have expressed great interest in the use of market-based policy instruments as a more efficient way of providing environmental quality and other environmental services. In 2006, USDA outlined its role in "market-based environmental stewardship." USDA is seeking to broaden the use of markets for environmental goods and services to "...encourage competition, spur innovation, and achieve environmental benefits..." (USDA, Natural Resources Conservation Service, 2006b). Some of the approaches that can be used to promote markets include credit trading, mitigation banking, and ecolabeling. To emphasize USDA's growing role, the Food, Conservation, and Energy Act of 2008 includes a provision directing USDA to facilitate the participation of farmers, ranchers, and forest landowners in environmental services markets. The U.S. Environmental Protection Agency (EPA) is promoting emissions trading as a way of reducing the cost of meeting air and water quality goals. The Organisation for Economic Co-Operation and Development is also promoting the use of market mechanisms for the provision of environmental services (Organisation for Economic Co-Operation and Development, 2005).

Creating markets for environmental services is no simple task. A key measure of a well-functioning market is how well it facilitates interaction between consumers and producers, which involves much more than simply the *sale* of environmental services.[1] A sustainable market should be based on more-or-less direct interaction between demanders and suppliers without constant government intervention when unanticipated changes occur.

Table 1. Matrix of Federal agricultural conservation/environmental policy instruments and problems.

	Participation							
	Involuntary			Voluntary				Facilitative
	Regulation	compliance	Taxes	Land retirement	Cost sharing	Incentive payments	Markets (Trading/ offsets/ labeling)	Education/ technical assistance
Problem:	Instrument							
Erosion: Soil productivity		Sodbuster/ compliance (1985)		Soil Bank (1956-60) CRP (1985)	ACP (1936-96) EQIP (1996)	CSP (2002) EQIP (1996)		CTA (1936) CEP (1914)
Erosion: sedimentation	CZARA (1990)	Sodbuster/ compliance (1990)		CRP (1990)	ACP (1936-96) EQIP (1996)	WQIP (1990-96) EQIP (1996) CSP (2002)		CTA (1936) CEP (1914)
Erosion: airborne dust	Clean Air Act	Sodbuster/ compliance (1990)		CRP (1996)	ACP (1936-96) EQIP (1996)	WQIP (1990-96) EQIP (1996) CSP (2002)		CTA (1936) CEP (1914)
Wetlands	CWA Section 404 (1972)	Swampbuster (1985)		Water Bank (1970-95) CRP (1988) WRP (1990) EWRP (1993)			Mitigation banking (1995)	CTPA (1936) CE (1914)
Water quality: nutrients	CWA Section 402 (2003)			CRP (1996)	EQIP (1996)	WQIP (1990-96) EQIP (1996) CSP (2002)	CWA (1990)	CTA (1936) CEP (1914)
Water quality: pesticides	FIFRA (1947) CZARA (1990)			CRP (1996)	EQIP (1996)	WQIP (1990-96) EQIP (1996) CSP (2002)		CTA (1936) CEP (1914)
Wildlife habitat	ESA (1973)			CRP (1996) GRP (2002)	WHIP (1996)	EQIP (1996) CSP (2002)	Conservaion banking (2003) Eco-labeling	CTA (1936) CEP (1914)

Acronyms:

ACP—Agricultural Conservation Program, CEP—Cooperative Extension, CRP—Conservation Reserve Program, CSP—Conservation Security Program, CTA—Conservation Technical Assistance, CWA—Clean Water Act, CZARA—Coastal Zone Act Reauthorization Amendments, EQIP—Environmental Quality Incentives Program, ESA—Endangered Species Act, EWRP—Emergency Wetland Reserve Program, FIFRA—Federal Insecticide, Fungicide, and Rodenticide Act, GRP—Grassland Reserve Program, WHIP—Wildlife Habitat Incentives Program, WQIP—Water Quality Improvement Program, WRP—Wetland Reserve Program.

Note: Year denotes first year Federal program authorized

Trading and offsets rely on regulatory measures to create a market. However, agriculture's participation is currently voluntary.

The purpose of this chapter is to explore the conditions under which markets for environmental services from agriculture might arise and when and how government intervention might help environmental service markets succeed. This chapter presents an extensive review of the types of environmental services farmers can produce, what is required for a market to form, and the problems these markets might face in functioning smoothly. We consider potential roles for government in creating and supporting a market, with a focus on reducing transaction costs.

The report also assesses the potential supply of environmental services to provide a perspective on the potential scale of such markets. By providing a clearer, stronger, more systematic motivation for government intervention in the development of environmental service markets, this chapter provides insight on ways in which government actions might link the public's demand for environmental services to agriculture's supply of these services and on conditions under which the formation of markets, despite government actions, is impracticable.

2. Environmental Services from Agriculture

Environmental services from agriculture are a subset of ecosystem services from agriculture. Ecosystem services are defined by the United Nations' Millennium Ecosystem Assessment as the "benefits people obtain from ecosystems" (Millennium Ecosystem Assessment, 2003, p. 3). These include a wide range of provisioning, regulating, cultural, and supporting services. Both unmanaged and managed ecosystems (such as agricultural lands) can provide these services.

Table 2.1. Some Environmental Services and Farm Management Options

Environmental service	Farm-level management option
Carbon sequestration in soils	Manage soil organic matter
Carbon sequestration in perennial plants	Convert cropland to grassland or forest
Methane emission reduction	Capture and destroy methane from animal waste storage structures
Water quality maintenance	Reduce agrichemical use, establish vegetative buffers, improve nutrient management
Erosion and sediment control	Manage soil conservation and runoff, increase soil cover
Flood control	Create diversions, wetlands, storage ponds
Salinization and water table regulation	Grow trees, manage water
Wildlife	Protect breeding areas and wild food sources, improve timing of cultivation, increase crop species/varietal diversity, reduce use of toxic chemicals

Source. Food and Agriculture Organization of the United Nations, 2007.

Farmers and ranchers constitute the largest group of natural resource managers in the world (Food and Agriculture Organization of the United Nations, 2007). Farms exist to produce food, fuel, and fiber and to sell them to consumers. However, farms also produce many other ecosystem services as externalities, in that they are unintended consequences of the primary production activities on the farm and those who are affected cannot infl uence their production. Farms can produce externalities as part of the production process (generally negative externalities, such as nutrient runoff or air pollution) or land use decisions (positive externalities, such as wildlife, wetland services, and water quality from farmland not planted to crops).

In this chapter, environmental services refer to positive externalities that result from stewardship on the farm (table 2.1). These externalities could include improved water quality from changes in crop management, carbon sequestration from converting cropland to forests, wetland services from preserving a wetland, and enhanced wildlife habitat by providing adequate food, cover, and nesting habitat.

Natural capital possesses the capacity of giving rise to the flow of environmental services (Boyd and Banzhaf, 2006; Costanza et al., 1997; Elkins, 2003). The natural capital that agricultural producers control is the land, water, air, and genetic resources on their farms. How these resources are managed affects the type and level of environmental services that can be produced.

Agriculture controls a large amount of natural capital in the United States. In 2002, private farms accounted for 41 percent of all U.S. land, including 434 million acres of cropland, 395 million acres of pasture and range, and 76 million acres of forest and woodland (USDA, National Agricultural Statistics Service, 2004). This capital can provide a host of environmental services, including water quality, air quality, flood control, wildlife, and carbon sequestration. These services can be consumed directly or combined by consumers with other goods to create final goods, such as sightseeing, fishing, wildlife viewing, or hunting. In this chapter, we focus on the provision of water quality, greenhouse gas reduction, wildlife, and wetland services. Markets have been developed for providing these services, and these are the ones specifically mentioned in the USDA policy on markets for environmental services (USDA, Natural Resources Conservation Service, 2006b).

Water Quality

The potential for agriculture to supply water quality improvement is defined largely by the significant *negative* impact that agriculture has historically had on water quality. Current production practices and inputs used by agriculture can result in a number of pollutants—including sediment, nutrients, pathogens, pesticides, and salts—entering water systems. Pollution from agriculture is generally exempt from regulations under the Clean Water Act, so agricultural producers have little incentive to address these largely offsite impacts.

Although no comprehensive national study of agriculture and water quality has been conducted, the magnitude of the impacts can be inferred from several water quality assessments. EPA's 2000 Water Quality Inventory reports that agriculture is the leading source of pollution in 48 percent of river miles, 41 percent of lake acres (excluding the Great Lakes), and 18 percent of estuarine waters that are impaired. The inventory shows these bodies of water to be water-quality impaired in that they do not support designated uses, such as swimming and aquatic life (U.S. EPA, 2002). The findings mean that agriculture is the leading source of impairment in the Nation's rivers and lakes and a major source of impairment in estuaries.

Agricultural producers can improve water quality by reducing the discharge of nutrients, pesticides, sediment, and other agricultural pollutants to water resources. The Natural Resources Conservation Service's (NRCS) technical field guide lists over 300 management practices that can improve water quality (USDA, NRCS, 2007b). These practices include conservation tillage, nutrient management, strip cropping, irrigation management, pesticide management, manure storage structures, vegetative buffer strips, fencing, and livestock watering facilities. Farmers can also retire cropland in sensitive areas and improve or restore wetlands to filter sediment and nutrients.

Air Quality

Agricultural production releases a wide variety of material into the air. Field operations produce windblown soil, nitrogen gases, and pesticides. Animal operations release hydrogen sulfide, ammonia, methane, volatile organic compounds, and odors. Internal combustion engines in field equipment and irrigation pumps and field burning produce fine particulates and nitrogen oxides. These pollutants may affect people's health, reduce visibility, and contribute to global warming or may simply be a nuisance. Agriculture can improve air quality by reducing the release of these materials through changes in soil, water, chemical, and manure management.

Greenhouse gases have been of particular recent interest due to their role in global climate change. Agriculture is both a source and a sink (storage in soil and in biomass) of greenhouse gases. It is a relatively small source of greenhouse gas emissions, accounting for about 8 percent of all U.S. greenhouse gas (GHG) emissions in 2005 (USDA, Office of the Chief Economist, 2007). The most important GHG emissions from agriculture are nitrous oxide (N_2O) and methane (CH_4). Agricultural soil management (60 percent), enteric fermentation (25 percent), manure management (13 percent), rice cultivation (2 percent), and agricultural residue burning (less than 1 percent) are the sources of agricultural GHG emissions.

Agriculture can sequester (store) carbon in soils and biomass, thus offsetting GHG emissions. Carbon entering the soil is stored primarily as soil organic matter. Agricultural soils sequestered an estimated 12.4 million metric tons carbon equivalent in 2004, less than 1 percent of U.S. emissions (U.S. EPA, Office of Atmospheric Program, 2006). Studies indicate that it may be *technically* possible to sequester an additional 89-318 million metric tons of carbon annually on U.S. croplands and grazing lands through various management practices, such as conservation tillage, crop rotations, and fertilizer management, or up to 16 percent of 2004 emissions (Lewandrowski et al., 2004). Shifting cropland to grasslands or forest could increase sequestration even more.

Wildlife

U.S. agriculture is in a unique position with respect to the Nation's wildlife resources. The historic development of U.S. agriculture required the development of large amounts of native grasslands, wetlands, and forests for agricultural purposes. Management of the

Nation's farms and ranches can play a major role in protecting and enhancing its wildlife. Because of the dominance of private land ownership in the United States, Federal and State governments cannot exercise effective responsibility for wildlife management without productive collaboration with private land managers (Benson, 2001b; Conover, 1998).

The quality of wildlife resources is a function of the amount, quality, and diversity of habitat. Grasslands and wetlands are two common types of habitat that can be protected, restored, or improved through conservation on agricultural lands.

Grassland Habitat

Grasslands constitute the largest land cover on America's private lands. These lands provide biodiversity of plant and animal populations and play a key role in environmental quality. Grasslands also improve the aesthetic character of the landscape, provide scenic vistas, open spaces, and recreational opportunities, and protect soil from water and wind erosion.

Large expanses of grassland acreage are annually threatened by conversion to other land uses, such as cropland and urban development. About half of all U.S. grasslands have been lost since settlement, much due to conversion to agricultural uses (Conner et al., 2001).

Wetland Habitat

Wetlands are complex ecosystems that provide many ecological functions valued by society. They take many forms, including prairie potholes, bottomland hardwood swamps, coastal salt marshes, and playa wetlands. Wetlands are known to be the most biologically productive ecosystems in temperate regions. More than a third of threatened and endangered species in the United States live only in wetlands, and nearly half use wetlands at some point in their lives (U.S. EPA, Office of Water, 1995a). Most freshwater fish depend on wetlands at some stage of their lives. Many bird species depend on wetlands for either resting places during migration, nesting or feeding grounds, or cover from predators. Wetlands are also critical habitat for many amphibians and fur-bearing mammals. Besides supporting wildlife, wetlands also supply water pollution control, flood control, water supply protection, and recreation.

When the country was first settled, there were 22 1-224 million acres of wetlands in the continental United States (Heimlich et al., 1998). Since then, about half have been drained and converted to other uses, nearly 85 percent for agricultural uses. Currently, there are about 111 million acres of wetlands on non-Federal lands (USDA, NRCS, 2004b). About 15 percent are on agricultural lands (cropland, pastureland, and rangeland).

Demand for Environmental Services

The existence of a market for an environmental service requires that potential consumers are willing to pay a price for those services. Numerous studies have found that people are in fact willing to pay for environmental services from agriculture (Environmental Valuation Reference Inventory, 2007). These findings do not mean a market should exist, but they are a prerequisite for a market to exist.

Another indication that demand for environmental services exists is that State and Federal governments have developed many programs to supply them, implicitly reflecting public demand. Conservation programs, such as the Conservation Reserve Program, Wetland Reserve Program, Environmental Quality Incentives Program, and Farm and Ranch Protection Program, provide financial and technical incentives to agricultural producers to retire land, adopt management practices that protect and enhance environmental quality, or preserve farmland. In recent years, USDA has spent over $4.5 billion per year on such programs (USDA, Economic Research Service, 2007a).

Environmental regulations are also used to ensure that environmental services are provided. Regulations in the Clean Water Act; Clean Air Act; Federal Insecticide, Fungicide, and Rodenticide Act; and Endangered Species Act keep harmful chemicals from water and air, prevent wetland loss, and protect habitat for endangered species. These and other regulations have been created because the public demands that environmental services be protected. Agriculture, however, is often exempt from these regulations, which leaves other mechanisms, such as financial assistance, to provide incentives for agriculture to increase its production of environmental services.

Demand for environmental services can also be expressed through private actions, such as the purchase of conservation easements by land trusts. Land trusts are one alternative mechanism by which individuals can choose to act privately to address the failure of both governments and private markets to provide environmental services (Sundberg, 2006). They preserve and increase environmental services, based on the perception of their members' interests, by obtaining fee title or conservation easements of land they want to protect. The United States had over 1,600 land trusts, protecting open space and habitat on over 37 million acres of land, in 2005 (Land Trust Alliance, 2006). About 1.2 million acres per year are added to the rolls of privately conserved land.

Summary

Agriculture controls natural capital that can provide environmental services. There is much evidence that people value these services, yet there is longstanding concern over their continued loss. Government programs and nongovernment efforts have been developed to motivate agricultural producers to provide environmental services. These efforts raise the question of why landowners do not market and sell these services as they do agricultural commodities, thereby attaining an additional stream of income. The next chapter reviews the function of markets and the reasons that markets fail to develop Markets are institutions through which potential buyers and sellers of goods and services deal with each other in the process of exchange. In a perfect world of competitive markets, resources move to their highest valued use (see box, "Value of Markets"). With market failure—that is, when markets do not operate properly—resources are not allocated to their highest valued use. Addressing market failure is one of the roles of government.

3. Market Basics

Markets are institutions through which potential buyers and sellers of goods and services deal with each other in the process of exchange. In a perfect world of competitive markets, resources move to their highest valued use (see box, "Value of Markets"). With market failure—that is, when markets do not operate properly—resources are not allocated to their highest valued use. Addressing market failure is one of the roles of government.

Markets for Environmental Services

Few well-functioning markets have developed for environmental services, even though evidence is strong that consumers are willing to pay for them (see chapter 2). The lack of markets has important consequences in the allocation of resources on farms. Without well-defined markets for environmental services, landowners are not rewarded financially for supplying them. For example, without a market for environmental services, a farmer with native vegetation on his or her land has no economic incentive to preserve the cover and the environmental services it provides. The farmer's land-use decision will be based on the potential return from agricultural commodities. If the value that society places on environmental services could be captured by the farmer, he or she would more likely keep a larger fraction of his or her land in a natural state.

Keep in mind that agricultural producers' motivations are more complex than simply profit maximization. Most agricultural producers value environmental services and may sacrifice some potential income to enjoy them on their farms. Without markets, however, agricultural producers' provision of environmental services is based on their own personal preferences, rather than the value society places on them. The result is likely to be an underprovision of those services.

Why Do Markets Fail?

Before exploring how markets for environmental services might be created, it is important to understand why markets fail. Markets for environmental services rarely exist because one or more of the following factors apply (Murtough, Aretino, and Matysek, 2002; Ruhl, Kraft, and Lant, 2007):

- Public good characteristics.
- Market burdens, such as large transaction costs and uncertainty.
- Institutional barriers.

Public Goods

Because environmental services are the product of complex ecosystem processes and delivered through a variety of landscape settings, they nearly always take on characteristics of public goods; they are nonexcludable and nonrival. With a private good, a producer can prevent someone who has not paid for it from obtaining it; it is excludable. For a public good, a provider cannot exclude someone who has not paid a price from obtaining it. For example, a farmer contemplating the sale of improved water quality by establishing vegetative buffers on his or her farm cannot exclude downstream users from benefiting; the downstream users are "free riders." In this situation, the farmer does not have an economic incentive to provide the good.

VALUE OF MARKETS

Markets are driven by individuals and firms striving to maximize their own well-being. Relative prices determined by the interaction of supply and demand satisfy the necessary conditions for maximizing social welfare. The producer combines price information from product markets with price information from input markets to determine how much to produce and how many inputs to purchase. The market supply curve for the product represents the production decisions made by all producers over a range of prices. The consumer participates in various product markets based on prices, income, and personal preferences. The market demand curve for a product represents the purchasing decisions made by all consumers over a range of prices.

If the market functions properly, factors of production move to those uses where they earn the highest return; resources are used most efficiently, and both producers and consumers enjoy maximum benefits from production and consumption.

Prices in a perfectly operating market tell participants how valuable one good or input is relative to another, making them the most important piece of infor-mation driving decisions about production and consumption. However, markets rarely operate perfectly. Various factors can affect the interplay of supply and demand so that prices no longer convey the true values of goods and services. A market for a good may also fail to form entirely. Under these conditions, factors of production do not move to those uses where they have the greatest value; resources are misallocated, and overall social welfare is lower than if markets operated perfectly.

The figure depicts the production possibility frontier (PPF) for a farm, or the marginal tradeoff between production of a commodity and an environmental service. The shape of the curve is a function of the farm's resource base and technology set and the farmer's management skills. The mix of commodities and environmental services provided by the farmer depends on the prices received for each. If no market exists for the environmental services and only the commodity has a price, then the farmer maximizes income by producing at point A; no envi-ronmental services are produced. Alternatively, if a market for environmental services could be created, then a price for that service would exist. The farmer would maximize net returns by producing at a point such as B, where the slope of the PPF equals the ratio of prices. Fewer commodities and more environmental services are produced.

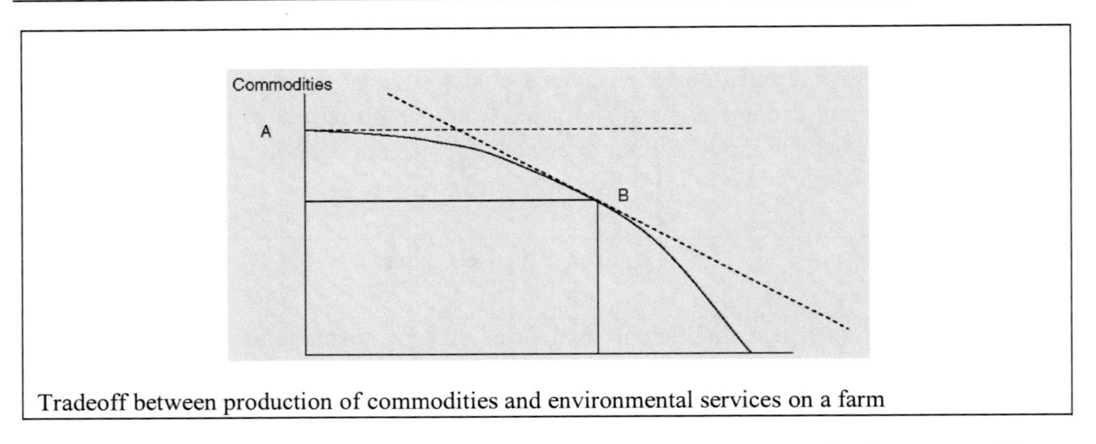

Tradeoff between production of commodities and environmental services on a farm

Furthermore, when a good is nonrival—that is, exclusive ownership is not possible—a buyer's purchase of a good will also benefit other individuals. Thus, the value to society of the good (say, improved water quality) is the sum of everyone's enjoyment. However, when individuals consider how much they will pay, they will not consider this sum; instead, they consider only their own personal values. Thus, even if a willing seller existed, the net price the producer could receive would be too low; it would reflect one individual' s value rather than the sum of the values of all individuals.

The point is that prices tell market participants how valuable one good or input is relative to another, making prices the most important piece of information driving decisions about production and consumption. If prices for environmental services under-represent their true value, fewer resources will be directed toward the production of environmental services than is socially optimal. The public-good nature of most environmental services is the primary reason that markets for them do not develop. Consequently, the price for most environmental services is zero.

Transaction Costs

Transaction costs are the costs of doing business. Parties must find one another and exchange information. They may also have to inspect or measure the good, draw up contracts, and consult with lawyers or regulators (Stavins, 1995). These actions require inputs of time or resources, costs that reduce the overall benefits expected from the transactions. If transaction costs are high relative to the value of the good, then exchange may have no benefits and a market could fail to develop.

Transaction costs associated with potential markets for environmental services are likely to be high. One issue with environmental services from agriculture is that they are often secondary to a farmer's primary activity of producing agricultural commodities; they are produced as externalities of agricultural production. It may be too costly for a farmer to learn about potential demand for an environmental service, develop a business plan, keep the necessary records, and integrate the new business into the traditional farming operations.

Environmental services, such as water quality and carbon sequestration, are difficult to measure. The monitoring necessary to measure these services is often expensive and may require intrusive visits to the farm.

Traditional farm commodities already have established systems for collection and distribution. Farm commodities are generally homogeneous, prices are established in centralized markets, and agricultural producers do not have to negotiate with each potential final buyer. On the other hand, environmental services tend to be unique for each farm, with no standard form of transac-tion. A farmer wishing to sell an environmental service may have to negotiate with each potential buyer, a potentially costly process. The same would be true for a buyer of environmental services, who may have to negotiate with many farmers.

Uncertainty

The performance of conservation practices in the production of environmental services is one of the most important sources of uncertainty in environmental markets. Uncertainty about the quantity and quality of services a farmer can produce affects both the demand and supply side of markets. Markets function best when information on the commodity is complete and readily available to all potential market participants. Environmental services, however, are often difficult to observe, such as the nutrient-filtering capacity of wetlands or the sequestration of greenhouse gases from adopting conservation tillage. Determining the quantity of services a farm can produce is, therefore, often left to estimation, based on farming practices and location. When information is missing, or otherwise inaccessible, potential customers may be reluctant to enter the market, or they may trade less. Uncertainty can also affect producers. Agricultural producers are reluctant to adopt management practices if potential returns are uncertain. Not knowing the quantity of environmental services that can be produced and sold would be a major impediment to entering a market for environmental services. Determining the amount and nature of the services a farm can produce can be costly, especially given their complex nature.

Institutional Barriers

Institutional barriers may prevent agricultural producers from selling an environmental service in existing markets. Agricultural producers may be unable to sell environmental services either by rule or because the rules that govern participation limit the supply of services a farm can provide in a market. For example, participants in the Wetlands Reserve Program are prohibited from selling some environmental services created by wetland restoration paid for by taxpayers, including carbon sequestration, open space, and wetland services (for the purposes of mitigation) (USDA, Natural Resources Conservation Service, 2007a).

Some markets do not allow environmental services from agricultural sources because of a high level of uncertainty about the amount actually produced or about their long-term supply. For example, some markets for greenhouse gas reduction do not allow credits from sequestration in agricultural soils because of the risk of future carbon emissions due to changes in management (known as the permanence issue) (Ecosystem Marketplace, 2007a). One could argue that uncertainty would reduce the demand for such credits in a market anyway and be reflected in price, but some markets have chosen to take away the choice entirely.

Some water quality trading programs require agricultural producers who wish to sell credits to be practicing a minimum level of stewardship. Requiring a minimum level of stewardship to participate in the trading program prevents the lowest cost credits from being marketed, raising the overall price of credits for point sources. The requirement is also a barrier for some producers, discouraging them from entering the market. A producer may be unwilling to bear the cost of achieving the minimum level of stewardship before being eligible to sell credits.

While not necessarily a barrier, government programs can sometimes compete for producers' investment in environmental stewardship. Government conservation programs and markets for environmental services sometimes have common objectives and outcomes. For example, conservation programs and trading programs may compete with each other for pollution reductions from agriculture. If a farmer enrolls in a conservation program to reduce nitrogen runoff, the marginal cost of making additional environmental gains (beyond those funded by the conservation program) is higher. If the farmer then wishes to participate in a trading program, the cost of abatement credits is higher than it would have been otherwise. Agricultural producers with a history of heavy involvement in conservation programs may have a more difficult time competing in a market than if they had not been as involved. While the environmental service is still being provided, market forces are not guiding the allocation of resources.

Summary

The public-goods nature of environmental services is the most important reason that markets for environmental services have not developed on their own. Transaction costs, uncertainty, and institutional barriers are also factors inhibiting markets. Government can use a variety of policy tools, including market mechanisms, to create incentives for farms to provide environmental services. The following chapter presents some examples of how market mechanisms have been used to spur the provision of environmental services, as well as steps government can take to promote the creation of sustainable markets.

4. What Can We Learn from Current Markets?

The previous chapter showed how characteristics of environmental services from agriculture prevent well-functioning markets from developing and hinder market function. As a result, the prices that convey information about the relative values of goods and services in well-functioning markets either do not exist in markets for environmental services or convey fl awed information. Can anything be done to "fix" the system so appropriate information is conveyed to landowners who provide environmental services?

To obtain a clearer understanding of how markets can be used to help provide environmental services, this chapter takes a close look at five different markets. For each market, we describe the "good" that is being bought and sold, impediments to demand, impediments to supply, and steps taken by government and/or market participants to overcome those impediments. Since we are also interested in the extent to which markets for environmental services might become a significant source of financial resources for stewardship on farms, we also explore the potential size of these markets. The five markets examined are water quality trading, carbon trading, wetland mitigation, wildlife, and eco-labels.

Water Quality Markets

Agriculture significantly affects water quality (chapter 2). Farmers and ranchers, for the most part, have little incentive to improve water quality. The primary U.S. water quality law, the Clean Water Act (CWA), regulates pollution only from point sources (for example, factories, sewage treatment plants, and large confined animal feeding operations). Voluntary approaches for controlling pollution from agriculture are the mainstay of Federal and State water quality improvement efforts. But benefits from water quality improvements occur mostly off the farm, and since they are public goods, few producers would voluntarily incur the costs of adopting management practices that improve water quality. How can a market for water quality be created?

One approach is emissions trading. Emissions trading is organized around the creation of discharge allowances, which is a time-limited permission to discharge a fixed quantity of pollutant into the environment. A discharge allowance has characteristics of a private good; it is rival and exclusive. Property rights are enforced by the regulatory agency managing the program.

A discharger (assumed to be a profit-maximizing firm) must own allowances to legally release pollutants. A regulatory agency creates demand for discharge allowances (and reduces pollution in regulated waterways) by restricting the number of allowances in a market. The regulatory agency first determines the maximum amount of discharge of a particular pollutant a watershed can absorb and still meet environmental quality goals. This becomes the emissions cap for the watershed. The cap is used to set discharge limits for each regulated firm operating within the watershed. Discharge allowances equal to the emissions cap are allocated to all regu-lated dischargers through an auction or some other means. By enabling allowances to be traded, a market is created that allocates discharges among regulated firms.

If a firm discharges more pollution than its holding of allowances during the year, it would be subject to fines and penalties. If a firm does not have enough discharge allowances, it can either reduce discharges or purchase allowances from other firms. If a firm discharges less than its holding of allowances, it can sell the excess. A firm will purchase allowances in the market if the price is less than its cost of reducing a unit of discharge. If a firm can reduce discharges at a cost lower than the price of an allowance, it will reduce emissions below its permit requirements and sell the excess allowances and earn a profit. If the market operates smoothly, it can achieve environmental goals at a lower cost than command and control regulations alone (Tietenberg, 2006). Firms with low pollution control costs will provide proportionately more pollution control, reducing total pollution control costs. A market allows maximum flexibility for firms in that a firm can meet its obligations by installing pollution control technology, adopting more efficient production technology, rearranging production processes, or purchasing credits (Ribaudo, Horan, and Smith, 1999). Emissions trading has been very successful in reducing the cost of regulations on sulfur dioxide emissions to the atmosphere from power plants (see box, "Trading Can Reduce the Cost of Lowering Emissions"). This program is estimated to have exceeded environmental goals at a savings of over $1 billion compared with a regulatory approach that does not allow trading (Stavins, 2005).

In the textbook example of emissions trading, all market participants are regulated under the cap. In water quality trading programs, EPA allows regulated point sources to purchase credits from unregulated nonpoint sources, such as agriculture. Sources of credit outside the cap are known as offsets.

Water quality trading markets must meet some basic conditions in order for demand for credits from nonpoint sources to develop. Units of trade must be clearly defined, defensible ecologically and economically, consistently measured, and enforced by the regulatory agency (Boyd and Banzhaf, 2006). The commodity to be traded must be a single pollutant in a common form that is understood by market participants. The discharge point of purchase and sale must be environmentally equivalent to ensure that expected water quality gains are achieved. The timeframes for buyers and sellers of credits must be aligned, in that purchased reductions in discharge must be produced during the same period that a buyer was required to produce them. The supply of nonpoint credits must be in balance with the point sources' demand for credits, in that there are enough potential nonpoint credits to satisfy the needs of potential purchasers. Otherwise, trading with nonpoint sources would not be able to generate pollution control savings.

Experience with water quality trading programs highlights the problems with nonpoint-source-created credits and some of the steps that can be taken to address those problems. Since 1990, 40 water quality trading programs have been started in the United States; 15 include production agriculture as a potential source of credits for regulated point sources (table 4.1) (Breetz et al., 2004). To date, trades between point and agricultural nonpoint sources have occurred in only four programs: Piasa Creek (Illinois), Red Cedar River (Wisconsin), Southern Minnesota Beet Sugar, and Rahr Malting (both Minnesota). These trades appear to be cost effective. For example, in the trading program established for Rahr Malting, four nonpoint-source projects controlled phosphorus runoff at a cost of about $2.10 per pound (Breetz et al., 2004). Rahr Malting would have had to pay an estimated $4-$ 18 per pound of phosphorus reduced if it had installed pollution control equipment. However, supply-side and demand-side impediments seem to be preventing trades in most trading programs. Simply creating a private good related to water quality by itself is insufficient for generating market activity.

Trading Can Reduce the Cost of Lowering Emissions

Without trading, the regulated firm reduces discharges by 500 pounds at a cost of $25,000 (500 pounds at $50 per pound), and the farm does nothing.

With trading, the firm reduces discharges by 400 pounds at a cost of $20,000 (400 pounds at $50 per pound). The farm is willing to reduce discharges for a price of $15 per pound. The firm purchases 100 pounds of reduction from the farm at a cost of $1,500 (100 pounds at $15 per pound). The firm's costs have been reduced to $21,500 (a savings of $3,500). The farm reduces discharges by 100 pounds at an actual cost of $1,000 (100 pounds at $10 per pound). The farmer receives a payment of $1,500 from the firm, so he or she actually realizes a profit of $500 for trading with the firm.

The total cost of reducing pollution (not considering profit to the farmer) has been reduced from $25,000 to $21,000.

Example: Firm discharge limit, no trading

Example: Firm discharge limit, with trading

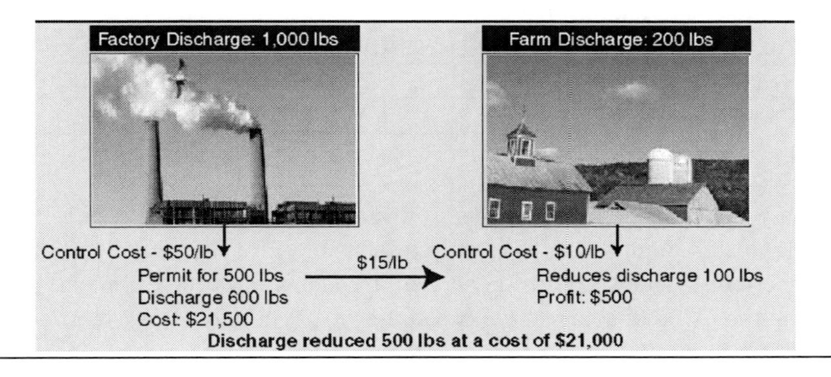

Issues In Demand For Credits From Agriculture

The source of demand for credits in any trading program is a regulation that establishes a cap on discharges that is below current levels. In the case of water quality, the Total Maximum Daily Load (TMDL) provision of the Clean Water Act is the legal mechanism that establishes a cap on pollution discharges in impaired watersheds. Without an effective or binding cap, regulated sources have no reason to seek credits in a market. Ineffective caps on point-source (regulated) dischargers are cited as the reason for lack of demand for nonpoint-source credits in three trading programs and may be a problem in others (Breetz et al., 2004).

Tab One of the requirements of trading is the equivalency of credits; ideally, point-source purchases of credits in a market have the same impact on water quality as if the firm reduced discharges itself. This equivalency ensures that water quality goals are actually met. Establishing equivalency between point and nonpoint sources must account for two factors—agricultural practice effectiveness and location relative to the point source.

The effectiveness of a best management practice (BMP) depends on site-specific conditions, implementation, and how well it is maintained (Mid-Atlantic Regional Water Program, 2006).

Table 4.1. Water Quality Trading Programs That Include Agriculture

Project	Pollutant traded	Trades
		Number
Cherry Creek, CO	Phosphorus	0
Lower Boise River, ID	Phosphorus	0
Piasa Creek, IL	Sediment	1
Acton, MA	Phosphorus	0
Massachusetts Estuaries Project	Nitrogen	0
Kalamazoo River, MI	Phosphorus	0
Rahr Malting, MN	Phosphorus	4
Southern Minnesota Beet Sugar, MN	Phosphorus	579
Tar-Pamlico, NC	Nitrogen, phosphorus	0
Clermont County, OH	Nitrogen, phosphorus	0
Great Miami River, OH	Nitrogen, phosphorus	0
Conestoga River, PA	Nitrogen, phosphorus	0
Fox-Wolf Basin, WI	Phosphorus	0
Red Cedar River, WI	Phosphorus	22
Chesapeake Bay Watershed	*Nitrogen, phosphorus*	0

Source. Breetz et al., 2004.

Uncertainty about such performance is a major stumbling block with point-nonpoint trading. If a regulated point source is legally responsible for achieving a particular discharge goal, the uncertainty about credits generated by nonpoint sources may make them an unattractive option. A point source's control strategy is generally a long-term decision, and it may be unwilling to rely on an uncertain source of credits because of the decision's inherent irreversibility (McCann, 1996). These factors may push point sources toward providing their own internal emission controls or trading with other point sources, rather than relying on nonpoint credits. Measurement problems were cited as obstacles in several existing trading programs (Breetz et al, 2004).

Uncertainty about practice performance can be addressed in three ways. One is to conduct research on the performance of practices under different conditions. USDA's Agricultural Research Service (ARS) conducts extensive research on the environmental performance of production practices and could provide information that reduces uncertainty in trading programs. Some water quality trading programs use simulation models to predict the performance of practices. Research and model development are costs that are generally borne by the public.

A second approach is to address the liability issue. A number of trading programs have created a reserve pool of credits that can be used by regulated point sources when an offset project fails to produce the expected number of credits. This pool could increase the willingness of point sources to trade with nonpoint sources. Rules that grant some leeway for point sources that purchase nonpoint-source offsets would also encourage pointnonpoint trading.

A third approach for addressing practice uncertainty is an uncertainty ratio. An uncertainty ratio is a type of trading ratio that generally requires more than one unit of nonpoint-source discharge reduction to offset one unit of point-source discharge. Uncertainty ratios in water quality trading programs generally range from 2:1 to 5:1, which means that a point source would have to purchase up to five units of pollutant reduction from a nonpoint source in order to ensure that its single unit of discharge is "covered" (Conservation Tillage Information Center, 2006). While providing assurance that the nonpoint-source reduction provides the expected gain in water quality, a trading ratio increases the effective price of nonpoint credits, thereby reducing point sources' demand for them. Research on practice performance could reduce this ratio, making nonpoint-source credits less costly to point sources.

Establishing equivalency between nonpoint offsets and point-source discharges also must take into account the location of nonpoint sources relative to the point source. Since equivalency is measured at the point source, the fate of pollutants when they leave a field must be considered as they move downstream. Take two fields, one close to the point source and the other much farther upstream. Identical reductions in nitrogen runoff at the two fields would affect water quality differently, as measured at the point source, due to biophysical activity along the way: the closer the source, the greater the effect. This difference must be accounted for when potential trades are constructed. A delivery or location ratio is another type of trading ratio, accounting for the location in the watershed of the nonpoint source relative to the point source: the smaller the distance, the smaller the ratio. While providing assurance that the nonpoint-source reduction provides the expected gain in water quality, a delivery ratio increases the effective price of nonpoint credits from farms located farther from the point source, thereby reducing point sources' demand for them.

Another issue facing point sources' demand for nonpoint credits is the cost of finding trading partners. Because farms are generally widely distributed across a watershed and each may be capable of producing a relatively small number of discharge credits, the transaction costs for point sources of identifying enough willing trading partners to satisfy their permits may discourage them from seeking trades. Some markets have developed formal clearing-houses that assemble information from both buyers and sellers, making it easier for potential trading partners to find each other (Breetz et al., 2004). Third-party aggregators are also used in several markets to assemble credits from nonpoint sources. Aggregators then market the credits to potential purchasers. Both government and nongovernment organizations are playing roles of clearinghouse and aggregator.

Issues In Supply Of Credits From Agriculture

Some of the impediments to the formation of trading markets fall on the supply side. Farm runoff is not regulated under the Clean Water Act, so producers are not compelled to actively seek trading partners. The expected returns from trading may not adequately compensate for the type of inspection and scrutiny the farm may receive if it enters into a trading program. Evidence from existing programs suggests that producers may also avoid trading programs because of a fear that entering into a trade is an admission that their farms pollute, exposing them to citizen complaint or future regulation (King and Kuch, 2003; King, 2005; Breetz et al., 2004).

Farmers may be uncertain about the number of credits they can reasonably expect to produce, making it difficult for a producer to determine whether it is financially beneficial to enter a market. Models and other tools could help farmers reduce this uncertainty. An example of this type of information source is the World Resources Institute's NutrientNet

(World Resources Institute, 2007). This online tool can function as an information source for farmers. Configured to a specific watershed, NutrientNet allows registered users to evaluate different trading options and assesses the combination of practices that works best for a farm with a particular set of resource characteristics.

Another tool currently under development is the NRCS/EPA Nitrogen Trading Tool (NTT). NRCS developed the NTT, in cooperation with ARS and EPA, as an online tool to help farmers determine how many potential nitrogen credits they can generate on their farms and sell in a water quality trading program (Gross et al., 2008). It allows a farmer to enter geographic, agronomic, and land use information to estimate baseline nitrogen loadings and changes in management practices or land use to calculate nitrogen load reductions that are the basis for credits in a trading market. Tools such as NutrientNet and the NTT can also reduce uncertainty on the demand side, if the model results are found to be reliable estimates.

A trading program may specify a set of practices eligible for producing credits to those for which performance data are readily available (Conservation Tillage Information Center, 2006). While simplifying the programs' problem of evaluating potential trades, it limits the choices a farmer may make in supplying credits. If the list of practices does not appeal to a farmer, he or she may decide not to participate.

Another supply-side issue arises when producers also participate in conservation programs, such as USDA's Environmental Quality Incentives Program (EQIP). Most trading programs do not allow producers receiving finan-cial assistance for water-quality-protecting management practices through

Federal programs to sell the subsequent water quality improvements as credits to point sources. An additional payment from a point source would not improve water quality beyond what the Government has already paid for. If farmers pay part of the cost of the practice out of their own pockets, one solution might be to allow a portion of the credits to be sold.

Some trading programs require a minimum level of stewardship before credits can be generated. For farms without "acceptable" management practices, credits cannot be created until the base level of environmental performance is attained. This requirement prevents the lowest cost credits from farms that have not adopted acceptable practices from being sold on the market, unless the returns from selling credits is so high that both the initial investment to achieve the baseline and the subsequent management costs can be covered. The bottom line is that the supply of low-cost credits is reduced, which has the effect of increasing the price regulated firms must pay.

Coordination of conservation programs with trading programs is one solution. USDA conservation programs, such as EQIP, could be targeted to producers who are not meeting the minimum level of stewardship to encourage them to participate in a trading program. The number of producers likely to participate in the trading program would increase, raising the potential supply of credits. However, average costs of credits would still be higher than if a stewardship-based baseline had not been used.

Producers may also face high transaction costs when trying to find trading partners. A farmer has to consider the type, amount, and timing of pollutant reductions generated on the farm and determine if they match the type, amount, and timing of pollutant reductions needed by regulated dischargers (Conservation Tillage Information Center, 2006). Unfamiliarity with the regulated community and the negotiation process could discourage producers from participating in a trading program. Third-party aggregators can play a role in addressing this issue and are being used in several projects. Trading programs have also established outreach programs to educate farmers about the opportunities that trading might offer and how to participate.

Future Role For Agriculture In Trading Programs?

USDA's interest in water quality trading (and other markets for environmental services) is based largely on the potential level of financial resources from private sources for conservation on farms. A question we examine is the extent to which water quality trading could provide enough financial assistance to producers to address a significant amount of agricultural nonpointsource pollution in impaired watersheds, assuming that demand and supply impediments could be overcome. We use a simple screening procedure to identify watersheds where demand for water quality credits by point sources may be high and where agriculture can provide enough credits to meet that demand, assuming that nonpoint sources of pollution, such as agricultural producers, remain unregulated.

Data And Analysis

Our goal is to identify watersheds that could support an active trading market with agriculture as a supplier of credits. To do so, we first identified watersheds where nutrient loadings are identified as a problem. Our analysis includes the 2,111 eight-digit Hydrologic Unit Code (HUC) watersheds of the contiguous United States. Data on nutrient impairment were obtained from EPA's 303(d) list of State-reported impaired waters (U.S. EPA, 2007a). With these data, we identified 710 HUCs containing water bodies impaired by nutrients—i.e., either nitrogen (N) or phosphorus (P).

We then identified watersheds where agriculture is likely to be a credit supplier. Because point sources may be required to purchase three or more credits from nonpoint sources for each unit of discharge, we assume that only watersheds where agriculture contributes a large portion of total nutrients— greater than 50 percent —might develop an active credit market where significant revenue for water-quality-enhancing practices flows to the agricultural sector. Finally, to ensure sufficient demand for nonpoint-source credits, we consider only watersheds where point sources contribute at least 10 percent of loadings. Estimates of nutrient loadings from point sources, agricultural nonpoint sources, and other nonpoint sources in each HUC were obtained from the U.S. Geological Survey (2000).

An important aspect of the potential price of credits from agriculture is the level of nutrient management that is part of the baseline (from which created credits are calculated). As discussed earlier, the cost of supplying a water quality credit is likely to be lower in watersheds with a lower percentage of cropland under a nutrient management plan (NMP). Data on the amount of cropland already covered by a NMP implemented with assistance from USDA in each HUC during 2004-06 were obtained from the NRCS Performance Results System (USDA, NRCS, 2007c).

Results

Agriculture is the primary source of nutrient loadings in most of the 710 impaired HUCs. Agriculture is responsible for 91-99 percent of N loadings in 68 percent of the impaired HUCs (Figure 4.1). Similarly, agriculture is responsible for 9 1-99 percent of P loadings in 52 percent of HUCs (Figure 4.2). We expect relatively low demand for agricultural credits (as a share of total agricultural discharges) by point sources in these watersheds because of the predominance of nonpoint-source loadings. Even though point sources may benefit from a plentiful supply of credits, only a small percentage of agriculture' s contribution to pollution will be addressed through management practices funded by point sources.

Agricultural contributions of N and P ranging between 50 and 90 percent are found in 142 and 224 of the impaired HUCs, respectively. We believe that demand and supply of credits is more balanced in these watersheds, which is necessary for an active market. Figures 4.3 and 4.4 show the spatial distribution of HUCs that meet our screening criteria for nitrogen and phosphorus. There are about 322,000 farms (15 percent of all U.S. farms) in the watersheds where phosphorus trading markets may be viable, and about 175,000 farms (8 percent) in the watersheds where nitrogen trading markets may be viable (table 4.2).

In terms of the cost of credits that agriculture might supply, no HUC had more than 22 percent of its cropland under a NMP, and most had less than 5 percent, which suggests that the level of NMP adoption would infl uence the level of trading in very few HUCs.

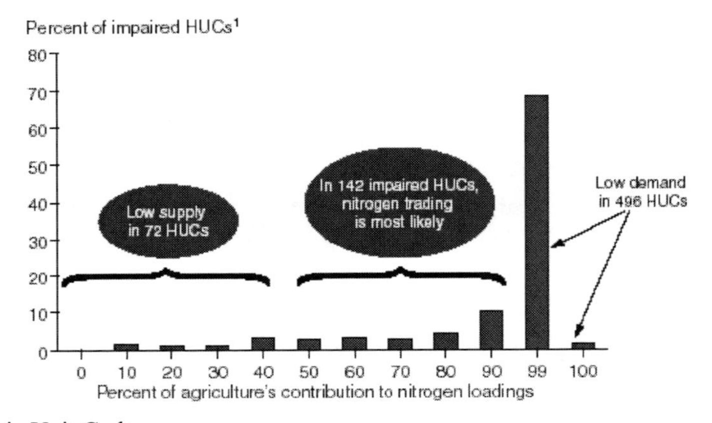

HUC = Hydrologic Unit Code.
[1] Number of impaired HUCs = 710.
Sources: USDA, ERS analysis of Environmental Protection Agency and U.S. Geological Survey data.

Figure 4.1. Agriculture's contribution to within-HUC nitrogen loadings Percent of impaired HUCsHUC = Hydrologic Unit Code

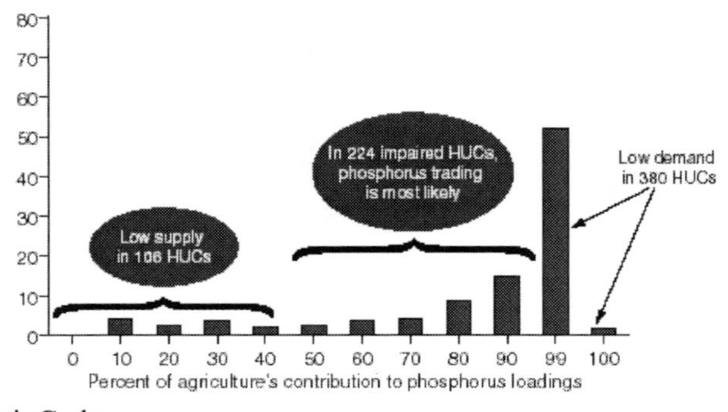

HUC = Hydrologic Unit Code.
[1] Number of impaired HUCs = 710.
Sources: USDA, ERS analysis of Environmental Protection Agency and U.S. Geological Survey data.

Figure 4.2. Agriculture's contribution to within-HUC phosphorus loadings

Table 4.2. Farms and Income in Watersheds Where Trading Most Likely

Indicator	Nitrogen		Phosphorus	
	<5% NMP	5-25% NMP	<5% NMP	5-25% NMP
Farms (number)	156,846	174,724	281,191	321,654
Crop sales ($1,000)	7,085,235	7,577,431	12,733,629	14,246,215
Livestock sales ($1,000)	4,997,249	5,718,705	10,455,075	12,411,055
Net cash income ($1,000)	2,632,522	2,864,816	5,315,743	6,048,930

NMP = Nutrient management plan.
Note. The United States has about 2.1 million farms.
Source. 2002 Census of Agriculture.

Figure 4.3
Nitrogen credit trading opportunities: Impaired watersheds where active trading is most likely

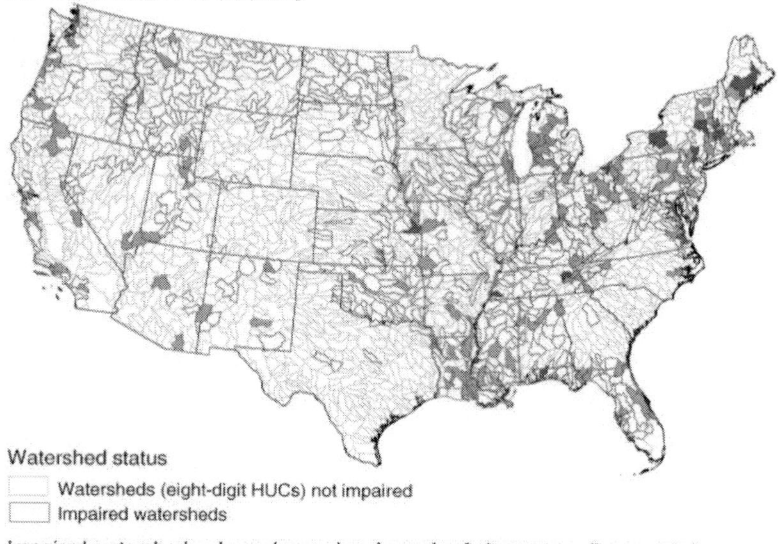

Watershed status
Watersheds (eight-digit HUCs) not impaired
Impaired watersheds

Impaired watersheds where demand and supply of nitrogen credits are most likely to be in balance (142 watersheds)
Greatest availability of low-cost credits (< 5 percent of cropland acres under nutrient management plan)
Somewhat lower availability of low-cost credits (5-25 percent of cropland acres under nutrient management plan)

HUC = Hydrologic Unit Code.
Note: None of the 142 watersheds have over 25 percent of cropland acres under nutrient management plan.
Source: USDA, ERS analysis of Environmental Protection Agency, U.S. Geological Survey, and Natural Resources Conservation Service data.

Trading could occur in any HUC where point sources are required to reduce nutrient loadings and are allowed to offset their discharges with reduction from farms. Most of trades that have actually occurred are single point sources that offset pollution through contracts with multiple producers. However, these results indicate that trading is not likely to be a major source of conservation assistance, even if impediments to trading are overcome. Relatively few impaired watersheds are in "balance," in that the potential demand for nonpoint-source credits is high enough to spur an "active" market with nonpoint sources. Although trading may represent an important source of conservation funding in some local areas, USDA will likely remain the primary source of financial assistance for water quality protection on farms, assuming that nonpoint sources remain unregulated.

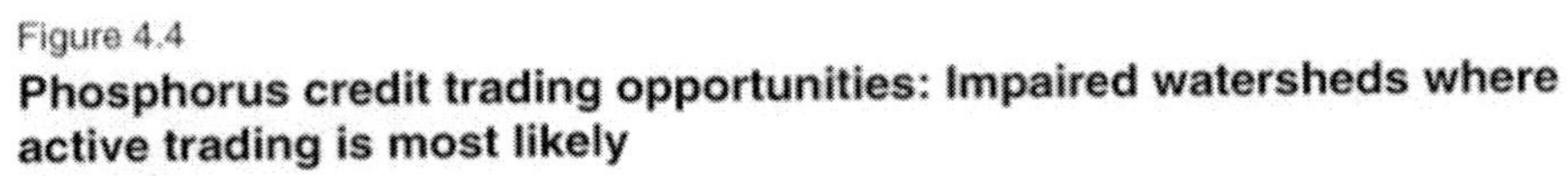

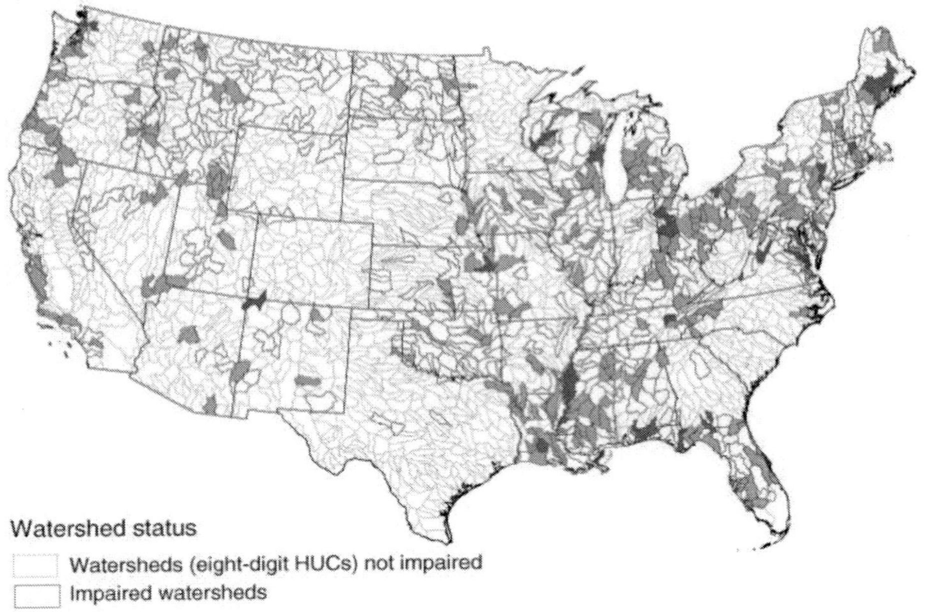

Watershed status

Watersheds (eight-digit HUCs) not impaired

Impaired watersheds

Impaired watersheds where demand and supply of phosphorus credits are most likely to be in balance (224 watersheds)

Greatest availability of low-cost credits (< 5 percent of cropland acres under nutrient management plan)

Somewhat lower availability of low-cost credits (5-25 percent of cropland acres under nutrient management plan)

HUC = Hydrologic Unit Code.

Note: None of the 224 watersheds have over 25 percent of cropland acres under nutrient management plan.

Source: USDA, ERS analysis of Environmental Protection Agency, U.S. Geological Survey, and Natural Resources Conservation Service data.

Greenhouse Gases and Agriculture

Concerns about global climate change have led to various strategies for reducing greenhouse gases (GHG) in the atmosphere. Most strategies combine reductions in emissions of GHG with sequestration (long-term removal of GHG from the atmosphere). One policy approach is to create markets for greenhouse gas reductions. As described in chapter 2, agriculture is both a source and sink for greenhouse gases, and producers might benefit in such markets by reducing greenhouse gas emissions, or by sequestering carbon in the soil or in biomass.

One factor in favor of developing an active market is the worldwide potential of such a market. Reducing greenhouse gas emissions or sequestering carbon have the same benefit no matter where they occur geographically, which means GHG reduction credits have many potential buyers and sellers, a necessary condition for an active market.

Issues In Demand For GHG Reductions

There are two primary scenarios for "creating" demand for GHG reductions (Council for Agricultural Science and Technology, 2004):

1. Regulatory cap and trade markets that set emission limits.
2. Voluntary markets driven by the following:
   - Consumer willingness to pay to reduce their carbon "footprint."
   - Firms wishing to show themselves as responsible environmental actors.
   - Firms wishing to gain control of low-cost alternatives that may be used to comply with future emission limitations (speculation) (Butt and McCarl, 2004).

Regulatory Markets

Regulatory markets create a property right for GHG reductions, in the form of tradable credits, much like the discharge allowance in water quality markets. The European Union Emissions Trading Scheme (EU ETS) is the world's largest market in greenhouse gas emissions. It was established primarily to help the 25 EU member states achieve their Kyoto Protocol targets (Ecosystem Marketplace, 2007b). The program established a mandatory cap and trade program for carbon dioxide in 2005 that did not include carbon sinks.

Several State and regional cap and trade programs have recently been approved in the United States to reduce GHG emissions. The Oregon CO_2 Standard, established in 1997, is the only State-level program currently underway. It requires new power plants to reduce emissions to 17 percent below those of the most efficient plant (Ecosystem Marketplace, 2007c). Affected companies have the option of meeting this requirement by financing carbon offset projects through the Climate Trust, a nongovernmental organization established to seek and finance offset projects and to verify offset credits. While the Oregon program rules do not place any limitations on the geographic location or types of CO_2 offset projects, Climate Trust does not accept offsets from sequestration in agricultural soils (Climate Trust, 2007).

Member States of the Regional Greenhouse Gas Initiative (RGGI) (consisting of 10 Northeastern and Mid-Atlantic States), Washington, and California are developing cap and trade programs for reducing GHG emissions, primarily from power plants. Rules are still being developed for these programs, but Climate Trust is already seeking carbon offset projects to meet future demand from RGGI (Climate Trust, 2007).

Voluntary Markets

Voluntary markets are currently the greatest source of demand for GHG reduction credits in the United States. The Chicago Climate Exchange (CCX) is a voluntary cap and trade program covering emission sources from the United States, Canada, and Mexico and offset projects from these countries and Brazil. While joining CCX is voluntary, members make a legally binding commitment to meet annual greenhouse gas emission reduction targets (Ecosystem Marketplace, 2007a). Members agree to annual reductions that will reduce their emissions of greenhouse gases by 6 percent below the average of their 1998-200 1 emissions baseline by 2010. Each member can meet its commitment through internal reductions, by purchasing allowances from other members, or by purchasing credits from emissions reduction projects. CCX issues tradable Carbon Financial Instrument contracts to owners or aggregators of eligible projects on the basis of sequestration, destruction, or displacement of GHG emissions. Eligible projects include agricultural methane, landfill methane, coal mine methane, agricultural and rangeland soil carbon, forestry, and renewable energy. As of July 2007, the price of a CO_2 equivalent[2] (CO_2e) was $3.25 per ton (Ecosystem Marketplace, 2007d). In contrast, carbon offsets in the European Union's Emissions Trading Scheme were trading at $30.60 per ton CO_2e (Ecosystem Marketplace, 2007d). This difference reflects the fact that the CCX is voluntary, while the EU ETS is not, and agricultural soil sinks are not recognized as a source of permanent carbon reductions by EU ETS. Since its inception in 1997, the CCX has traded almost 24 million metric tons of CO_2e.

The CCX addresses the issue of the cost of finding potential offsets from a geographically dispersed sector through the use of third-party aggregators. Aggregators create, aggregate, register, and trade certified carbon credits to buyers in the CCX. Fifty-three offset aggregators are members of the CCX and include farm groups (e.g., Iowa Farm Bureau) as well as private corporations (Chicago Climate Exchange, 2007).

An important question is why a firm would voluntarily enter into a legally binding commitment to reduce its carbon emissions. Some of the benefits include the following (Chicago Climate Exchange, 2007):

- Capture gains and manage risks in the growing carbon market.
- Acquire cutting-edge measurement and trading skills that will be needed as markets develop.
- Demonstrate strategic vision on climate change to shareholders, rating agencies, customers, and citizens.
- Gain leadership recognition for taking early, credible, and binding action to address climate change.

A purely voluntary retail market for carbon offsets has developed for individuals, businesses, and other institutions that find the concept of being "carbon neutral" an attractive one. Approximately 35 retail offset providers currently offer "carbon neutrality" for a fee to consumers and businesses. These retailers fund projects that are intended to offset GHG emissions from cars, airplanes, and special events, such as concerts and weddings. Offset projects include a variety of actions, including methane capture from animal feeding operations and landfills, reforestation, developing renewable energy, and improving energy efficiency. Some retailers purchase reductions on the CCX rather than directly funding projects. Retailers are currently charging from $4 to $35 per ton of CO_2e.

Demand in this retail market is largely unknown. Relatively little information is available regarding the volume of trades or the composition of volume by project type (Trexler, Koslof, and Silon, 2006). In addition, little research has been conducted on consumers' willingness to pay to be carbon neutral. The good being sold does not have private-good characteristics, so free riding is an issue for market development.

Another implication of the newness of these markets is that they currently have no accepted standards for what qualifies as an offset for making consumers carbon "neutral." Because the commodity is intangible, it is very difficult for consumers to differentiate between high-quality and low-quality offerings based on the information provided by retailers (Trexler Climate + Energy Services, 2006). In addition, there are no industry quality standards for offsets, no reliable certification process for retailers, and no effective disclosure and verification protocols. Consumers who are willing to pay may be reluctant to enter such markets because of this uncertainty. Such uncertainty reduces overall demand, keeps prices low, and stifles market growth.

A related issue affecting demand in all carbon markets is whether nonagricultural entities will purchase carbon credits from agriculture. The quantity and permanence of carbon sequestration on agricultural soils is less certain than for other types of GHG reductions (Zeuli and Skees, 2000). The amount of carbon sequestered in agricultural soils is determined by the interaction of soils, climate, land use, crop rotation, fertilizer management, and other management practices (Council for Agricultural Science and Technology, 2004). An accurate measurement of sequestration rates has to be made locally rather than relying on regional estimates. Demand by nonagricultural interests in this regard is not clear (McCarl and Schneider, 2000; Zeuli and Skees, 2000). Emphasis in many offset projects in both regulatory and voluntary markets is on permanent, easy-to-measure offsets, such as methane capture and destruction.

Research can address uncertainty issues surrounding potential sequestration of soil carbon. GRACEnet is an ARS project for estimating net GHG emissions of current agricultural systems and the impacts of alternative management (USDA, ARS, 2007). It will reduce uncertainty about how agricultural management might alter the amount of GHG emitted to the atmosphere by identifying the best regionally specific management practices for increasing soil carbon and reducing the net global warming potential of greenhouse gases emitted by agriculture. It will also provide a scientific basis for possible carbon credit and trading programs.

Issues In Supply Of Carbon Sequestration

Carbon sequestration by agriculture is enhanced under management systems that (1) minimize soil disturbance and erosion, (2) maximize the amount of crop-residue return, and (3) maximize water and nutrient use efficiency in crop production (Council for Agricultural Science and Technology, 2004). Changes in cropland management include adopting conservation tillage and residue management, improving crop rotations and cover crops, eliminating summer fallow, improving nutrient management, using organic manure and byproducts, and improving irrigation management (Lewandrowski et al., 2004). Land use changes include converting cropland to forests, perennial grasses, conservation buffers, and wetlands.

The potential supply is dictated partly by standards set by individual markets. A trading program's success depends on the agricultural sinks' ability to offer management practices that are visible and have a predicted effectiveness within acceptable degrees of certainty (McCarl and Schneider, 2000). The Chicago Climate Exchange limits cropland eligibility for carbon credits from conserva-tion tillage to soils that have been evaluated for that purpose (Chicago Climate Exchange, 2007). The CCX also limits rangeland eligibility to regions where research on soil sequestration is available. Research programs, such as the one associated with GRACEnet, could reduce uncertainty about the potential for soils and management to sequester carbon, increasing the potential supply of offsets. Such research also makes it easier for farmers to estimate expected returns from entering a market for carbon offsets.

A program that can help reduce uncertainty and transactions costs in greenhouse gas mitigation markets is the U.S. Department of Energy's revised Voluntary Greenhouse Gas Reporting Registry. The revised Registry, also known as the 1605b program, is a voluntary program for reporting GHG emissions to the Federal Government.[3] Participants can establish a record of emissions and emissions reductions that will be deemed "credible" over the widest possible range of potential uses (U.S. Department of Energy, 2007). The Registry provides guidance, tools, and standardized methodologies for estimating greenhouse gas emissions and removals. Possible benefits of the Registry include enhancing participants' ability to take advantage of future Federal climate policies in which emission reductions have value and helping agriculture and forest entities take advantage of State- and private-sectorgenerated opportunities to trade emission reductions and sequestered carbon.

We use a 2004 study by Lewandrowski et al., to get an idea of how farmers might respond to a price of $3.35 per ton of CO_2e (price as of June 14, 2007, on the CCX) that is paid for gross reduction sequestration (not accounting for potential increases in GHG emissions elsewhere on the farm). Based on the study results, farmers in the 48 coterminous States would shift about 2.3 million acres of cropland to forest, about 11.5 million acres of grazing land to forests, and about 80 million acres of conventional tillage to conservation tillage. Net farm income for all farms would increase about 0.9 percent. These results assume no transaction costs or uncertainty and adequate demand to purchase all the credits farmers can sell at that price.

That the estimated levels of land-use changes have not occurred is due to a limited number of purchasers, transaction costs, uncertainty, and higher commodity prices than are used in the analysis.

What happens if payments are based on *net* sequestration rather than on gross sequestration? Unlike many commodities, what matters is not the flow of product (such as consumable ears of corn) but the stock (such as the amount of CO_2 in the atmosphere). GHG trades outside the United States often require that emission reductions on a project site not lead to increases elsewhere (Butt and McCarl, 2004). For example, if a farmer retires 1 acre of cropland for production simply to switch his or her production efforts to an idle acre, there is no net benefit on the "stock" of carbon in the atmosphere (known as leakage). In the analysis described above, price increases for farm commodities provide an incentive to bring more land into production. Some of the benefits from sequestration are lost because of increased emissions on the farm. For the same price of $3.35 per ton of CO_2e, if farmers are debited for changes in land uses and production practices that increase carbon emissions, the acres shifting to forests are the same but far fewer acres shift to conservation tillage—only 7 million acres compared with 80 million if payments are based on gross sequestration (Lewandrowski et al., 2004). Payments based on net sequestration are less attractive to farmers.

Supply of GHG reduction credits could be affected by how carbon markets and conservation programs are coordinated. Land retirement programs, such as the Conservation Reserve Program and Wetlands Reserve Program, and working lands programs, such as the Environmental Quality Incentives Program and Conservation Security Program, provide financial incentives for management practices that could be eligible for producing credits on the CCX. Unlike water quality trading programs, which often disallow credits produced through conservation programs, the CCX does allow credits produced via projects subsidized by conservation payments. Producers have an extra incentive to enroll in USDA conservation programs if they can also sell credits to the CCX. However, future carbon trading programs may not allow this "double dipping" because of questions about additionality.[4]

Farmer participation in carbon markets is also infl uenced by their willingness to accept market requirements. A shift to reduced- or no-till conservation practices could increase variations in net returns and discourage participation. The price of a carbon credit would have to be high enough to account for the increased uncertainty for a farmer to participate. Required practices may have management characteristics that do not mesh well with the rest of the farm, discouraging participation in the market (McCarl and Schneider, 2000). Producers may also be unwilling to make the long-term commitment that sequestration projects often require to be effective. This unwillingness may be exacerbated by tenure arrangements, which are generally for shorter periods and do not support long-term planning.

The way projects address uncertainty can affect the supply of credits. For example, each CCX project must place 20 percent of eligible carbon offsets in a reserve pool to provide coverage in the event the project fails to produce projected offsets (Chicago Climate Exchange, 2007). Offsets in the pool are held by the farmer and, if not needed, can be released for sale at the end of the accounting period. On the one hand, such a pool enhances trades by reducing uncertainty on the demand side. However, the uncertainty that this requirement imparts on expected income could discourage some producers from initiating a Project.

Issues In The Supply Of Methane Capture

Animal feeding operations are a potential source of methane emission reductions that are eligible in all carbon markets as offsets. Capturing methane and burning it, with or without the production of energy, reduces net GHG emissions. Eligible agricultural methane collection/combustion systems include covered anaerobic digesters, complete-mix, and plug-fl ow digesters. Methane emission reductions on animal feeding operations have been encouraged by assistance programs, such as EPA's AgSTAR, but the emergence of this market has improved the benefits to farmers of installing such systems.

Methane recovery systems are most effective for confined livestock facilities that handle manure as liquids or slurries, such as dairy and swine. EPA estimates that about 6,900 dairy and swine operations could benefit financially by installing anaerobic digesters (U.S. EPA, AgSTAR, 2006). These include dairy operations with more than 500 head, swine operations with more than 2,000 head if using a system other than deep pit, and swine operations with more than 5,000 head if using deep-pit storage. Financial benefits include sale of carbon offsets to the various carbon markets and/or sale of electricity generated on the farm from captured methane. EPA estimates that it is technically possible for anaerobic digesters to reduce GHG emissions by 30 million metric tons CO_2e per year (U.S. EPA, AgSTAR, 2006). For the sake of comparison, managed livestock waste emits about 51 million metric tons CO_2e per year in the United States (USDA, Office of the Chief Economist, 2007). About 90 operations have installed systems and are reducing GHG emissions by about 30,000 metric tons CO_2e per year. How much of this is being sold on credit markets is not known. A number of factors affect whether animal operations would enter the GHG market. Digesters are very expensive, require a high level of management skill, and should be customized for each farm. Additionally, maintenance costs can be high.

Wetlands' Environmental Services and Agriculture

Agriculture has traditionally had a profound effect on the supply of wetland services. Wetlands are rich ecosystems that provide a multitude of environmental services (see chapter 2). The number, mix, and quality of services in each bundle vary across wetlands, depending on their size and type, weather and climatic conditions, surrounding environment, and other factors.

As reported in chapter 2, wetland losses, primarily to agriculture, have been extensive (Figure 4.5). Wetlands were drained without landowners considering their value; wetland services are public goods for which markets do not exit. Also, until the 1980s, USDA provided financial support for draining and filling wetlands, further tilting economic incentives toward reducing wetland services.

Demand for wetland services is expressed, indirectly, through purchases by government and nongovernment entities trying to preserve wetlands. Federal, State, and local governments act on the public's demand for wetland services by implementing programs and regulations that create and preserve wetlands and restrict wetland losses. One program, USDA's Wetlands Reserve Program (WRP), has restored more wetlands than any other public effort. The WRP restores wetlands on agricultural lands and purchases easements on these lands. By the end of 2005, the WRP had enrolled over 1.9 million acres. WRP easements are found in every State (Figure 4.6).

Figure 4.5
County-level estimates of converted wetland acreage

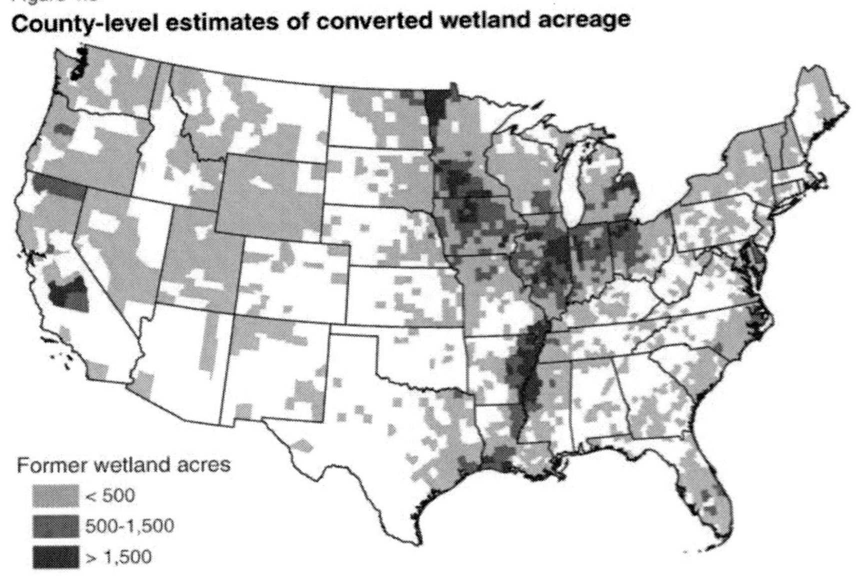

Source: USDA, ERS analysis of USDA's National Resources Inventory data (USDA, Natural Resources Conservation Service, and Iowa State University, 2000).

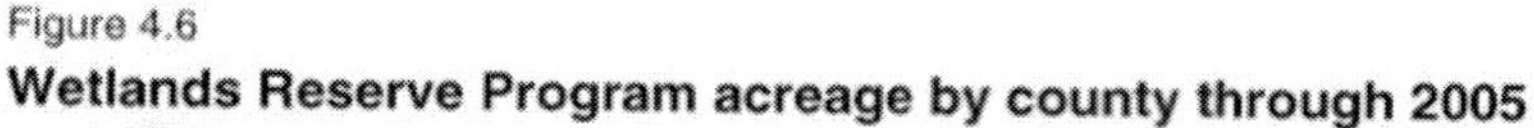
Figure 4.6
Wetlands Reserve Program acreage by county through 2005

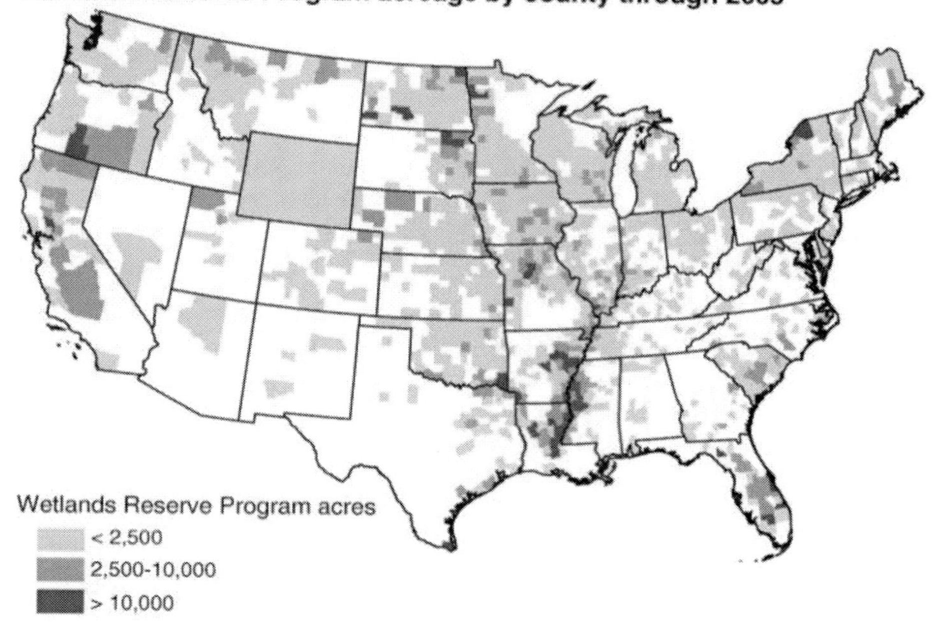

Source: USDA, ERS analysis of Wetlands Reserve Program contract data.

Nongovernment entities purchase wetland easements to preserve and increase the availability and quality of wetland environmental services. The magnitude of these purchases is difficult to gauge. Two of the more sizable organizations involved are The Nature Conservancy and Ducks Unlimited. The Nature Conservancy (TNC) uses a wide range of approaches to meet its mission, including the purchase of land and easements. Wetlands are included in over 4,000 projects encompassing more than 2.5 million acres and valued at nearly $2.6 billion.[5] In 2005, Ducks Unlimited controlled nearly 222,000 acres under easements or deed restrictions for the purpose of restoring and improving wetlands critical for waterfowl (Ducks Unlimited, 2007).

Wetland Mitigation Markets

Producers have some opportunities to sell wetland services directly to consumers, but functioning markets are rare. When able to control access to a wetland, landowners may have opportunities to sell rights to hunt, fish, and view wildlife. Limited data suggest that some producers are marketing fishing, hunting, and wildlife viewing on wetlands, but the extent is not known.

A better opportunity for producers to sell wetland services may exist in the offset markets created by Section 404 of the Clean Water Act. As in the case of water quality trading, a regulation is used to create demand for a private good (mitigation credits) that is closely linked to a public good (wetland environmental services).

The Act creates demand by requiring that any loss in wetland services be offset by a new or improved wetland that offers similar services (known as mitigation). Anyone wishing to drain or fill a wetland must first take all appropriate and practicable steps to avoid and then minimize harmful effects to wetland environmental services (U.S. Government Accountability Office, 2005). To offset any subsequent impacts, the firm or individual can either create wetland offsets (or credits) or purchase wetland credits from a mitigation bank.

The number of credits needed to mitigate lost wetland services is determined by a Mitigation Bank Review Team (MBRT), chaired by the U.S. Army Corps of Engineers (USACE). Besides the bank sponsor and wetland developer, participants typically include representatives from the EPA, Fish and Wildlife Service, National Marine Fisheries Service, and USDA's NRCS, as appropriate. Other Federal, State, and local governmental regulatory and resource agencies may participate, as well as tribal and other entities. Proposals must also be open to public review and comment.

Under Section 404, MBRT is to base its estimates of the number of credits a bank has available at a given time on the observed level of services and not on expectations of future additional services. That is, credits must be created before being sold. In theory, transactions involve no loss in wetland environmental services. The bank sponsor is responsible for creating, oper-ating, and managing the bank; preparing and distributing monitoring reports, conducting compliance inspections of the mitigation, and securing funds for the long-term operation and maintenance of the bank (U.S. EPA, Office of Water, 1995b).

Banks may be sited on public or private lands (wildlife management areas, national or State forests, public parks, etc.). Federally funded wetland conservation projects undertaken via separate authority and for other purposes, such as the WRP, cannot be used in banking arrangements.

Banks must be located in an area (e.g., watershed, county) where it can reasonably be expected to provide comparable environmental services to offset the impact of wetland drainage. Data suggest that wetlands have been drained and mitigation banks created in both urban and nonurban counties (Figure 4.7). Urban development pressure is a commonly cited reason for wetland loss, but clearly nonurban factors, such as highway construction and expansion, play a role.

The use of mitigation banks has increased steadily. In the 1980s and early 1990s, most wetland offsets were done through onsite restoration by developers. Few mitigation banking permits were approved. But in the mid-1990s, the number of bank approvals increased substantially (Figure 4.8). Over the past decade, 30-50 mitigation banks have been approved annually. In total, over 600 mitigation banks have been approved or are under consideration for approval. Thirty-four States have at least one mitigation bank, but 80 percent are concentrated in 10 States (table 4.3).

Farmland owners should be in a good position to supply mitigation services. Nearly 60 percent of all mitigation counties have agricultural lands that were once wetlands. Prior wetland acreage is not a necessity—wetlands can be and are created on lands that have not previously been wetlands. But wetland resto-ration tends to be less costly on converted wetland acreage because soil type, topology, and other factors are favorable to wetland development. However, evidence suggests that agricultural landowners have played a small role in mitigation markets, although they may have sold lands to mitigation bank owners. In 2004, the first mitigation bank owned by a farmland owner was approved (USDA, NRCS,

2004a). Agricultural producers have not played a more direct role in wetland mitigation markets for a number of reasons.

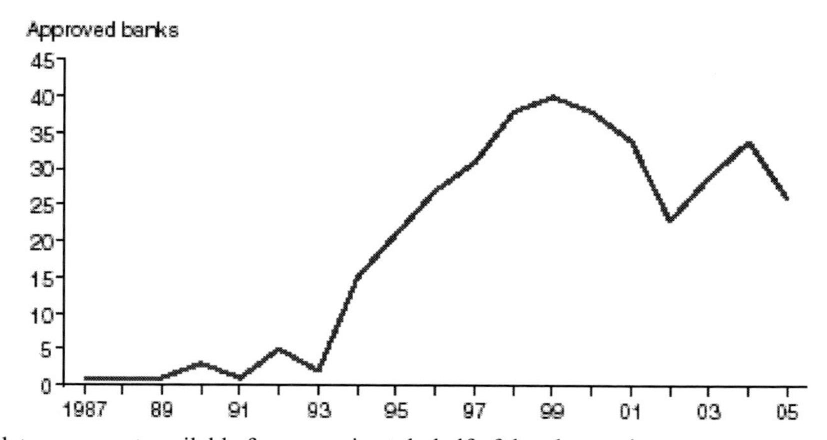

[1]Approval dates were not available for approximately half of the observations.
Source: USDA, ERS analysis of Environmental Law Institute mitigation banking data.

Figure 4.8. Mitigation banks approved[1]

Table 4.3. Number of approved mitigation banks by State

Rank	State	Banks	Rank	State	Banks
		Number			*Number*
1	Louisiana	100	18	Alabama	8
2	Georgia	74	19	Arkansas	8
3	California	61	20	Tennessee	8
4	Florida	55	21	Utah	7
5	Virginia	47	22	Idaho	5
6	Illinois	40	23	Indiana	5
7	Texas	22	24	Kentucky	5
8	Oregon	19	25	New York	5
9	North Carolina	16	26	Iowa	4
10	South Carolina	15	27	Nebraska	4
11	Colorado	14	28	Michigan	3
12	Mississippi	14	29	Delaware	2
13	Ohio	13	30	Maryland	2
14	Wisconsin	12	31	Kansas	1
15	Missouri	11	32	Oklahoma	1
16	New Jersey	9	33	South Dakota	1
17	Washington	9	34	West Virginia	1

Source. ERS analysis of U.S. Army Corps of Engineers data assembled by Environmental Law Institute.

Issues In Supply

Producers considering whether to become a mitigation banker could be infl uenced by several factors. Success of mitigation depends on, among other things, the permanence of the compensatory services. Mitigation permits must include a mechanism that guarantees long-term support of the wetland services. Bank sponsors are responsible for long-term costs of monitoring the wetlands, reporting banks' compliance, and maintaining the wetlands to ensure that banks' operations continue to provide the agreed-upon level of wetland services.

Uncertainty imposes an additional cost on suppliers. An entity wishing to produce and sell mitigation credits will not know the level of credits that a wetland will provide until after an MBRT's evaluation. Being uncertain of the credits a wetland may produce makes estimating income potential difficult and may discourage investment in mitigation projects on the farm.

The long lag time between a wetland's restoration, the recovery of the wetland's environmental services, and the approval to sell credits can be a major issue (Shabman and Scodari, 2005). The loss in output from the land used to produce the bank and capital construction costs are certain, whereas income from the sale of credits is uncertain. Individual producers may not be comfortable taking on such a risk. Furthermore, the mitigation banker faces the risk of rule changes in the CWA or other legislation that might reduce or eliminate demand.

Issues In Demand

We assume that farmers would enter the mitigation market with the goal of maximizing profits. However, not all mitigation bankers have that goal. Nearly 20 percent of all mitigation banking credits are supplied by nongovernmental organizations (NGO), that may be willing to absorb economic losses in exchange for increased wetland services. Farmers, therefore, may be at a competitive disadvantage and face a reduced demand for their wetland services in areas where NGOs are actively supplying wetland services (Shabman and Scodari, 2004). In addition, approximately 19 percent of all mitigation banking credits are supplied by government agencies, such as USDA and the Fish and Wildlife Service. The objectives of public agencies may or may not be profit maximization. In some cases, agencies have created more wetland credits than they need to mitigate their own actions, and have offered the excess in mitigation markets at reduced prices (Wilkinson and Thompson, 2006).

Demand for mitigation credits from producers may also be lost to in-lieu-fee mitigation. In-lieu-fee allows credits to be sold or accepted as offsets before being created, which eliminates many transaction costs and uncertainties faced by mitigation bankers and enables credits to be sold at a lower price than from a traditional mitigation bank. Only government agencies and NGOs are allowed to be in-lieu-fee bankers. Proposed changes to mitigation regulations could eliminate the cost advantage given in-lieu-fee banks (Kenny, 2007).

Balancing Supply And Demand

The success of the mitigation system depends on regulators' and arbitrators' abilities to recognize the quantity of services lost through development and provided by a mitigation bank. Participants in the MBRTs—mitigation bankers, developers, public agencies, NGOs, local communities—negotiate an agreement on the value of services lost and gained (U.S. Government Accountability Office, 2005). When participants have different goals, as is often the case (i.e., profit maximization versus ensuring protection of environmental services), negotiating trades can be time consuming and costly.

Figure 4.9

Counties most likely to have an agricultural mitigation bank

Source: USDA, ERS analysis.

The Competitiveness Of Mitigation Markets

We are interested in the extent to which producers might be able to benefit by participating in the wetland mitigation market. Data are too limited to allow us to estimate supply functions, but the available data allow us to compare wetland restoration costs. Based on WRP data from 1995 through 2007, county-level wetland restoration costs averaged $73-$525 per acre across counties with mitigation banks, with a maximum of about $2,500 per acre. Conversely, restoration costs of mitigation banks, in most cases, exceeded $5,000 per acre and, in some cases, exceeded $125,000. Assuming that the mitigation banks were successful financially, agricultural producers in those same counties would have also benefited if they had established mitigation banks on their land. While our analysis does not explain why we see such a difference, it gives us reason to believe that farmland owners may have a competitive advantage in wetland restoration.

The likelihood that a county will have a new mitigation bank can be estimated from historical data and measures of land characteristics. We estimated a probability model to predict the likelihood that a county will have at least one mitigation bank created in the future, given current development pressures. Factors that are likely to have an impact on the likelihood of a bank being developed include urban development pressure, wetland acreage (together with urban development, the source of demand), and total agricultural land (the likely source of supply). (For details of the model and the analysis, see Appendix: Predicting the Location of New Mitigation Banks.)

Based on the locations of current mitigation projects and the results of the model, 326 counties are predicted to be the most likely to see new mitigation banks in the near future (270 counties currently with mitigation banks and 56 additional counties with high development pressure and available wetlands). The likelihood for future mitigation projects is greatest in the coastal and Gulf States and parts of the Corn Belt, Lake States, and Mississippi Delta regions (Figure 4.9). These counties contain 260,000 farms (12 percent of all farms). Farmers in these counties with the appropriate soils would have the greatest opportunity to create mitigation banks for the purpose of selling wetland services to developers. However, farmers may continue to decide not to accept the risk of becoming a mitigation banker but instead sell or lease land to mitigation banks.

Market Incentives for Wildlife

Hunting is a popular recreation activity in the United States. Private lands are an important source of hunting opportunities. While wildlife residing on the land is a public good, the right to hunt on private lands is a private good controlled by landowners, one that can be sold to hunters willing to pay a fee. While some producers market hunting opportunities on their land, most do not. Thus, producers may have substantial opportunities to increase fee hunting, which could increase both producers' income streams and opportunities available to hunters. What's more, any increase in fee hunting may provide an economic incentive to producers to improve wildlife habitat that benefits both game and nongame species. In this case study, we identify factors that hinder the supply of and demand for fee hunting. We also consider the implications of using the Conservation Reserve Program

—a conservation program that pays farmers to retire land—to promote habitat improvement as well as access to hunting areas.

Background

Hunting in the United States is shaped by two fundamentals: (1) wildlife is owned in common by all citizens, and (2) most of the Nation's wildlife habitat is on private land (Benson, Shelton, and Steinbach 1999).[6] Due to the dominance of private land ownership, Federal and State governments cannot exercise effective responsibility for wildlife management without productive collaboration with private land managers (Benson, 2001b; Conover, 1998). Although wildlife is a public resource and individuals cannot claim ownership over it, private property access rights give landowners de facto control over wildlife residing on their land (Butler et al., 2005).

Reflecting this pattern of land ownership, the U.S. Fish and Wildlife Services' 2001 Fishing, Hunting, and Wildlife Associated Recreation survey (FHWAR2001) found that almost 75 percent of hunting days occurred on private land, 57 percent of all hunters hunted only on private lands, and nearly two-thirds hunted at least part of the time on private land.

Thus, private provision of the "hunting" environmental service is common. However, most of this provision does not rely on markets, where access is controlled by price. A 1993 national survey indicated that, while 77 percent of farmers allowed hunting, only 5 percent charged a fee (Conover, 1998). Several State studies report similar results.[7] Farm survey data from USDA indicate that only 1 to 2.5 percent of farms received income from recreation activities each year from 2000 to 2005 (USDA, ERS and National Agricultural Statistics Service).

While hunting on private lands is common, in recent years, many observers perceive that gaining access to private land for hunting has become more difficult (Larson, 2006; Bihrle, 2003). This observation is suggested by national data indicating that participation in hunting has dropped about 7 percent between 1996 and 2001 (U.S. Department of Interior, Fish and Wildlife Service and U.S. Department of Commerce, Bureau of the Census, 2002). This reduction may be partially explained by difficulties in obtaining access to land. For example, a more urbanized population is less likely to have personal connections to rural landowners from whom they can easily obtain hunting access. Similarly, liability and other concerns seem to have driven an increase in the fraction of land that is posted (for no hunting).[8]

Yet, given the perceived decreased in supply, why is fee hunting so uncommon? Hunter surveys have consistently found that at least half would be willing to pay for hunting access (Benson, Shelton, and Steinbach, 1999). So why has fee hunting not expanded to satisfy demand, especially since income from recreation activities can be substantial?[9] Average gross revenue from fee-based recreation activities ranged between $13,000 and $18,000 per farm offering these activities between 2000 and 2005 (USDA, ERS and National Agricultural Statistics Service).[10]

Supply Issues

Farmer decisions about whether to market wildlife, such as through selling hunting access, hinge on several factors. These factors include attitudes about wildlife and permitting access to hunters, expected economic returns, and personal opinions about the marketing of wildlife.

Many producers apparently value having wildlife on their lands (Conover, 1998). A survey of producers' perceptions about wildlife on their farms found that 51 percent purposely managed their farm for wildlife. However, wildlife can also be seen as a problem to some farmers. The same survey found that 80 percent of the surveyed farmers experience wildlife-caused damage on their farms, and 53 percent stated that damage exceeded their tolerance levels. About a quarter indicated that wildlife damages reduced their willingness to enhance wildlife habitat. Farmers experiencing damage were more willing to allow hunting on their land, probably as a means of reducing wildlife damage, but this willingness to allow hunting does not mean producers would be willing to *encourage* wildlife by investing in habitat improvement.

One obstacle that limits greater use of markets may be the asymmetric distribution of costs and benefits. Wildlife does not respect property boundaries, which can limit the incentives for landowners to invest in habitat enhancements because some of the return will accrue to owners of adjacent parcels or even to other States (migratory waterfowl) (Lewandrowski and Ingram, 2001).

A fee hunting enterprise is not without cost. Setting up a fee hunting enterprise involves the time and expense of advertising, handling contracts, and addressing liability concerns on the farm. The latter is particularly important. The property must be inspected for abandoned wells, fences, dead trees, and other potential hazards that could lead to a liability suit if a hunter was injured. Also, a certain amount of compromise is necessary between the production of agricultural commodities and wildlife-related recreation in land management decisions in order to optimize income on all the land on the farm (Pierce, 1997). Management practices that can enhance game populations include brush control, planting perennial grasses, tillage practices, choice of crops, crop harvest, weed control, haying, grazing methods, stocking rates, fencing, and fertilization. Knowing how these practices affect game on the farm is critical to efficient management, and many landowners do not have the training to be effective wildlife managers (Butler et al., 2005). The higher the quality of the hunting or wildlife-viewing experience a farmer can offer, the greater the fee that can be charged.

Fee hunting does not always mean improved wildlife habitat. Many landowners who charge a fee are not increasing the provision of environmental services by managing their land for wildlife (Butler et al. 2005; Wiggers and Rootes, 1987; Benson, 2001a; Jones et al., 1999). In Mississippi, Jones et al. found that only 19 percent of farmers offering fee hunting actively managed their lands for wildlife.

Demand Issues

One of the largest issues in potential demand for fee hunting on private land is the belief that access to hunting areas should not be restricted by price. Many hunters and even landowners dislike the concept of fee hunting. For example, a North Dakota survey reports that over 50 percent of North Dakota farmers and over 60 percent of North Dakota hunters were "philosophically opposed to charging hunters for access" (Bihrle, 2003). A number of States actively promote open-access programs to counter fee hunting. In Washington, for example, a State program to open more private lands to hunters was instituted to "...combat the proliferation of fee hunting on private land...." (Washington Department of Fish and Wildlife, 2004). These programs pay a small fee, averaging about $5 per acre to participating landowners. In exchange, hunters are granted free walk-in access to the lands during hunting season without the need to obtain personal permission from the landowner. For most of these programs, the State also publishes (in print or on-line) land atlases that list all lands in the program. In several programs, participating landowners are covered under State liability insurance. Although such programs may compete with landowners who wish to charge a fee for access, they are likely to provide a "lower quality" recreational experience than is typical on fee hunting operations, which offer a wider range of services to hunters (Butler et al., 2005).

A Policy Simulation

Overall, the use of market mechanisms to provide wildlife-related environmental services, while not unusual, is not widespread. Obstacles include the public-goods nature of wildlife, inadequate education of potential private benefits from developing a fee hunting business, the complications of operating a hunting and farming business on the same land, and the transaction costs of bringing potential demanders and suppliers together. In this section, we consider how a Federal conservation program might be used to increase the willingness of producers to invest in improved habitat and supply hunting opportunities on their land for a financial gain—in other words, using an existing program to kick-start the market.

USDA's Conservation Reserve Program (CRP) is an example of a payment-for-environmental-services program. Established by the Food Security Act of 1985, the program uses contracts with agricultural producers and landowners to retire over 34 million acres of highly erodible and environmentally sensitive cropland and pasture from production for 10-15 years. When first started, the CRP's primary goal was soil conservation. However, it has evolved beyond soil conservation, with greater weight given to wildlife habitat and air and water quality. The CRP has successfully provided a variety of environmental services, including significant reductions in soil erosion, hence, cleaner waterways, and large increases in wildlife populations. However, the question remains whether market mechanisms can be harnessed to further increase the benefits of land retirement and the quality of wildlife habitat.

The CRP could provide for more improved habitat than fee hunting alone, bring together buyers and sellers, and provide economic opportunities for landowners in areas where there are cultural objections to fee hunting. A number of States with walk-in access hunting programs specifically target land enrolled in CRP (table 4.4). The use of CRP to promote both wildlife habitat enhancement and hunting would also have implications for the distribution and rental rates of enrolled acres. Enrollment decisions by USDA could favor landowners who allow public access to their lands. To gauge how an aggressive policy of using the CRP to market hunting services would affect enrollment, several scenarios are examined using a simulation model combined with a measure of potential hunting demand.

The CRP uses an Environmental Benefits Index (EBI) to determine what lands to accept from among all farmer offers (USDA, ERS, 2007b). The EBI has a wildlife component that could be modified to reward offers that permit public access, even if the landowner charges a fee. Such a modification could substantially increase hunting access in some States. However, it might have little net impact in regions where alternatives (public land or an active market for leases) are available.

Table 4.4. States with walk-in hunting-access programs that enroll CRP Acreage

State	CRP acreage in program	2005 CRP acreage in State	Notes
Colorado	135,000	2.2	Total program size is 160,000 acres
Kansas	534,000	2.8	Total program size is about 1 million acres http://www.kdwp.state.ks.us/
Nebraska	180,000	1.2	Complements 250,00 acres of Park and Game Commission land http://www.ngpc.state.ne.us/hunting/
South Dakota	About 333,000	1.5	About 1,000,000-acre program (private communication, SD Division of Wildlife, Bill Smith)
North Dakota	About 180,000	3.3	About 425,000 acres in ND Private Land Initiative. http:// www.nodakoutdoors.com/ valleyoutdoors10.php

CRP = Conservation Reserve Program.

Source. As noted and from Helland (2006). Several other States have walk-in hunting-access programs that do not use significant CRP acreage. Payments to landowners may depend on the quality of the wildlife habitat and on habitat-improving practices installed by the landowner.

We used the USDA Farm Service Agency's Likely To Bid (LTB) model to predict what lands would be enrolled in the CRP if landowners were provided an additional incentive to be more open to selling the "hunting" environmental service (say, through a federally operated program). Data from the 2001 Fishing, Hunting, and Wildlife Associated Recreation Survey (FHWAR2001) and 2000 National Survey of Recreation and the Environment (NSRE2000) are used to identify the potential demand for hunting:

- The NSRE2000 is a midsized dataset (several thousand respondents) that asks questions about participation in wildlife-related activities. About 1,600 respondents provided a distance (from their residence) to the location they visited on a wildlife-related trip and a direction (i.e., 120 miles to the north).
- The FHWAR2001 is a large dataset (about 25,000 respondents) that asks extensive questions about hunting, including States in which the respondents hunted.

The FHWAR2001 provides accurate measures of how many hunting trips were made to each State. The NSRE2000 provides a relative measure of where hunters went within each State. Combining the two provides an estimate of total trips hunting, a measure of "hunting pressure," for all U.S. counties. For this simulation, we ranked all counties by this "hunting pressure" measure and classified the upper 50 percent as "hunting counties." Figure 4.10 displays the results of combining these two datasets, with counties in green being "hunting counties" and urbanized areas in orange.[11]

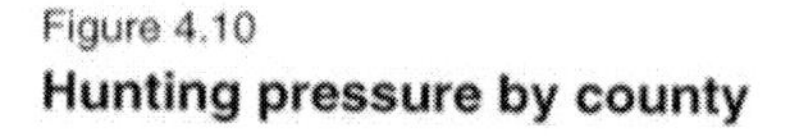

Figure 4.10

Hunting pressure by county

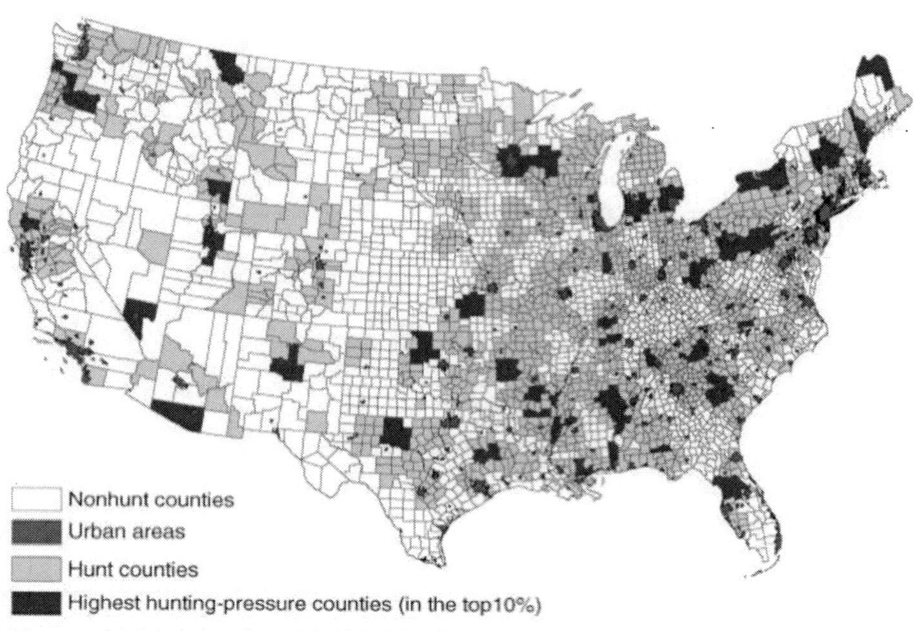

Source: USDA, ERS, using FHWAR2001 and NSRE2000 data.

We specify a baseline scenario under current policy and compare this baseline to several alternatives. In all the alternatives, only acreage in "hunting counties" responds to the scenarios' postulated changes.[12] The scenarios examined are as follows:

1. Where demand is high, farmers recognize their potential to sell hunting leases. They lower their offer prices (the amount they ask for to enroll their land in the CRP) in response to potential income from retired land. This scenario assumes a reduction of $5 per acre in the offer price ($5 per acre is an upper-end value for several of the State "walk-in" programs).
2. Where demand is high, the government successfully encourages applicants to maximize the N1 "wildlife points" in the EBI. In practice, this might be achieved by fully cost-sharing wildlife-enhancing practices (rather than the standard 50-percent cost share).
3. Combination of 1 and 2: Farmers lower offer prices, and the government subsidizes wildlife practices.
4. Similar to 3, but landowners do not lower their bids, although they still assume they can earn $5 per acre from hunting leases. While this hunting lease income does not infl uence the EBI scores of submitted bids, it does increase the acreage offered to the program (since farmers will receive both government and hunter payments for their CRP land).

Table 4.5 summarizes the results of the policy experiments and of a base-line that uses the current CRP rules. The model is calibrated for 2006, so it does not reflect current high commodity prices. However, comparisons of scenarios to the baseline should be roughly accurate. Note that, in all scenarios, a 35-million-acre program is simulated.[13] These results are best used as indicators of the range and types of changes to the CRP rather than of specific predictions.

The most general results are not surprising: The average bid decreases when expected, and the wildlife score (N1) increases. However, a few points are worth noting:

1. Acreage shifts from nonhunt counties to hunt counties can be substantial: In the fourth scenario, over 3 million acres shift, leading to about a 20-percent increase in hunt-county CRP acreage.[14]
2. The average bid does not decrease by $5 per acre in hunt counties, which is due to the heterogeneity of land types and the increased likelihood of accepting land as bid rates drop. Thus, decreasing the bid of a previously rejected “environmentally desirable but expensive” offer may result in its acceptance, even though its offer price may still be greater than average.
3. N1 scores increase by about 25 percent. Although occurring largely in hunt counties, wildlife scores also increase in nonhunt counties. The increase in average EBI scores causes marginal offers in nonhunt counties (those that often have relatively low N1 scores) to be dropped, increasing the overall average.
4. The fraction of landowners willing to make offers to the program increases when they are assumed not to lower their bids (scenario 4).

Table 4.5. General Results of Scenarios

Item	Baseline	Lowered bids	Increased wildlife points	Lowered bids and increased wildlife points	Increased income and increased wildlife points
Acreage enrolled (million acres):					
All counties	35.0	35.0	35.0	35.0	35.0
In nonhunt counties[1]	19.7	18.8	17.2	16.1	16.5
In hunt counties	15.3	16.2	17.8	18.9	18.5
Average bid of enrolled acres ($):					
Across all counties	24	23	25	23	26
In nonhunt counties	21	21	21	21	21
In hunt counties	28	25	28	25	30
Average N1 score of enrolled acres: [2]					
Across all counties	74	74	82	83	83
In nonhunt counties	74	74	75	75	75
In hunt counties	73	73	90	90	90

[1]“Hunt counties” are counties identified using the hunting-pressure index (all counties with a hunting-pressure index score greater than the median score).

[2]The maximum value of the N1 component of the Environmental Benefits Index is 100.

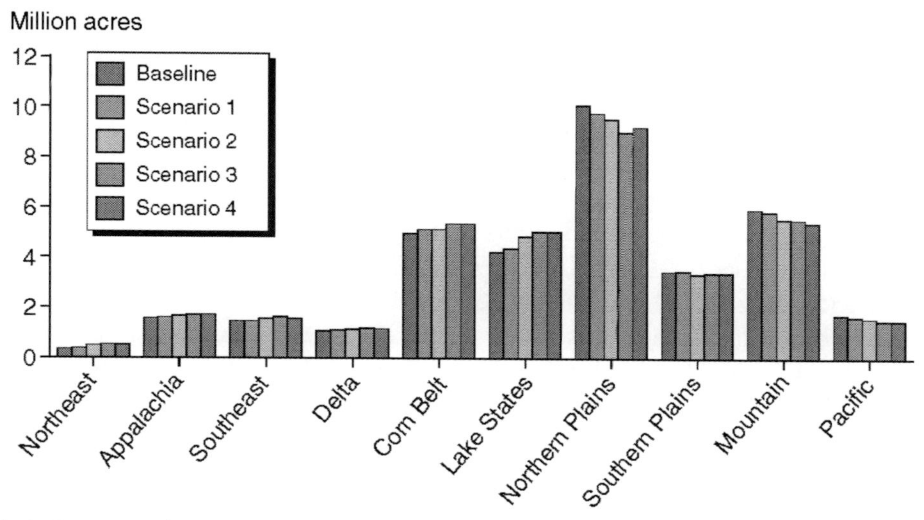

Scenario 1. Farmers reduce CRP offers by $5.

Scenario 2. Farmers maximize wildlife points in Environmental Benefits Index and receive full-cost share.

Scenario 3. Farmers reduce CRP offers by $5, maximize wildlife points, and receive full-cost share.

Scenario 4. Farmers do not lower CRP offers, maximize wildlife points, and receive full-cost share.

Notes: The scenario descriptions summarize several possible per acre changes. The possibility of leasing their CRP land to hunters may lead landowners to reduce their bids, to improve the wildlife habitat on their land, or to factor in lease income when deciding whether or not to offer their land to the program.

Source. USDA, ERS.

Figure 4.11. Enrollment in Conservation Reserve Program (CRP) by region for fee hunting scenarios

In addition to national impacts, the regional distribution of CRP land may change under different scenarios. Changes across the 10 USDA Farm Production Regions are summarized in figure 4.11. The most striking result is the acreage reduction (compared with the baseline) in the Northern Plains, Pacific, and Mountain States. These acres are reallocated to the other regions, especially the Corn Belt and Lake States, which are largely driven by the greater hunting pressure east of the Mississippi.

Scenarios where fee hunting opportunities cause landowners to reduce their offers (hence increasing the likelihood of an offer acceptance) yield similar results to scenarios where offers are not reduced (hence increasing the likelihood of an offer being made). There are a few differences. For example, in the Northeast, acreage is slightly higher in the "reduce-the-bid" scenario than in the "do-not-reduce-bid" scenario. Conversely, in the Northern Plains, the opposite is observed.

Implications

Recreational hunting primarily occurs on private lands. While most of this access is through informal mechanisms, landowners have a long history of charging willing hunters to access their land. However, for a variety of reasons, the marketing of the "hunting" environmental service is still rela-tively small. And although some of these reasons (such as landowner reluctance to give strangers with guns access to their property) are unlikely to change, others (such as the difficulty of connecting landowners to hunters) may be amenable to institutional solutions. However, even with greater farmer participation, evidence suggests that fee hunting does not always lead to improved wildlife management, which is a major reason for promoting the creation of markets.

One institutional solution for improving wildlife habitat is via government policy vis-à-vis agricultural conservation programs, such as the CRP. While only suggestive, the alternative scenarios for coordinating the CRP with hunting access indicate that such programs could reduce CRP costs (with a 10-percent reduction in offer price in some scenarios) and increase the quality of wildlife habitat (with a 25-percent increase in one measure of wildlife habitat in some scenarios). These scenarios also suggest that CRP acreage may shift toward more populated areas of the country (where there are more hunters). An indirect benefit of this could be increased values from the provision of other environmental services, such as open space and water quality.

Overall, current trends suggest greater restriction on casual access to hunting lands, with continued urbanization further weakening the link between nonrural hunters and rural landowners. Thus, the prospects of using private provisions of hunting services are likely to increase.

"USDA Organic" and Other Eco-Labels in Agriculture

One way that a farmer could benefit financially by providing an environmental service is to link the provision of the service to the sale of a private good. Eco-labeling is a way of informing consumers of the process used to produce the private good and, concurrently, its impact on environmental services.

Consumers who care about environmental services may be willing to pay a higher price for products produced in a way that provides those services.

Starting with the organic label in the 1950s, eco-labels have been used to tout reduced pesticide use, wildlife protection, and other environmental services tied to specific agricultural production systems. Food that has an organic or other eco-label is fundamentally a "credence good"—it cannot be distinguished visually from conventional food—and consumers must rely on labels and other advertising tools for product information. Many consumers associate enhanced food safety and nutrition, environmental protection, and other qualities with eco-labels. We examine experience with the organic label for lessons on how this approach could be expanded to a wider set of environmental goals.

National Organic Standards Define An Ecological Production System

The organic label is the most prominent eco-label in the United States, reflecting decades of private-sector development and subsequent initiation of a government regulatory program. Congress passed the Organic Foods Production Act of 1990 (OFPA) to establish national standards for organically produced commodities in order to facilitate domestic marketing of organically produced fresh and processed food and to assure consumers that such products meet consistent, uniform standards.

The program establishes: (1) national production and handling standards for organically produced products, including a national list of substances that can and cannot be used; (2)

certification requirements for organic growers; (3) a national-level accreditation program for State and private entities, which must be accredited as certifying agents under the USDA national standards for organic certifiers; (4) requirements for labeling products as organic and containing organic ingredients; and (5) civil penalties for violations of these regulations.

In setting the soil fertility and crop nutrient management practice standard, USDA requires the producer to use practices that maintain or improve the physical, chemical, and biological condition of soil and minimize soil erosion. The producer is required to manage crop nutrients and soil fertility through rotations, cover crops, and the application of plant and animal materials and is required to manage plant and animal materials to maintain or improve soil organic matter content in a manner that does not contribute to contamination of crops, soil, or water by plant nutrients, pathogenic organisms, heavy metals, or residues of prohibited substances.

Environmental benefits that can be attributed to organic production systems include the following:

- *Reduced pesticide residues in water and food.* Organic production systems virtually eliminate synthetic pesticide use, and reducing pesticide use has been an ongoing U.S. public health goal as scientists continue to document their unintentional effects on nontarget species, including humans.
- *Reduced nutrient pollution, improved soil tilth, soil organic matter, and productivity, and lower energy use.* A number of studies have documented these environmental improvements in comparing organic farming systems with conventional systems (USDA Study Team on Organic Farming, 1980; Smolik et al., 1993; Mäder et al., 2002; Marriott and Wander, 2006).
- *Carbon sequestration.* Soils in organic farming systems (which use cover crops, crop rotation, fallowing, and animal and green manures) may also sequester as much carbon as soils under other carbon sequestration strategies and could help reduce global warming (Lal et al., 1998; Drinkwater et al., 1998). (See "Issues in Supply of Carbon Sequestration" for more detail.)
- *Enhanced biodiversity.* A number of studies have found that organic farming practices enhance the biodiversity found in organic fields compared with conventional fields (Mäder et al., 2002; Altieri, 1999) as well as improving biodiversity in field margins (Soil Association, 2000).

Issues In Supply: Major Farm Sectors Lag In Adopting Organic Systems

U.S. farmland under organic management has grown steadily for the last decade as farmers strive to meet consumer demand in both local and national markets. U.S. certified organic crop acreage more than doubled between 1992 and 1997 and doubled again between 1997 and 2005 (USDA, Economic Research Service, 2007c). Organic fruit and vegetable crop acreage, along with acreage used for hay and silage crops, expanded steadily between 1997 and 2005. However, most of the acreage increase for organic grain and oilseed crops took place early in this period, and organic soybean acreage has declined substantially since 2001.

California had more certified operations than any other State, with just over 1,900 operations in 2005, up 20 percent from the previous year. Wisconsin, Washington, Iowa,

Minnesota, New York, Vermont, Oregon, Pennsylvania, and Maine rounded out the top 10. Many of these States have a high proportion of farms with fruits and vegetables and other specialty crops. Also, some of these States, particularly in the Northeast, have relatively little cropland but a large concentration of market gardeners.

While adoption of organic farming systems showed strong gains between 1992 and 2005 and the adoption rate remains high, the overall adoption level is still low: Only about 0.5 percent of all U.S. cropland and 0.5 percent of all U.S. pasture were certified organic in 2005 (Figure 4.12). About 8,500 operations are certified organic (out of over 2 million farms).

One of the biggest obstacles to adoption is the cost of converting a farm to organic production. The transition from conventional production systems to organic systems typically involves high managerial costs and risks (Oberholtzer, Dimitri, and Greene, 2005). Production costs may be higher because of more intensive use of labor, use of substitutes for synthetic chemicals, longer crop rotations for disease and pest control, reduced yields, and increased recordkeeping.

Another issue is access to production and market information. The infrastructure for extension, marketing, handling, and transport is far less developed than for conventional production systems (Lohr and Salomonsson, 2000). A lack of publicly funded organic farm advisors and relatively little government-funded research on organic production and marketing systems has hindered adoption in the recent past (Lipson, 1997). Most organic information is disseminated by farmers and private organizations (Lohr and Salomonsson, 2000).

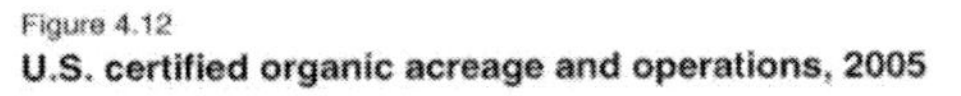

Figure 4.12

U.S. certified organic acreage and operations, 2005

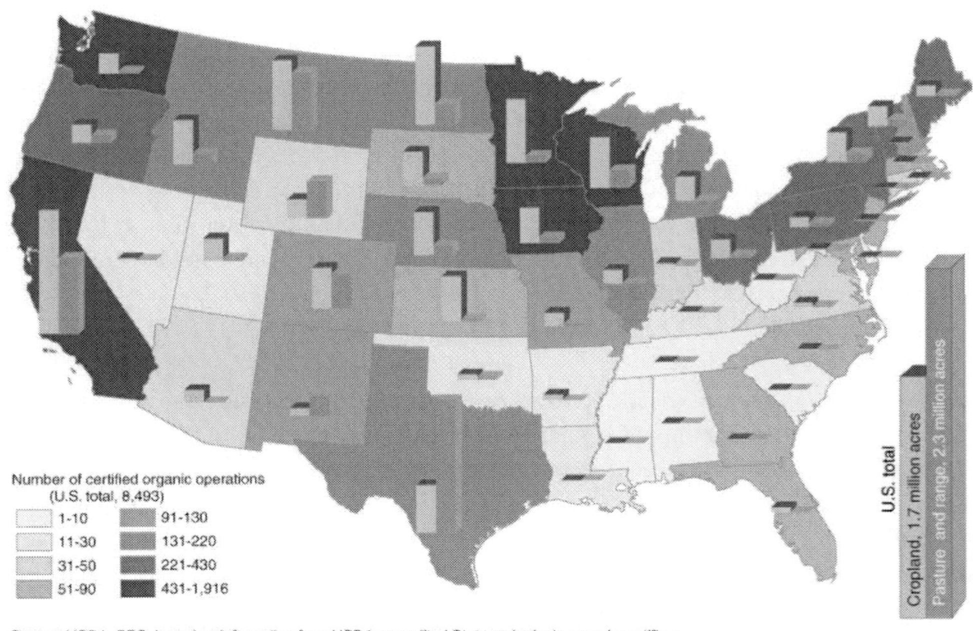

Source: USDA, ERS, based on information from USDA-accredited State and private organic certifiers.

Issues In Demand

Farmers, food processors, and other businesses that produce and handle organically grown food have a financial incentive to advertise that information because consumers have been willing to pay a price premium for these goods. Academic research studies in the 1980s and early 1990s found that consumers were purchasing organic products in response to environmental concerns, such as the impacts of pesticide use on the environment, ground-water, wildlife, and agricultural workers, as well as personal safety concerns (Bruhn et al., 1991; Weaver, Evans, and Luloff, 1992; Cuperus et al., 1996; Goldman and Clancy, 1991; Davies, Titterington, and Cochrane, 1995; Morgan, Barbour, and Greene, 1990).

Although eco-labels enable consumers who value environmental services to pay for the services through the purchase of certain goods, environmental services are still public goods, which presents an opportunity for some of those who value these services to free ride. To the extent that free riding occurs, the prices that organic farmers receive do not reflect the full value that society places on the environmental services provided by this approach to farming.

A negative feature of a proliferation of eco-labels and other labels related to such issues as social justice is that label effectiveness may diminish because multiple, competing label claims may cause consumer confusion (U.S. EPA, 1998). On the other hand, many consumers may be savvy enough to see the differences between unregulated labeling terms that indicate the use of some alternative production practices and a government-regulated, fully defined, independently certified product label like "USDA organic." Also, some of these labels are complementary, not competitive. Organic certifying entities, both State and private, already certify producers and processors to a number of other standards—including food safety standards and international organic standards that already incorporate a social justice component. A product might easily carry both an organic label, denoting the ecologically based production system used, and a locally grown logo, denoting the number of food miles to deliver the product to the consumer.

Emerging Eco-Labels

In addition to the organic label, a number of other eco-labeling programs have emerged in the food and agricultural sector for a broader group of farm-related characteristics (table 4.6). Some of these programs use private third-party certification to enhance consumer confidence, but none has a government regulatory program similar to the organic program.

Several process-based labels have emerged, which have a regional focus. In 1998, the World Wildlife Federation collaborated with another nonprofit, "Protected Harvest," to initiate a label for potato farmers in Wisconsin that would reduce the use of some toxic pesticides and encourage other environmentally beneficial production practices. About the same time, a nonprofit in the Pacific Northwest developed a "Salmon Safe" label that recognizes the adoption of "ecologically sustainable agricultural practices that protect water quality and native salmon." This label encourages restoration of riparian habitat adjacent to fields, as well as improved cropping system practices. Fewer than 50 farmers were using these programs in 2005/06. About 100 growers in New York, using a variety of production systems, were using a "Pure Catskills" promotional label to indicate their participation in a watershed protection program.

Table 4.6. Eco-Labels Used in U.S. Food and Agriculture Sectors

Label	Private standards	Government standards	Program certification	Certified operations, 2005/06 *Number*	Farmland 2005/06 *Acres*	Retail sales 2006 *$ million*
Process-based						
USDA organic (organic production and food processing systems)	International (IFOAM; Codex)	USDA-AMS, Federal Register, National Organic Program, fi nal rule, Dec. 21, 2000, pp. 80548-80684	Private-1971 State-1980 Federal-2000	50 States; Farmers-8,493 Processors—3,000	Cropland—1,723,271 Pasture—2,331,158	16,000
Healthy Grown (alternative pest management)	World Wildlife Fund-protected harvest	No	1998	Wisconsin; Farmers—11	Cropland—5,823	—
Salmon Safe (alternative production and salmon habitat restoration practices)	Salmon safe	No	1997	Northwest (4 States); Farmers—39	Cropland and pasture—30,000	—
Pure Catskills (alternative production practices)	Watershed Agricultural Council	No	Promotion only	New York; Farmers—102	Cropland—	—
Responsible Choice (alternative apple production, packing, and shipping practices)	Stemult Growers, Inc. (1989)	—	1989	Farmers—250 Processor—1	Cropland—	—
Product-based						
Natural (minimal processing, no artificial ingredients, additives, or coloring)	—	Definition, but no standards (FTC, 1970s; USDA, 1982)	Promotion only	—	—	5,140

Table 4.6. (Continued)

Label	Private standards	Government standards	Program certification	Certified operations, 2005/06 *Number*	Farmland 2005/06 *Acres*	Retail sales 2006 *$ million*
Location-based						
State logos (such as Jersey Fresh, Minnesota Grown, Pride of New York, and Virginia's Finest)	—	State departments of agriculture	Promotion only (first logo, 1983)	44 States	—	—
Food miles emissions) (CO_2 emissions	Iowa State University Leopold Center (pilot)	No	—	—	—	—
Social Justice						
Food Alliance Certified (standards for working conditions and alternative production practices)	Food Alliance	No	1998	10 States; Farmers—159	Cropland—156,001 Pasture—4,148,467	82
Just Organic (farmers' rights, farm workers' rights, fair trade and indigenous peoples' rights)	Florida Certified Organic Growers and Consumers (pilot)	No	—	,	—	—

IFOAM=International Federation of Organic Agriculture Movements; AMS=Agricultural Marketing Service; FTC=Federal Trade Commission.

Sources. USDA-ERS, www.ers.usda.gov/briefi ng/organic; Wyman, 2006; Food Alliance, www.foodalliance.org; and Saam, 2007.

Location-based labels may have had limited use in this country, but a "food miles" label may emerge as interest in reducing the energy costs and environmental impacts of food transportation increases (Leopold Center, 2003). States have been developing promotional logos to appeal to consumer interest in helping to protect their State's farmland from development since the early 1980s, and 44 States now have their own agricultural logo—i.e., "Jersey Fresh" and "Virginia's Finest." Many consumers may associate environmentally friendly production practices with local production and local labels. However, these labels address product freshness and the energy used in transportation during the food distribution process but not necessarily environmentally friendly production practices.

Summary

The organic label is the most important eco-label in the United States. It has benefited from consumer demand, a clearly defined set of standards, a strong certification system, and a system of enforcement. However, the adoption of organic production systems is still fairly low. Obstacles to adoption by farmers include high managerial costs and risks of shifting to a new way of farming, limited awareness of organic farming systems, uncertainty over expected yields and returns, lack of marketing and infrastructure, and inability to capture marketing economies (Greene, 2001). These factors are likely to be issues for other types of eco-labels as well.

The proliferation of other local and national eco-labels for a variety of environmental services and labels for other causes may pose a challenge to consumers. Many of these labels do not come with the standards and certification of the organic label, raising the uncertainty of the label claims. Even if consumers are willing to pay a premium to support the supply of environmental services on farms, too much information may make deciding between competing goods difficult. However, careful development of new production standards and labeling regulations, along with consumer education, production research, and other policy initiatives, can mitigate consumer confusion and address the obstacles to adoption.

Even if price premiums for eco-labels can be maintained, however, the public-goods nature of environmental services, such as biodiversity and water quality, implies that they do not reflect the true social value of these services. Eco-labels alone do not provide a socially optimal level of environmental services.

5. Lessons Learned and Potential Roles for Government

Producers have opportunities to sell environmental services in a number of well-functioning markets (table 5.1). EPA is encouraging States to use the Clean Water Act's permit program to establish markets for pollution discharge allowances and to include agriculture in these markets. Producers can sell credits for greenhouse gas reductions on the Chicago Climate Exchange and in a growing number of retail carbon markets. Wetland mitigation markets are operating in many States, and the concept has been expanded to protect endangered species habitat. Fee hunting operations are commonplace in a few States and demonstrate that producers can earn substantial income that could be used to support wildlife habitat. Organic labeling is well established, and food labels are expanding to include information related to the provision of a wider set of public goods on farms.

Table 5.1. Summary of Existing Markets for Environmental Services and Some Important Characteristics

Market	Water quality trading	Chicago Climate Exchange	Retail carbon market	Wetland mitigation banking	Organic labeling	Fee hunting
Environmental service	Water quality	Reductions in net green-house gas emissions	Reductions in net green-house gas emissions	Wetland services	Various (water quality, biodiversity, air quality)	Wildlife
Good traded	Discharge allowance	Carbon credit	Carbon credit	Qualified wetland acreage	Agricultural food, fiber, and other products	Access to land
Source of property right	Regulatory agency	CCX rules	Retail carbon provider	Regulatory agency	Private good	Private good
Source of demand	Regulatory dis-charge cap on point sources	Legally binding discharge cap on member firms	Private sentiment	Legally binding no-net loss rules	Private sentiment	Private sentiment
Standards?	Yes	Yes	No	Yes	Partial	No
Steps being taken to reduce uncertainty	Research on perfor-mance of conserva tion practices, flexible rules for point sources, verification, enforcement	Research on performance of conservation practices, verification	None	Research on measuring and verifying wetland services	Uniform national standards, mandatory certification, Federal enforcement	Research on improving habitat, outreach
Steps being taken to reduce transactions costs	Third-party aggregator, clearing house, outreach, models	Third-party aggregator, models, Voluntary Green-house Gas Reporting Registry	Online decision aids	Third-party arbitrators	Reduction in multi-ingredient certification disputes	Outreach, clearing house operated by State, liability coverage
Remaining impediments or issues	Producer reluctance, lack of binding caps, interactions with conservation programs	Lack of national binding cap, interactions with conservation programs	Lack of standards and verification	Up-front costs and market uncertainty, interactions with conservation programs	Information overload, free-riding on environmental benefits	Public sentiment, free-riding on wildlife services

CCX=Chicago Climate Exchange.

Overall, however, farmer participation in these markets has been limited. Part of the reason is that many of the markets themselves are limited in scope. Experiences with these markets have also identified a number of impediments that limit producers' participation. Many of these impediments are unlikely to be overcome without direct involvement by government, including USDA. USDA has already identified some of the actions it can take to assist in the development of markets and to increase farmer participation (see box, "USDA Commitments to Markets for Environmental Services"). Economic theory and experience with the markets described in the case studies highlight a number of issues that are of primary importance in the successful creation of markets for environmental services.

Issue: Performance of Management Practices

One of the biggest issues facing producers who wish to participate in markets for environmental services is uncertainty about the environmental performance of conservation practices, such as conservation tillage, riparian buffers, and nutrient management. In emissions trading and offset markets, uncertainty about the quantity of credits that can be supplied reduces demand for environmental services from agriculture. Markets often try to account for this uncertainty by requiring that a lost unit of wetland services or a point- source unit of pollution discharge be replaced or mitigated with two or more units of services (credits) from farms. This practice essentially increases the price of mitigation to buyers and reduces overall demand for farmer- produced credits.

Uncertainty of practice performance also affects the potential supply of environmental services. Uncertainty about the quality or quantity of the environmental services a farm can produce makes it difficult for producers to decide the long-term economic benefit of investing in a wetland mitigation bank, to make wildlife habitat improvements for a fee hunting business, to enter an emissions trading market, or to enter the organic market. Uncertainty about the impact of a new practice on crop yields can also affect a farmer's decision to implement a practice in order to enter a market. In the case of the Chicago Climate Exchange, lack of scientific evidence about a soil's ability to sequester carbon can prevent a farm from entering the market.

USDA can play a role in providing research on the effectiveness of different conservation practices for producing environmental services. USDA already provides farmers and ranchers with information on the impact of conservation practices on air, water, and wildlife habitat through sources like the NRCS Field Office Technical Guide. However, much more detailed information is needed to estimate the number of credits that might be produced for sale in emissions trading markets or the wetland services that can be sold by a mitigation bank.

USDA supports the development of tools and methods for quantifying how changes in farming practices affect environmental services (USDA, Natural Resources Conservation Service, 2006b). For example, the Nitrogen Trading Tool and GRACEnet can help reduce uncertainty in water quality and carbon trading markets, respectively.

Another broader effort is the Conservation Effects Assessment Project (CEAP). The goal of CEAP is to quantify the environmental benefits of conservation practices used by private landowners participating in USDA conservation programs. Field-level sampling, monitoring, and modeling are being used to estimate the impacts of conservation practices on water quality, wildlife, and soil quality. In addition, collaborative regional assessments are developing models for estimating environmental services from wetlands, including carbon storage, sediment, and nutrient reduction, flood water storage, wildlife habitat, and biological sustainability (USDA, NRCS, 2006a). CEAP also includes watershed assessment studies that are to provide a framework for evaluating and improving the performance of water quality assessment models. Such models are critical for estimating the equivalency of water quality credits that are produced in different parts of a watershed. Models that can predict the movement of chemicals carried in runoff with a degree of certainty sufficient to allow agricultural credits to be traded would make it easier for producers to participate in trading programs. Models would also allow uncertainty ratios (trading ratios that specifically reflect practice uncertainty) to be lowered, reducing the cost of agricultural credits and making them more attractive to point sources. In addition, research sponsored by the USDA Cooperative State Research, Education, and Extension Service is also addressing practice performance in a variety of settings, as well as supporting the development of assessment tools.

USDA Commitments to Markets for Environmental Services

In 2006, USDA released a departmental regulation defining its policy on markets for environmental services. This policy stated that USDA would do the following:

- Cooperate with other Federal, State, and local governments to establish a role for agriculture in environmental markets.
- Find ways to make USDA policies and programs support producers wanting to participate in such markets.
- Conduct research and develop tools for quantifying environmental impacts of farming practices.

A partnership agreement between EPA and NRCS to collaborate on efforts to establish viable water quality trading markets was signed in 2007. A goal is to develop a pilot water quality trading project in the Chesapeake Bay watershed.

The Food, Conservation, and Energy Act of 2008 contained a section in the Conservation Title outlining USDA's role in support for market-based conserva-tion. The provision required the following:

- The Secretary of Agriculture will establish technical guidelines for measuring environmental services from conservation and other land management activ-ities, and priority will be given to developing guidelines for participation in carbon markets.
- Guidelines will be established for a registry to collect, record, and maintain information on measured benefits.
- Guidelines will be established for a process to verify that a farmer has implemented the conservation or land management activities reported in the registry.

Uncertainty over the economic performance of practices implemented to produce environmental services can also be overcome through risk-management instruments, such as insurance (Zeuli and Skees, 2000). Private companies could provide such instruments, but government could also offer them if an active market for environmental services is an important conservation goal and private insurance is not available.

Issue: Standards and Verification

One of the requirements for a smoothly operating market is that the good being traded is of a consistent quality that is known to all. Organic agriculture and emissions trading markets have very specific standards for the services that are marketed, which is not the case for the retail carbon market and some of the newer eco-labels. Consumers may not know what they are buying or how the environmental services provided by one supplier differ from another. For example, what does "wildlife-friendly" agriculture really mean? What does it really take to eliminate the carbon footprint of an airline flight or a wedding? As long as labels and advertising are the only ways consumers have of discriminating between the ability of producers to provide environmental services, consumers are likely to be skeptical of suppliers' claims. Third-party certification is considered the only reliable way to signal product quality claims in organic markets (Cason and Gangadharan, 2002).

USDA is playing an important role in setting standards and providing certification for organic agriculture. Standards and certification provide the assurance to consumers that the claims on the label are believable and protect producers from dilution of price premiums due to less rigorous (and less costly) applications of organic standards. The department regulation outlining USDA's roles in "market-based stewardship" calls for USDA to cooperate with other Federal Departments and groups in developing accounting practices and procedures for quantifying environmental goods and services in other types of markets. Research on practice performance would help USDA contribute to such a role.

Verification that standards are being followed and that promised management practices are being implemented is a related issue. Many environmental services are not easily observed. Verification is based on the farming practices that have been implemented, and this often requires on-site visits. Particularly in markets created through regulation, such as water quality trading and wetland mitigation, the prospects of on-site visits by representatives of EPA or other regulatory agencies have been a deterrent to farmer participation (Breetz et al., 2004). In some markets, such as the CCX and some water quality trading programs, aggregators or other third-party service providers, rather than a government agency, verify that practices are in place. Although farmers may be less reluctant to deal with USDA-led verification for market services, such a role could put USDA at odds with its historical constituents. Experience with conservation compliance and Swampbuster (a compliance program to discourage the draining of wetlands) would seem to bear this out. The Government Accountability Office found that almost half of all NRCS field offices were not properly verifying that producers were meeting the requirements of the compliance and Swampbuster provisions (U.S. General Accounting Office, 2003). A reluctance to assume an enforcement role was cited as one of the reasons. Improved remote-sensing technology might provide more acceptable (less intrusive) means of verification, although this practice may not be applicable for all types of management options.

Verification almost always concerns management practices or land use, rather than the environmental services that are being produced. Measuring environmental services, such as water quality, carbon sequestration, wetland functions, and wildlife, is often extremely difficult and costly. Verifying practices is much less costly and is sufficient as long as market participants accept that the expected services are actually being produced.

Issue: Cost of Information

An important aspect of a market for environmental services is that participants have access to the information they need to make informed decisions. Producers need to know which markets they can participate in, how to produce the services demanded, and what the total cost to the farm business will be. Producers are not likely to have the time to research all the questions that need to be answered, given the time needed for managing the farm.

Government and other groups can reduce the costs of participating in a market by providing the necessary information. The USDA departmental regulation calls for USDA to conduct outreach, education, technology- transfer, and partnership-building activities with producers, using established institutional arrangements, to help producers participate in markets for environmental services. Many State cooperative extension offices have developed publications to help producers set up a fee-hunting business, with checklists to help identify business goals, the type of lease to offer (daily, long term, lease to a hunt club), other services to offer (bed and breakfast, guides, game cleaning), how to advertise, and how to manage risk (Chopak, 1992; Porter et al., 2007). Nongovernment organizations and private businesses that benefit by farmer participation in markets also have an incentive to reduce producers' information costs. NutrientNet and the Nitrogen Trading Tool are examples of tools that can reduce information costs, as well as uncertainty.

Educating the public presents an important step in increasing demand for environmental services. Raising the public's awareness of the potential threats from GHG emissions could increase their willingness to pay for GHG reductions in retail markets (Trexler, Kosloff, and Silon, 2006).

Issue: Bringing Together Buyers and Sellers

Environmental services are produced across a diverse landscape. It may be costly for individual buyers to find all potential suppliers and to discover what each is selling, especially when the demand from a single source is much greater than the supply from a single farm. For example, a single sewage treatment plant may require nutrient credits from multiple farms to meet its permit requirements. Similarly, it can be costly for producers to find potential buyers, many of whom may be residing some distance away.

One way that markets have addressed this issue is through formal clearinghouses that assemble information from both buyers and sellers, making it easier for potential trading partners to find each other and to gauge supply and demand. The Internet is an obvious tool that could be used to facilitate trades. For example, NutrientNet, World Resources Institute's on-line nutrient-trading tool, could play a clearinghouse role in water quality trading programs (Kramer, 2003).

Government is playing a clearinghouse role in some markets. State-operated clearinghouses make it easier for point sources and nonpoint sources to find each other in some water quality trading programs (Breetz et al., 2004). The Voluntary Greenhouse Gas Reporting Registry can help agriculture and forest entities take advantage of State- and private-sector-generated opportunities to trade emission reductions and sequestered carbon.

Third-party brokers and aggregators also play a more direct role of bringing buyers and sellers together by purchasing credits from producers and selling them to buyers. Aggregators play a critical role in the Chicago Climate Exchange and are present in some water quality trading programs. In some cases, government plays an aggregator role by purchasing credits from producers and selling them on the market (such as what North Carolina does in its Tar-Pamlico trading program). State agencies serve as third-party brokers in some wetland mitigation markets to reduce uncertainty and arbitration costs. A number of State programs purchase hunting access rights from landowners and make these available to the hunting public. Hunters can consult State-provided atlases to find hunter-accessible land, with no need to seek out the individual landowner.

Issue: Coordinating Conservation Programs with Markets

Federally funded conservation programs and markets for environmental services can interact in several ways. USDA for the most part does not claim any credits in markets for environmental services that are created through practices implemented with financial assistance from conservation programs, allowing landowners to sell them. However, the WRP does not allow environmental services (such as carbon sequestration) created by wetland restoration to be sold. Markets for environmental services and conservation programs can also compete with each other for the same natural capital, driving up costs to the possible detriment of market development. For example, the WRP may, in some areas, reduce the stock of lands most suited to wetland restoration, leaving mitigation bankers with higher restoration costs.

Rules of individual markets may present confl icts with conservation programs. Many water quality trading programs do not allow producers to sell pollution reductions from practices financed through a conservation program, arguing that these improvements would have occurred without trading. This restriction is similar to the WRP example above. On the other hand, the Chicago Climate Exchange has no such restriction and will pay producers for carbon sequestration from practices for which producers have already received payment (raising the question of additionality).

Coordinating conservation programs and environmental service markets can enhance the performance of both. In trading programs that establish a baseline on a minimum level of stewardship, targeting conservation programs, such as EQIP, at producers with the most serious environmental problems not only increases program performance, but could also increase the number of producers who are willing to enter a market. The policy simulation on pages 39-43 indicates that coordinating the CRP with fee hunting opportunities could benefit the program as well as producers and wildlife by reducing the rental rates landowners are willing to accept while increasing their efforts to improve wildlife habitat.

Of interest is the potential impact of participation in markets for environmental services on USDA's compliance programs. Conservation compliance requires farmers to meet particular soil conservation goals in order to receive program benefits. Similarly, Swampbuster requires that producers not drain wetlands as a condition for receiving program benefits. Compliance requirements may be less costly to producers if credits produced by adopting soil- conserving practices or maintaining wetlands could be sold in water quality, carbon, or other markets.

USDA has developed a partnership agreement with EPA to coordinate agency policies and activities that promote the effective use of water quality credit trading. To this end, USDA agrees to identify and remove program barriers that might impede the development of water quality trading markets. What these are, however, will depend on the rules adopted in each market. Similar agreements could be developed for other markets as well.

Issue: The Role of Policy

The design and eligibility requirements of markets for environmental services can greatly affect how attractive they are to potential participants. As discussed in the "Water Quality Markets" section of chapter 4, baseline requirements can greatly infl uence the cost and supply of credits. As shown in the greenhouse gas case study, basing credits on *net* sequestration rather than *gross* sequestration greatly affects potential returns to producers from trading.

Major expansions in some markets (i.e., wetland services, water quality, and greenhouse gases) come only with expanded or more stringent regulations on environmental quality. The low price for carbon credits in the CCX reflects the relatively low level of demand inherent in a voluntary program. A number of water quality trading programs cited lack of trades for discharge allowances because discharge caps were too high to stimulate demand (Breetz et al., 2004). Also, in a global sense, the demand for water quality improvements from producers is currently low because few impaired watersheds have opted to implement a water quality trading program and nonpoint sources are not capped.

Increased demand for environmental services from agriculture could occur when regulations change or trading programs are expanded into new areas. Requiring agricultural sources to also meet an emissions cap in a carbon market would greatly enhance demand for sequestration and result in a much larger market. Regulating all emission sources would also address the problem of leakage that occurs when payments are based on gross sequestration rather than net sequestration. Similarly, a more vigorous water quality trading market would be realized if nonpoint sources were included under a cap just as point sources are. This practice would spur nonpoint-nonpoint trading, as well as point-nonpoint.

Because of program requirements, producers considering whether to enter the wetland mitigation market face a relatively long period between starting wetland restoration and being able to sell wetland credits. A Government interested in promoting producer participation in mitigation banks could reduce these startup costs by working with lending institutions to construct loans that provide capital in increments, negotiate flexibility on loan repayment dates (perhaps delaying loan payments until wetland credits are marketed), and guarantee loans so that producers could receive a lower interest rate.

Markets Are Not Always The Answer

We have shown that markets for environmental services rarely develop without some type of outside intervention. Government and other groups can reduce supply and demand impediments through regulation, market design, program coordination, education, verification, certification, and research. One of the features of working markets is the incentive to reduce transaction costs. While transaction costs may be high initially, and require Government assistance to reduce them to get the market started, costs tend to decrease over time as new institutions and mechanisms are developed by those who benefit most from them.

What the ultimate scale of markets for environmental services might be is difficult to say. For fee hunting, which is not a new concept, attitudes of both landowners and hunters may prevent much expansion. Both the water quality trading and wetland case studies indicate that the combination of factors required for markets to develop may be limited to a relatively few areas, given the current regulatory regime. On the other hand, the market for greenhouse gas reductions could be greatly expanded if a national discharge cap is implemented and producers across the country could participate in the global market. Organic agriculture and other labels are relatively new, and increased concerns over the environment could raise demand for foods produced in such a way as to provide environmental services.

Even though government can take a number of actions to promote markets for environmental services, such actions may not always be advisable. The costs of setting up and supporting a market may outweigh the benefits.

The uncertainties associated with nonpoint-source pollution from farms may never be overcome sufficiently enough to allow water quality trading markets to develop on a wide scale. Government may have to use alternative approaches, such as regulation or financial incentives, to reduce pollution from nonpoint sources and to improve water quality. Similarly, difficulties in measuring wetland services that are being lost through development or gained through restoration could relegate mitigation banking to a seldom-used tool and increase the demand for regulation or other approaches for meeting the national goal of no-net loss of wetlands. Free riding will continue to limit demand for foods covered by an eco-label, reducing the economic incentive to expand eco-friendly agriculture. Fee hunting may never become widespread because of long-ingrained attitudes about access to land for hunting.

It is probably safe to say that markets for environmental services will never supplant the need for traditional conservation programs, which will continue to play a major role in providing environmental services. Where markets do develop, government can play a role in advising market managers on the potential tradeoffs between different design and eligibility options, in providing outreach and information to reduce transactions costs and uncer-tainty for market participants, and in establishing standards and certification that provide consumer confidence in environmental services produced by farmers and ranchers.

APPENDIX: PREDICTING THE LOCATION OF NEW MITIGATION BANKS

If the forces that drove the demand for and, subsequently, the creation of mitigation banks continue, history can provide a perspective of where future mitigation banks might be located. Data from the USACE (edited by the Environmental Law Institute) identify counties with mitigation banks (both approved and waiting for approval). Currently, banks are dispersed across and within multiple States and in rural and urban areas, suggesting that opportunities have been widespread (see Figure 4.7).

To predict the probability that a county might have a mitigation bank in the near term, a probit model is estimated using data on existing mitiga-tion banks, county population demographics, land use, and other factors. As discussed above, one would expect that new banks will be created in counties with higher urban pressure and greater wetland acreage. Based on these and other factors, the probability model is expressed as:

Bank = f(urban pressure (low, medium, high), wetland acres, wetland acres squared, total agr land)

Where:

1. *Bank* equals 1 if the observation county has a mitigation bank or application and zero otherwise;
2. *low (medium, high)* equals 1 if urban pressures (as defined by county PIZA scores) are low (medium, high) and zero otherwise. Population-interaction zones for agriculture (PIZA) codes are derived from a classification scheme that indexes small geographic areas according to the size and proximity of population concentrations (akin to a gravity model). The codes are discrete values ranging from 1 (rural) to 4 (high population interaction);
3. *wetland acres* is a county's wetland acreage;
4. *wetland acres sq* is wetland acres squared;
5. *total agr land* is the total agricultural land within the county.

Data on wetland acreage are from the National Resources Inventory (USDA, NRCS, 2000). Total agricultural land is from the 2000 Agricultural Census, and the qualitative measures of urban pressure are from ERS.

Results of the analysis are statistically significant and consistent with expectations (app. table 1). Urban pressure variables are significant determinants of the probability of a mitigation bank, and the sizes of their coefficients indicate that, as expected, the probability of a mitigation bank increases with greater urban pressure. The positive coefficient on *wetland acres* and negative coefficient on *wetland acres sq* indicate that greater wetland acreage is likely to increase the need for a mitigation bank (e.g., without wetlands, mitigation is not necessary) but at a decreasing rate as wetland acreage increases. Mitigation may be avoided if alternative lands—lands without wetlands—are available for development. The negative and significant coefficient on t*otal agr land* supports this proposition.

Appendix Table 1. Regression results: Probit model of the probability of a county's having an application for a new mitigation bank

County variables	Parameter standard		
	Estimate	**Error**	**Pr>ChiSq**
Intercept	-1.829	0.0735 0.092	<0.0001*
Low	0.388		<0.0001*
Medium	0.801	0.104	<0.0001*
High	0.904 11.586	0.107 1.339	<0.0001*
Wetland acres			<0.0001*
Wetland acres sq	-27.313 0.420	5.189	<0.0001*
Total agr lands	3,101	0.138	<0.0023*
Number of observations		NA	NA

1) USACE district mitigation banking data as edited by the Environmental Law Institute: Dependent variable (yes/no mitigation bank).
2) USDA, NRCS, National Resources Inventory: wetland acres, and former wetlands.
3) 2000 Census of Agriculture: Net farm income, farms, and government payments, USDA, ERS (2005): Low, Medium, and High urbanization measures based on PIZA scores, www.ers.usda .gov/Data/PopulationInteractionZones/discussion.htm

To provide a perspective of its reliability, we use the probit model to "back forecast"—that is, see how well the model predicts observed values. We found that the model correctly predicted counties with mitigation banks only 12 percent of the time. Counties without mitigation banks were correctly predicted 98 percent of the time. Although we have confidence that the variables in the model are appropriate, additional data related to the economics and landscape characteristics of wetland mitigation are needed to estimate a more robust prediction model.

Number of counties that have:	Number of counties predicted to have		
	Banks	**No Banks**	**Percent correct**
Banks	33	267	12%
No Banks	56	2,740	98%
Percent correct	37%	89%	90%

Note. 52 counties have no agricultural land (they are not included in the table).

REFERENCES

Altieri, M. (1999). "The Ecological Role of Biodiversity in Agroecosystems," *Agriculture, Ecosystems & Environment*, *74*, 19-31, June.

Antle, J. M. (1999). "The New Economics of Agriculture," *American Journal of Agricultural Economics*, *81(5)*, 993-1010, December.

Benson, D. E. (2001a). "Survey of State Programs for Habitat, Hunting, and Nongame Management on Private Lands in the United States," *Wildlife Society Bulletin*, *29(1)*, 354-58, Spring.

Benson, D. E. (2001b). "Wildlife and Recreation Management on Private Lands in the United States," *Wildlife Society Bulletin*, *29(1)*, 359-71, Spring.

Benson, D. E., Shelton, R. & Steinbach. D. W. (1999). *Wildlife Stewardship and Recreation on Private Lands*, College Station, TX: Texas A&M University Press.

Bihrle, C. (2003). "Perceptions and Realities: Game and Fish Surveys Provide Insight into Current Issues," *ND Outdoors*, 16-20, August.

Boyd, J. & Banzhaf, S. (2006). "What are Ecosystem Services?: *The Need for Standardized Environmental Accounting Units*," Discussion Paper 06-02, Resources for the Future, Washington, DC.

Breetz, H. L., Fisher-Vander, K., Garzon, L., Jacops, H., Kroetz, K. & Terry, R. (2004). "Water Quality Trading and Offset Initiatives in the U.S.: A Comprehensive Survey," Hanover, NH: Dartmouth College, August, Accessed at www.dartmouth.e du/~kfv/waterqualitytradingdatabase.pdf.

Bruhn, C. M., Diaz-Knauf, K., Feldman, N., Harwood, J., Ho, G., Ivans, E., Kubin, L., Lamp, C., Marshall, M., Osaki, S., Stanford, G., Steinbring, Y., Valdez, I., Williamson, E. & Wunderlich, E. (1991). "Consumer Food Safety Concerns and Interest in Pesticide-Related Information," *Journal of Food Safety*, *12(3)*, 253-62, October.

Butler, M. J., Teascher, A. P., Ballard, W. B. McGee. & B. K. (2005). "Commentary: Wildlife Ranching in North America—Arguments, Issues, and Perspectives," *Wildlife Society Bulletin*, *33(1)*, 381-89, April.

Butt, T. A. & McCarl, B. A. (2004). "Farm and Forest Carbon Sequestration: Can Producers Employ it to Make Some Money?" *Choices*, *19(3)*, 27-31, 3rd Quarter.

Cason, T. & Gangadharan, L. (2002). "Environmental Labeling and Incomplete Consumer Information in Laboratory Markets," *Journal of Environmental Economics and Management*, *43(1)*, 113-34, January.

Chicago Climate Exchange. (2007). *Soil Carbon Management Offsets*, August, Accessed at www.chicagoclimatex.com/docs/offsets/CCX_Soil_Carbon_ Offsets.pdf.

Chopak, C. (1992). *Promoting Fee-Fishing Operations as tourist Attractions*, ID: E-2409, Michigan State University Extension, June, Accessed at web1.msue.msu.edu/imp /modtd/33809023.html.

Climate Trust. (2007). Window #1: Offsets for use in the Regional Greenhouse Gas Initiative, Accessed at www.climatetrust.org/solicitations_RGGI.php.

Conner, R., Seidl, A., VanTasssell, L. & Wilkins, N. (2001). "United States Grasslands and Related Resources: An Economic and Biological Trends Assessment," Texas A&M University, *Institute of Renewable Natural Resources*.

Conover, M. R. (1998). "Perceptions of American Agricultural Producers about Wildlife on Their Farms and Ranches," *Wildlife Society Bulletin*, *26(3)*, 597-604, Autumn.

Conservation Tillage Information Center. (2006). *Getting Paid for Stewardship: An Agricultural Community Water Quality Trading Guide*, West Lafayette, IN.

Costanza, R., d'Arge, R., deGroot, R., Farber, S., Grasso, M., Hannon, B., Limburg, K., Naeem, S., O'Neill, R. V., Paruelo, J., Raskin, R. G., Sutton, P. & van den Belt, M. (1997). "The Value of the World's Ecosystem Services and Natural Capital," *Nature*, *387*, 253-60, May 15.

Council for Agricultural Science and Technology. (2004) *Climate Change and Greenhouse Gas Mitigation: Challenges and Opportunities for Agriculture*, Task Force Report No. 141, Ames, IA.

Cuperus, G., Owen, G., Criswell, J. T. & Henneberry, S. (1996). "Food Safety Perceptions and Practices: Implications for Extension," *American Entomologist*, *42*, 201-03, Winter.

Davies, A., Titterington, A. J. & Cochrane, C. (1995). "Who Buys Organic Food? A Profile of the Purchasers of Organic Food in Northern Ireland," *British Food Journal*, *97(10)*, 17-23.

Drinkwater, L. E., Wagoner, P. & Sarrantonio, M. (1998). "Legume-Based Cropping Systems Have Reduced Carbon and Nitrogen Losses," *Nature*, *396*, 262, November 19.

Ducks Unlimited. (2007). Accessed at www.ducks.org/Conservation/PriorityAreas/1599/PriorityAreasHome.html, Ecosystem Marketplace. *Backgrounder: Chicago Climate Exchange (CCX)*, Katoomba Group, Accessed at ecosystemenvironmentalmarketplace.com/pages/marketwatch.backgrounder.php?market_id=13&is_aggregate=1.

Ecosystem Marketplace. (2007b). *Backgrounder: European Union Emissions Trading Scheme (EU ETS)*, Katoomba Group, Accessed at ecosystemmarketplace.com/pages/marketwatch.backgrounder. php?market_id= 10&is_aggregate=0.

Ecosystem Marketplace. (2007c) *Backgrounder: Non-Kyoto*, Katoomba Group, Accessed at ecosystemmarketplace.com/pages/marketwatch.backgrounder.php? market_id= 11&is_aggregate=0.

Ecosystem Marketplace. (2007d). *Marketwatch*, Accessed at ecosystemmarketplace.com/pages/static/marketwatch.php.

Elkins, P. (2003). "Identifying Critical Natural Capital: Conclusions About Critical Natural Capital," *Ecological Economics*, *44*, 277-92, March.

Environmental Valuation Reference (2007). Inventory. www.evri.ca/, Accessed September 30. *Food Alliance.* Accessed at foodalliance.org.

Food and Agriculture Organization of the United Nations. (2007). *The State of Food and Agriculture: Paying Farmers for Environmental Services,* FAO Agriculture Series No. *38*, Rome.

Goldman, B. & Clancy, K. L. (1991). "A Survey of Organic Produce Purchases and Related Attitudes of Food Cooperative Shoppers," *American Journal of Alternative Agriculture*, *6(2)*, 89-92.

Greene, C. (2001). *U.S. Organic Farming Emerges in the 1990s: Adoption of Certified Systems*, Agriculture Information Bulletin No. 770, U.S. Department of Agriculture, Economic Research Service, June.

Gross, C. M., Delgado, J. A., McKinney, S. P., Lal, H., Cover, H. & Shaffer. M. J. (2008). "Nitrogen Trading Tool to Facilitate Water Quality Credit Trading," *Journal of Soil and Water Conservation*, *63(2)*, 44A-45A, March-April.

Heimlich, R., Claassen, R., Wiebe, K. D., Gadsby, D. & House. R. M. (1998). *Wetlands and Agriculture: Private Interests and Public Benefits*, Agricultural Economic Report No. 765, U.S. Department of Agriculture, Economic Research Service, August, Accessed at. www.ers.usda. gov/publications/aer765/.

Helland, J. (2006). *Walk-In Hunting Programs in Other States, Information Brief,* Minnesota House of Representatives Research Department, St. Paul, MN.

Jones, W., Munn, I., Grado, S. & Jones, J. (1999). *Fee-Hunting and Wildlife Management Activities by Non-industrial, Private Landowners in the Mississippi Delta*, Article No. FO 123, Forest and Wildlife Research Center, Mississippi State University, Accessed at sofew.cfr. msstate.edu/papers/0114jones.pdf

Kenny, A. (2007). "Bankers, Developers and Environmentalists Weigh in on New Wetlands Regulations," in *Banking on Conservation 2007: Species and Wetland Mitigation Banking*, Ecosystem Marketplace.

Kieser and Associates. (2004). "Ecosystem Multiple Markets: A White Paper," Draft paper produced on behalf of The Environmental Trading Network, Accessed at www.envtn.org/docs/EMM_WHITE_PAPERApril04.pdf.

King, D. M. (2005). "Crunch Time for Water Quality Trading," *Choices*, *20(1)*, 7 1-76, 1st Quarter.

King, D. M. & Kuch, P. J. (2003). "Will Nutrient Credit Trading Ever Work? An Assessment of Supply and Demand Problems and Institutional Obstacles," *Environmental Law Reporter*, *33*, 10352-68.

Kramer, J. (2003). *Lessons from the Trading Pilots: Applications for Wisconsin Water Quality Trading Policy*, Resource Strategies, Inc., Madison, WI.

Lal, R., Kimble, J. M., Follett, R. F. & Cole. C. V. (1998). *The Potential of U.S. Cropland to Sequester Carbon and Mitigate the Greenhouse Effect*, Chelsea, MI: Ann Arbor Press.

Langner, L. (1987). "Hunter Participation in Fee Access Hunting," *Transactions of the 52nd North American Wildlife Natural Resources Conference, Wildlife Management Institute*, 475-82.

Land Trust Alliance. (2006). *2005 National Land Trust Census Report*, Washington, DC, November, Accessed at www.lta.org/aboutus/census.shtml.

Larson, C. (2006). "The End of Hunting? How Only Progressive Government Can Save a Great American Pastime," *Washington Monthly*, January-February.

Leopold Center. (2003). "Ecolabel Value Assessment: *Consumer and Food Business Perceptions of Local Foods*," Iowa State University, Ames, IA.

Lewandrowski, J. & Ingram, K. (2001). "Agricultural Resources and Environmental Indicators: Wildlife Resources Conservation," *Agricultural Resources and Environmental Indicators*, Agriculture Handbook No. 722, U.S. Department of Agriculture, Economic Research Service, April, Accessed at www.ers.usda.gov/publications/arei/ ah722/arei3_3/DBGen.htm

Lewandrowski, J., Peter, M., Jones, C., House, R., Sperow, M., Eve, M. & Paustian, K. (2004). *Economics of Sequestering Carbon in the U.S. Agricultural Sector*, Technical Bulletin No. 909, U.S. Department of Agriculture, Economic Research Service, April.

Lipson, M. (1997). *Searching for the O-word*, Organic Farming Research Foundation, *Santa Cruz*, CA.

Lohr, L. & Salomonsson, L. (2000). "Conversion Subsidies for Organic Production: Results From Sweden and Lessons for the United States," *Agricultural Economics*, *22(2)*, 133-46, March.

Mäder, P., Fliebach, A., Dubois, D., Gunst, L., Fried, P. & Niggli. U. (2002). "Soil Fertility and Biodiversity in Organic Farming," *Science*, *296(5573)*, 1694- 97, May 31.

Marriott, E. & Wander, M. M. (2006). "Total and Labile Soil Organic Matter in Organic and Conventional Farming Systems," *Soil Science Society of America Journal*, *70*, 950-59.

McCann, R. J. (1996). "Environmental Commodities Markets: 'Messy' versus 'Ideal' Worlds," *Contemporary Economic Policy*, *14(3)*, 85-97, July.

McCarl, B. A. & Schneider, U. A. (2000). "U.S. Agriculture's Role in a Greenhouse Gas Emission Mitigation World: An Economic Perspective," *Review of Agricultural Economics*, *22(1)*, 134-59, June.

Mid-Atlantic Regional Water Program. (2006). *A Primer on Water Quality Credit Trading in the Mid-Atlantic Region*, Agricultural Research and Cooperative Extension, Pennsylvania State University, University Park, PA.

Millenium Ecosystem Assessment. (2003). *Ecosystems and Human Well-being: A Framework for Assessment*, Washington, DC: Island Press.

Morgan, J., Barbour, B. & Greene, C. (1990). "Expanding the Organic Produce Niche: Issues and Obstacles," *Vegetables and Specialties: Situation and Outlook Report*, VGS-263, U.S. Department of Agriculture, Economic Research Service.

Murtough, G., Aretino, B. & Matysek, A. (2002). *Creating Markets for Ecosystem Services*, Productivity Commission Staff Research Paper, AusInfo, Canberra.

Oberholtzer, L., Dimitri, C. & Greene, C. (2005). *Price Premiums Hold on as U.S. Organic Produce Market Expands*, VGS-308-01, U.S. Department of Agriculture, Economic Research Service, May.

Organisation for Economic Co-Operation and Development. (2005). *Multifunctionality in Agriculture: What Role for Private Initiatives?* Paris.

Pierce, R. A. (1997). II. *Lease Hunting: Opportunities for Missouri Landowners*, University Extension, University of Missouri-Columbia.

Porter, M. D., Masters, R., Bidwell, T. G. & Hitch, K. L. (2007). *Lease Hunting Opportunities for Oklahoma Landowners*, Bulletin T-5032, Oklahoma Cooperative Extension Service, Oklahoma State University, Stillwater, OK.

Ribaudo, M. O., Horan, R. D. & Smith, M. E. (1999). *Economics of Water Quality Protection from Nonpoint Sources: Theory and Practice*, Agricultural Economic Report No. 782, U.S. Department of Agriculture, Economic Research Service, November.

Ruhl, J. B., Kraft, S. E. & Lant, C. L. (2007). *The Law and Politics of Ecosystem Services*, Washington, DC: Island Press.

Saam, H. (2007). *Food Alliance, Personal communication,* September 21.

Shabman, L. & Scodari, P. (2005). "The Future of Wetland Mitigation Banking," *Choices*, *20(1)*, 65-70, 1st Quarter.

Shabman, L. & Scodari, P. (2004). *Past, Present, and Future of Wetlands Credit Sales*, Discussion Paper 04-48, Resources for the Future, Washington, DC, Accessed at www.rff.org/Documents/RFF-DP-04-48.pdf

Smolik, J. D., Dobbs, T. L., Rickerl, D. H., Wrage, L. J., Buchenau, G. W. & Machacek, T. A. (1993). *Agronomic, Economic, and Ecological Relationships in Alternative (Organic), Conventional, and Reduced-Till Farming Systems*, B718, Agricultural Experiment Station, South Dakota State University, September.

Soil Association. (2000). *The Biodiversity Benefits of Organic Farming*, Bristol House, UK, May.

Stavins, R. N. (2005). "Lessons Learned from SO_2 Allowance Trading," *Choices*, *20(1)*,53-58, 1st Quarter.

Stavins, R. N. (1995). "Transaction Costs and Tradeable Permits," *Journal of Environmental Economics and Management*, *29(2)*, 133-48, September.

Sundberg, J. O. (2006). "Private Provision of a Public Good: Land Trust Membership," *Land Economics*, *82(3)*, 353-66, August.

Tietenberg, T. H. (2006). *Emissions Trading: Principles and Practice*, Washington, DC: Resources for the Future, Washington, DC.

Trexler Climate + Energy Services, Inc. (2006). *A Consumers' Guide to Retail Offset Providers*, Prepared for Clean Air-Cool Planet, December.

Trexler, M. C., Kosloff, L. H. & Silon, K. (2006). *EM Market Insights: Carbon— Going Carbon Neutral: How the Retail Carbon Offsets Market Can Further Global Warming Mitigation Goals*, Ecosystem Marketplace.

U.S. Department of Agriculture, Agricultural Marketing Service. (2000). "National Organic Program; Final Rule, 7 CFR Part 205," *Federal Register*, December 21, Accessed at www.ams.usda.gov/nop.

U.S. Department of Agriculture, Agricultural Research Service. (2007). "Gracenet: An Assessment of Soil Carbon Sequestration and Greenhouse Gas Mitigation by Agricultural Management," Project 5402-11000-008-00, December, Accessed at: www.ars.usda.gov/research/projects/projects.htm?accn_no=4 11610

U.S. Department of Agriculture, Economic Research Service. (2007a). "Conservation Policy: Background," April, Accessed at www.ers.usda.gov/ Briefing/ConservationPol icy/background.htm.

U.S. Department of Agriculture, Economic Research Service. (2007b). "Conservation Policy: Land Retirement Programs," April, Accessed at www.ers. usda.gov/Briefi ng/ConservationPolicy/retirement.htm.

U.S. Department of Agriculture, Economic Research Service. (2007c). "Organic Production," Accessed at www.ers.usda.gov/Data/Organic/ on October *5*.

U.S. Department of Agriculture, Economic Research Service. (2005). "Population-Interaction Zones for Agriculture (PIZA): Discussion," May. Accessed at www.ers.usda.gov/Data/PopulationInteractionZones/ discussion.htm.

U.S. Department of Agriculture, Economic Research Service and National Agricultural Statistics Service. *USDA Agricultural Resource Management Survey*, multiple years.

U.S. Department of Agriculture, National Agricultural Statistics Service. (2004). *2002 Census of Agriculture, Vol. 1*: Part 51, Chapter 2, AC-02-A-51, United States Summary and State Data, June.

U.S. Department of Agriculture, Natural Resources Conservation Service. (2004a). "Chief Knight Tours Nation's First Ag Wetland Mitigation Bank," *NRCS This Week*, April 28, Accessed at www.nrcs.usda.gov/news/thisweek/2004/040428/ moknightwetlandb ankearthday.html.

U.S. Department of Agriculture, Natural Resources Conservation Service. (2006a). *Conservation Effects Assessment Project (CEAP)*. December, Accessed at www.nrcs.usda.gov/technical/nri/ceap/ceapgeneralfact.pdf

U.S. Department of Agriculture, Natural Resources Conservation Service. (2007a). eDirectives, Title 440, Pat 514 – Wetland Reserve Program, Accessed at policy.nrcs.usda.gov/viewerFS.aspx?id=2192

U.S. Department of Agriculture, Natural Resources Conservation Service. (2007b). *"Electronic Field Office Technical Guide,"* Accessed at www.nrcs.usda. gov/technical/efotg/ in October.

U.S. Department of Agriculture, Natural Resources Conservation Service. (2004b). *National Resources Inventory 2002 Annual NRI*, Accessed at www.nrcs.usda.gov/technic al/land/nri02/nri02wetlands.html.

U.S. Department of Agriculture, Natural Resources Conservation Service. (2007c). *Performance Results System*, Accessed at ias.sc.egov.usda.gov/prshome/.

U.S. Department of Agriculture, Natural Resources Conservation Service. (2006b). *USDA Roles in Market-Based Environmental Stewardship*, Departmental Regulation 56000-003, December 20.

U.S. Department of Agriculture, Natural Resources Conservation Service, and Iowa State University, Statistical Laboratory. (2000). *Summary Report: 1997 National Resources Inventory (revised December 2000)*, December, Accessed at www.nrcs.usda.gov/technical/NRI/1997/summary_report/

U.S. Department of Agriculture, Office of the Chief Economist, Global Change Program Office. (2007). *U.S. Agriculture and Forestry Greenhouse Gas Inventory: 1990-2005.*

U.S. Department of Agriculture, USDA Study Team on Organic Farming. (1980). *Report and Recommendations on Organic Farming*, U.S. GPO No. 620-220/3641, July.

U.S. Department of Energy, U.S. (2007). Energy Information Administration. *Voluntary Reporting of Greenhouse Gases Program*, Accessed at www.eia.doe.gov/oiaf /1605/Brochure.html.

U.S. Department of Interior, Fish and Wildlife Service. (2007). "National Wetlands Inventory: Wetland Plants," 2007, Accessed at www.fws.gov/nwi/plants. htm on August.

U.S. Department of Interior, Fish and Wildlife Service, and U.S. (2002). Department of Commerce, Bureau of the Census. *2001 National Survey of Fishing, Hunting, and Wildlife-associated Recreation.*

U.S. Environmental Protection Agency, AgSTAR. (2006). *Market Opportunities for Biogas Recovery Systems: A Guide to Identifying Candidates for On-Farm and Centralized Systems*, EPA-430-8-06-004.

U.S. Environmental Protection Agency, Office of Atmospheric Program. (2006). *Inventory of U.S. Greenhouse Gas Emissions and Sinks: 1990-2004*, EPA 430-R-06-002.

U.S. Environmental Protection Agency, Office of Water. (1995a). *America's Wetlands: Our Vital Link Between Land and Water*, EPA 843-K95-001.

U.S. Environmental Protection Agency, Office of Water. (1995b). "Federal Guidance for the Establishment, Use, and Operation of Mitigation Banks," *Federal Register* 60(228) :58605-614, November 28, Accessed at www. epa.gov/owow/wetlands

U.S. Environmental Protection Agency, Office of Water. (2007a). *National Section 303(d) List Fact Sheet*, Accessed at oaspub.epa.gov/waters/national_rept. control#IMP_STATE.

U.S. Environmental Protection Agency, Office of Water. (2002). *National Water Quality Inventory: 2000 Report to Congress*, EPA-841-R-02-001, August.

U.S. Environmental Protection Agency, Pollution Prevention Division. (1998). "Environmental Labeling: Issues, Policies, and Practices Worldwide," EPA Contract No. 68-W6-0021, December.

U.S. Geological Survey. (2000). *SPARROW Surface Water-Quality Modeling Nutrients in Watersheds of the Conterminous United States*, Accessed at water.usgs.gov/nawqa/spa rrow/wrr97/results.html.

U.S. General Accounting Office. (2003). *Agricultural Conservation: USDA Needs to Better Ensure Protection of Highly Erodible Cropland and Wetlands,* Publication No. GAO-03-418, Washington, DC, April.

U.S. Government Accountability Office. (2005). *Wetlands Protection: Corps of Engineers Does Not Have an Effective Oversight Approach to Ensure that Compensatory Mitigation is Occurring*, Report GAO-05-898, Washington, DC, September.

Washington Department of Fish and Wildlife. (2007). *Private Lands Wildlife Management Area (PLWMA): Private Land Partnerships for Hunter Access*, Discussion paper, August 2004, Accessed at wdfw.wa.gov/wlm/ plwma/plwma_accessprogram.htm on February *16.*

Weaver, R. D., Evans, D. J. & Luloff. A. E. (1992). "Pesticide Use in Tomato Production: Consumer Concerns and Willingness-to-Pay," *Agribusiness*, *8(2)*, 131-42.

Wiggers, E. P. & Rootes, W. (1987). "Lease Hunting: Views of the Nation's Wildlife Agencies," *Transactions Of the North American Wildlife and Natural Resources Conference*, *52*, 525-29.

Wilkinson, J. & Thompson, J. (2006). *2005 Status Report on Compensatory Mitigation in the United States,* Environmental Law Institute, April, Accessed at www.elistore.or g/reports_detail.asp?ID=11137

World Resources Institute. (2007). "About NutrientNet," Accessed at www.nutrientne t.org/about.cfm,.

Wyman, J. (2006). "The Wisconsin Health Grown Potato Story," presentation, IPM Symposium, St. Louis, April 4.

Zeuli, K. A. & Skees, J. R. (2000). "Will Southern Agriculture Play a Role in a Carbon Market?" *Journal of Agricultural and Applied Economics*, *32(2)*, 235-48, August.

End Notes

[1] Payments to agricultural producers for the production of environmental services are fairly common. USDA currently supports the production of environmental services through conser-vation programs, such as the Environ-mental Quality Incentive Program, Wildlife Habitat Incentive Program, Conservation Reserve Program, and Wetland Reserve Program. Land trusts, such as the Nature Conservancy and Ducks Unlimited, purchase land or easement to land in order to protect the fl ow of environmental services, primar-ily wildlife or biodiversity.

[2] A carbon equivalent is an internation-ally accepted measure that expresses the global warming potential of greenhouse gases in terms of the amount of carbon dioxide that would have the same global warming potential.

[3] The Registry was first created by paragraph 1605(b) of the Energy Policy Act of 1992. In February 2002, President Bush directed the Secretaries of Energy, Agriculture, and Commerce and the Administrator of the Environ-mental Protection Agency to recom-mend improvements to the program. The revised program stresses compre-hensive (all GHG sources and sinks) and continuous (yearly) reporting, transparency in estimating emissions, and use of standardized estimation methods.

[4] Additionality refers to emission reductions that are in addition to business-as-usual. They would not have occurred without the program.

[5] ERS analyzed TNC project data. The contract data do not identify the wetland acreage involved, whether the projects involved land or easement pur-chases, or whether the projects involve existing or restored wetlands.

[6] As stated in chapter 2, in 2002, private farms accounted for 41 percent of all U.S. land.

[7] The level of fee hunting tends to be higher in regions with smaller propor-tions of public lands, such as the South and Plains States (Langner, 1987; Con-over, 1998). Jones et al. (1999) reported that, in Mississippi, 11-14 percent of landowners charged a fee for hunting in 1996-98, with gross revenues from hunting averaging about $3,300 per landowner in 1997 and 1998.

[8] For example, Benson (2001) reports that 43 percent of State wildlife manag-ers reported a decrease in hunting ac-cess between 1985 and 1994, whereas 8 percent reported an increase. In North Dakota, between 1992 and 2001, the share of landowners who posted their land increased from about 61 percent to over 68 percent (Bihrle, 2003).

[9] Some farms obtain more income from their hunting operations than from the crops they produce (Benson, Shelton, and Steinbach, 1999).

[10] These activities include hunting, fishing, petting zoos, tours, and onfarm rodeos.

[11] The NSRE has about 100 observa-tions that could be classified as "hunt-ing trips to CRP-like lands." Thus, in order to get a reasonable national cov-erage, all observations (for all wildlife-related trips) were used, which

may introduce bias because hunting trips are probably to locales that systematically differ from trips for other wildlife-related recreation. Nevertheless, the NSRE does capture the distribution of population and does relate to wildlife-associated recreation.

[12] Demand for hunting is relatively low in "nonhunting counties"; thus, in these nonhunting counties, the alternatives assume no changes in the factors that infl uence which acres are offered to the CRP.

[13] The size of the CRP will be re-duced to 32 million acres over the next several years, as specified in the Food, Conservation, and Energy Act of 2008.

[14] An extension to scenario 4, that classifies the Northern Plains States with "walk-in hunting access pro-grams" (North Dakota, South Dakota, Kansas, and Nebraska) as "hunting pressure," yielded similar results, although the acreage in the Northern Plains increases substantially, largely at the expense of the Mountain, Southern Plains, and Corn Belt regions.

INDEX

A

abatement, 75, 124, 135, 196
access, 11, 51, 91, 115, 120, 153, 169, 184, 213, 219, 220, 221, 225, 226, 228, 237, 238, 240, 250
accounting, 73, 124, 141, 144, 158, 159, 174, 175, 178, 189, 201, 210, 211, 236
accreditation, 227
accuracy, 156
acid, 103, 135
acidic, 103
adaptation, 76
advantages, 50, 115, 143
adverse effects, 36, 65
aesthetic, 3, 12, 16, 17, 103, 138, 143, 145, 176, 190
aesthetic character, 190
aesthetics, 21, 22, 23, 29, 48, 121
affective experience, 61, 174
age, 78, 79
agencies, 3, 6, 7, 9, 10, 11, 24, 25, 52, 53, 54, 60, 72, 80, 96, 114, 115, 126, 128, 134, 139, 145, 147, 151, 152, 153, 154, 208, 214, 217, 236, 238
aggregation, 33, 34, 49, 92, 123, 164, 178
agricultural producers, 185, 188, 191, 192, 195, 203, 218, 221, 249
agricultural sector, 203, 229
agriculture, 100, 103, 172, 176, 183, 185, 186, 187, 188, 189, 190, 191, 194, 196, 197, 198, 199, 201, 203, 204, 206, 209, 210, 212, 231, 232, 234, 235, 236, 238, 239, 240, 241
air pollutants, 96, 103, 172
air quality, 16, 46, 47, 48, 61, 76, 96, 122, 124, 160, 166, 170, 188, 189, 233
airplanes, 208
algorithm, 87
altruism, 19
ambient air, 96
American Educational Research Association, 169
ammonia, 100, 101, 189
amphibians, 117, 190
anaerobic digesters, 211
anoxia, 100
anthropologists, 137
APA, 180
aquaculture, 96, 98, 102
aquatic life, 188
aquatic systems, 51, 52
aquifers, 119, 157
arbitration, 238
ARC, 159
arsenic, 132, 135, 139
arson, 179
assessment models, 51, 235
assessment tools, 235
assets, 176
assimilation, 119
asthma, 46
asthma attacks, 46
atmosphere, 43, 100, 197, 206, 209, 210
authority, 12, 98, 116, 126, 214
average costs, 202
avian, 124
avoidance, 106, 160, 170, 177
awareness, 232, 237

B

background information, 177
bankers, 217, 238
banking, 185, 186, 214, 215, 217, 233, 240, 242
bankruptcy, 132
banks, 214, 215, 216, 217, 218, 239, 241, 242
bargaining, 183
bargaining XE "bargaining" costs, 183
barriers, 166, 192, 195, 196, 239
base, vii, 3, 8, 20, 32, 37, 57, 68, 69, 104, 106, 112, 115, 123, 146, 148, 150, 155, 173, 178, 193, 202, 214, 223
behavioral change, 98
behavioral intentions, 59
behavioral sciences, 16

benzene, 133
benzo(a)pyrene, 132
bias, 22, 26, 66, 68, 111, 167, 250
biodiversity, 4, 12, 17, 20, 51, 57, 71, 72, 115, 116, 117, 118, 120, 121, 123, 124, 125, 145, 159, 160, 162, 163, 168, 170, 171, 174, 175, 179, 190, 227, 232, 233, 249
biological activities, 15
biological control, 21
biomass, 20, 42, 47, 71, 189, 207
biosphere, 31, 74, 160
biotic, 50, 118, 119, 135, 164, 177
birds, 12, 117, 124, 133
bonds, 116, 117, 123
Bureau of Labor Statistics, 179
Bureau of Land Management, 72
Bush, President, 249
businesses, 25, 208, 229, 237
buyer, 194, 195, 198
buyers, 183, 191, 192, 198, 201, 207, 208, 221, 234, 237, 238

C

cadmium, 135, 139
calibration, 78
candidates, 80, 101
capital account, 73, 174
carbon, 42, 43, 44, 49, 71, 124, 157, 170, 174, 188, 189, 194, 195, 196, 207, 208, 209, 210, 211, 227, 232, 233, 234, 235, 236, 237, 238, 239, 249
carbon dioxide, 43, 207, 249
carbon emissions, 195, 208, 210
carbon monoxide, 124
carbon neutral, 208
case studies, 8, 10, 28, 54, 80, 100, 109, 111, 113, 115, 126, 129, 150, 152, 177, 184, 234, 240
case study, 28, 69, 101, 161, 163, 165, 167, 170, 182, 218, 239
cash, 205
casting, 67
categorization, 15, 16, 18
causal beliefs, 62, 177
causation, 51
Census, 179, 205, 219, 241, 242, 245, 247, 248
CERCLA, 129, 156, 172
certification, 183, 209, 227, 229, 230, 231, 232, 233, 236, 240
challenges, 6, 8, 17, 23, 28, 31, 38, 42, 43, 49, 55, 78, 80, 91, 92, 93, 98, 102, 104, 112, 138, 150, 156, 178
chemical, 12, 15, 31, 40, 44, 129, 132, 137, 138, 139, 140, 141, 177, 189, 227
chemicals, 44, 131, 132, 187, 191, 228, 235
Chicago, 2, 50, 76, 115, 116, 117, 118, 119, 120, 121, 122, 123, 124, 125, 132, 160, 165, 169, 179, 180, 207, 208, 209, 211, 232, 233, 234, 238, 243, 244
children, 17
chromium, 133
chronology, 116, 179, 180
citizens, 67, 115, 120, 162
City, 2, 62, 74, 124, 125, 167, 169
Civil War, 134
clarity, 92, 95, 149
classification, 18, 46, 118, 162, 241
Clean Air Act, 77, 96, 97, 102, 103, 111, 171, 175, 178, 186, 191
cleaning, 237
cleanup, 15, 24, 70, 130, 134, 136, 143, 150, 157
clients, 156
climate, vii, 3, 12, 16, 21, 45, 53, 62, 145, 161, 165, 189, 206, 208, 209, 210
climate change, 45, 62, 161, 165, 189, 206, 208
closure, 132, 134
CO2, 207, 208, 210, 231
coal, 208
coastal region, 51
cognitive psychology, 159
collaboration, vii, 3, 36, 117, 137, 143, 145, 190, 219
combustion, 189, 211
commercial, 31, 51, 63, 64, 102, 103, 104, 107, 125, 176
commodity, 185, 193, 195, 198, 209, 210, 224
communication, 14, 47, 62, 63, 83, 85, 89, 90, 91, 92, 93, 95, 121, 128, 130, 131, 137, 143, 157, 167, 171, 175, 222, 246
communities, 12, 15, 33, 51, 102, 115, 117, 118, 119, 120, 121, 123, 124, 127, 164, 217
community, 4, 10, 19, 20, 25, 31, 32, 44, 48, 50, 52, 59, 60, 67, 68, 69, 71, 79, 106, 117, 118, 120, 121, 127, 129, 130, 131, 133, 138, 140, 144, 152, 154, 163, 178, 202
compensation, 18, 19, 59, 63, 67, 74, 97, 136, 164
competing interests, 23, 136
competition, 41, 183, 185
competitive advantage, 218
competitive markets, 191, 192
competitiveness, 218
complement, 13, 35, 181
complex interactions, 98
complexity, 10, 38, 43, 44, 45, 49, 50, 52, 54, 93, 109, 140, 152
compliance, 8, 97, 107, 150, 186, 214, 216, 236, 239
complications, 221
composition, 103, 140, 208
compounds, 51, 100

computer, 59, 61, 63, 87, 144
computer simulation, 63
conception, 104, 134
conceptual model, 6, 8, 28, 29, 31, 35, 38, 39, 40, 41, 45, 53, 83, 105, 106, 107, 108, 109, 110, 112, 129, 137, 138, 139, 144, 147, 150, 177
conceptualization, 30
consensus, 18, 84, 115, 117, 154, 173, 177
consent, 57, 123, 164
conservation, 60, 68, 69, 71, 72, 116, 117, 118, 121, 123, 126, 134, 136, 142, 158, 163, 165, 166, 167, 170, 171, 181, 182, 183, 186, 187, 189, 190, 191, 195, 196, 202, 203, 205, 209, 210, 211, 214, 219, 221, 226, 233, 234, 235, 236, 238, 239, 240
conservation XE "conservation" programs, 181, 182, 183, 196, 202, 210, 226, 233, 234, 238, 240
constituents, 40, 52, 236
construct validity, 58
construction, 23, 70, 132, 166, 170, 177, 214, 216
consumer demand, 227, 232
consumer education, 232
consumer surplus, 101
consumers, 103, 185, 188, 190, 192, 193, 208, 209, 213, 226, 229, 232, 236
consumption, 60, 72, 193, 194
contaminant, 133
contaminated sites, 9, 24, 70, 128, 129, 131, 138, 140, 144, 151, 157, 161, 162
contamination, 101, 128, 129, 132, 135, 136, 138, 140, 166, 177, 227
content analysis, 46, 106
controversial, 66
cooperation, 202
coordination, 48, 240
copper, 135, 139
correlation, 49, 51
cost benefits, 104
cost saving, 74
covering, 207
credit market, 203, 211
crop, 44, 103, 165, 187, 188, 189, 209, 220, 227, 228, 234
crop production, 209
crop rotations, 209, 228
crops, 42, 103, 188, 209, 220, 227, 228, 249
CRP, 186, 221, 222, 223, 224, 225, 226, 238, 249, 250
CTA, 186
cultivation, 187, 189
curriculum, 169
customers, 195, 208
cycles, 52, 103
cycling, 15, 21, 41, 52

D

damages, iv, 59, 103, 122, 135, 136, 138, 139, 140, 220
data availability, 32, 38
data collection, 11, 85, 90, 94, 129, 153
data set, 51, 122, 177
data transfer, 53
database, 80, 82, 119
datasets, 52, 111, 126, 222
decision makers, 6, 7, 9, 14, 18, 23, 38, 45, 48, 50, 54, 83, 85, 89, 90, 91, 93, 94, 95, 120, 124, 125, 126, 148, 149, 151, 169, 177
deforestation, 12
degradation, 36, 51, 62, 63, 67, 68, 162
Delta, 218, 244
demand curve, 193
democracy, 68, 161
Department of Agriculture, 27, 103, 153, 170, 179, 181, 182, 185, 244, 245, 246, 247, 248
Department of Commerce, 219, 248
Department of Energy, 210, 248
Department of Health and Human Services, 179
Department of the Interior, 157
Department of Transportation, 180
deposition, 101, 103, 104, 114
depth, 30, 176
deregulation, 64, 174
destruction, 208, 209
developing countries, 66, 158
dichotomy, 136
dioxins, 103
direct measure, 48, 109, 142
directives, 4, 9, 146
discharges, 27, 98, 99, 100, 104, 176, 183, 197, 198, 199, 201, 203, 205
disclosure, 209
diseases, 3, 12, 145
displacement, 208
dissolved oxygen, 99, 102
distribution, 8, 11, 28, 85, 86, 87, 88, 108, 109, 111, 113, 114, 150, 153, 194, 204, 220, 221, 225, 232, 250
distribution function, 86
disturbances, 86
diversity, 71, 100, 158, 176, 179, 187, 190
dominance, 190, 219
double counting, 34, 47, 141, 142, 177
draft, 2, 27, 28, 45, 89, 172, 173, 175
drainage, 135, 156, 214
drawing, 6, 53, 112, 125, 150
drinking water, 47, 74, 101, 122, 139, 176, 177
drought, 86, 185

drugs, 98, 99
dynamic systems, 43
dynamism, 44

E

eco-labeling, 229
ecological data, 11, 52, 54, 76, 153
ecological indicators, 6, 92, 95, 147, 164, 169, 177
ecological information, 4, 10, 22, 38, 52, 72, 92, 95, 146, 152
ecological processes, 4, 31, 39, 44, 50, 114, 116, 126, 139, 177
ecological restoration, 134, 140
ecological systems, vii, 2, 3, 6, 12, 13, 14, 17, 18, 20, 21, 23, 25, 36, 39, 41, 43, 44, 45, 47, 49, 51, 55, 67, 72, 73, 90, 91, 92, 95, 101, 108, 116, 117, 118, 123, 125, 144, 147, 149, 150, 152, 154, 169, 173, 175, 177
Ecological Systems and Services, v, vii, 1, 2, 13, 156, 172, 173, 175, 176
ecology, 13, 17, 22, 38, 39, 43, 49, 73, 84, 127, 162, 163, 165, 168, 174, 175
economic assessment, 24
economic consequences, 50
economic incentives, 212
economic indicator, 33
economic losses, 217
economic performance, 235
economic systems, 72, 169
economic theory, 58, 108, 110, 169
economic values, 4, 9, 19, 20, 24, 32, 33, 37, 64, 67, 69, 71, 74, 107, 108, 155, 158, 175, 176
economics, 4, 13, 17, 22, 24, 41, 47, 64, 65, 72, 73, 84, 97, 103, 122, 124, 127, 146, 157, 158, 159, 160, 162, 163, 164, 166, 168, 170, 173, 174, 175, 176, 242
economy, 73, 143, 158, 159, 174
education, 4, 146, 162, 183, 184, 185, 221, 232, 237, 240
effluent, 40, 96, 98, 171
Elam, 1
electricity, 211
emission, 75, 187, 200, 207, 210, 211, 238, 239, 249
endangered species, 190, 191, 232
energy, 4, 15, 20, 60, 71, 72, 73, 103, 141, 145, 155, 157, 160, 163, 165, 169, 178, 208, 211, 227, 232
energy efficiency, 208
energy transfer, 20
enforcement, 25, 232, 233, 236
engineering, 13, 88, 131, 175
England, 114
enrollment, 221
environment, 3, 12, 15, 16, 36, 63, 64, 67, 68, 72, 114, 116, 122, 134, 138, 139, 145, 157, 162, 164, 167, 169, 171, 172, 173, 182, 184, 197, 212, 229, 240
environmental characteristics, 59, 101
environmental conditions, 63, 97
environmental contamination, 128
environmental degradation, 67, 68
environmental economics, 64, 65, 122, 162, 166, 168, 170
environmental effects, 65, 105
environmental factors, 32
environmental impact, 100, 129, 232, 235
environmental issues, 122
environmental organizations, 115
environmental policy, 19, 69, 104, 163, 186
environmental protection, 115, 226
Environmental Protection Act, 25
Environmental Protection Agency, v, 1, 12, 128, 161, 164, 165, 169, 171, 172, 173, 185, 204, 248
environmental quality, 46, 50, 66, 96, 128, 170, 182, 185, 190, 191, 197, 239
environmental resources, 67, 70, 163
environmental services, vii, 67, 77, 140, 181, 182, 183, 184, 185, 187, 188, 190, 191, 192, 193, 194, 195, 196, 203, 212, 213, 214, 216, 217, 220, 221, 226, 229, 232, 234, 235, 236, 237, 238, 239, 240, 249
environmental stress, 39, 98
environmental threats, 100
equipment, 41, 99, 189, 198
equity, 23
erosion, 21, 190, 209, 221, 227
ethics, 62, 159, 162
EU, 66, 158, 207, 208, 244
Europe, 168
European Union, 207, 208, 244
evaporation, 135
evapotranspiration, 49
evidence, 51, 66, 68, 78, 93, 159, 162, 169, 191, 192, 214, 226, 234
exclusion, 95
execution, 87
Executive Order, 24, 27, 28, 29, 96, 97, 104
executive orders, 3, 20, 150
exercise, 21, 26, 31, 39, 55, 94, 112, 120, 123, 176, 190, 219
expenditures, 59, 65, 74, 158
experiences, 36, 144
expertise, 9, 12, 25, 48, 91, 115, 116, 120, 124, 127, 151
exploration, 33, 93
exposure, 44, 51, 101, 114, 129, 139, 140, 172

external constraints, 7, 26, 148
externalities, 167, 174, 188, 194

F

farm income, 210, 242
farmers, vii, 181, 182, 183, 184, 185, 187, 195, 201, 202, 203, 210, 211, 217, 218, 219, 220, 223, 227, 228, 229, 231, 232, 234, 236, 239, 240
farmland, 165, 188, 191, 214, 218, 227, 232
farms, 100, 174, 183, 188, 190, 192, 196, 201, 202, 203, 204, 205, 210, 218, 219, 220, 228, 232, 234, 237, 240, 242, 249
federal government, 177
Federal Government, 183, 210
federal programs, 176
Federal Register, 66, 157, 175, 177, 230, 247, 248
fee hunting, 218, 219, 220, 221, 225, 226, 234, 238, 240, 249
feed additives, 99
feedback, 118, 120, 156
fencing, 189, 220
fermentation, 189
fiber, vii, 15, 181, 182, 188, 233
filtration, 74, 122
financial, 85, 94, 115, 116, 149, 184, 185, 191, 196, 203, 206, 211, 212, 221, 229, 238, 240
financial incentives, 211, 240
financial resources, 196, 203
financial support, 212
fires, 86
fish, 12, 48, 49, 51, 64, 69, 81, 100, 101, 103, 107, 135, 137, 139, 140, 142, 161, 164, 166, 177, 190, 213
Fish and Wildlife Service, 72, 139, 214, 217, 219, 248
fisheries, 51, 120, 168, 169, 176, 177
fishing, 64, 75, 79, 81, 99, 101, 102, 104, 107, 133, 135, 136, 137, 139, 142, 173, 177, 179, 188, 213, 249
flexibility, 197, 239
flight, 236
flooding, 119, 124, 137, 142
fluctuations, 52, 161
focus groups, 30, 32, 48, 59, 62, 66, 106, 123, 137
food, vii, 12, 15, 39, 99, 100, 135, 136, 142, 181, 182, 187, 188, 226, 227, 229, 230, 232, 233
food additives, 99
food chain, 100, 135
food safety, 226, 229
Food, Conservation, and Energy Act of 2008, 184, 185, 235, 250
forecasting, 157
forest ecosystem, 12, 53, 170
forest habitats, 120
formation, 16, 187, 201
foundations, 22, 169
framing, 22, 92, 136
free riders, 193
freedom, 115
freshwater, 160, 166, 190
funding, 116, 128, 206, 208
funds, 90, 116, 132, 214

G

GAO, 248
GDP, 19, 73, 156, 165
General Accounting Office, 236, 248
geography, 170
Georgia, 125, 216
global climate change, 62, 165, 189, 206
global scale, 3, 12, 145
global warming, 12, 189, 209, 227, 249
God, 123
goods and services, 3, 12, 18, 19, 21, 41, 59, 63, 64, 67, 97, 145, 174, 185, 191, 192, 193, 196, 236
goose, 69, 166
governance, 21
government intervention, 185, 187
government payments, 242
government policy, 177, 226
governments, 9, 24, 114, 115, 126, 151, 184, 185, 190, 191, 212, 219, 235
GPRA, 25, 156, 176
grasses, 209, 220
grasslands, 133, 182, 189, 190
grazing, 189, 210, 220
Great Lakes, 188
greenhouse, 174, 183, 188, 189, 195, 206, 207, 208, 209, 210, 232, 239, 240, 249
greenhouse gas emissions, 189, 207, 210
greenhouse gases, 183, 189, 195, 206, 208, 209, 239, 249
gross domestic product, 19
groundwater, 43, 44, 47, 119, 132, 133, 134, 135, 138, 139, 229
group processes, 89, 136
grouping, 92
groupthink, 89
growth, 12, 31, 43, 48, 62, 102, 103, 143, 159, 162, 209
guidance, 18, 26, 28, 30, 32, 45, 55, 92, 96, 97, 110, 118, 139, 167, 173, 210
guidelines, 35, 50, 54, 91, 95, 96, 98, 149, 158, 164, 171, 235
guiding principles, 176

H

habitat, vii, 16, 33, 42, 50, 51, 60, 69, 74, 75, 99, 114, 120, 124, 125, 126, 127, 136, 138, 142, 161, 165, 168, 176, 181, 182, 186, 188, 190, 191, 218, 219, 220, 221, 222, 225, 226, 229, 230, 232, 233, 234, 235, 239
habitat quality, 50
habitats, 119, 120, 134
harbors, 125
harmful effects, 214
harvesting, 137
hazardous substances, 135
hazardous waste, 3, 24, 74, 132, 150
headache, 77
health, 12, 17, 46, 48, 77, 100, 132, 133, 157, 159, 163, 164, 189, 227
Health and Human Services, 179
health effects, 28, 46
health insurance, 77
health status, 77
heavy metals, 98, 100, 132, 135, 139, 227
heterogeneity, 78, 111, 224
historical data, 218
history, 12, 52, 53, 132, 133, 134, 136, 196, 225, 241
homeowners, 101
homes, 76, 120, 132, 143
honey bees, 42, 163
hormones, 98, 100, 101, 177
host, vii, 181, 182, 188
House, 244, 245, 246
House of Representatives, 244
housing, 76, 101, 160
human dimensions, 44
human experience, 12
human health, 3, 12, 19, 23, 28, 29, 36, 46, 56, 89, 96, 100, 101, 124, 125, 129, 133, 134, 136, 138, 145, 148, 176
human values, 15, 36, 49, 78
human welfare, 26, 27, 33, 35, 71, 176
Hunter, 219, 245, 249
hunting, 142, 184, 188, 213, 218, 219, 220, 221, 222, 223, 224, 225, 226, 232, 233, 234, 237, 238, 240, 249, 250
hybrid, 163
hydroelectric power, 163
hydrogen, 100, 189

I

identification, 28, 30, 35, 46, 47, 48, 84, 104, 107, 147
impact assessment, 76, 79, 89, 96, 109
Impact Assessment, 25
impacts, 5, 9, 12, 28, 39, 85, 86, 88, 90, 95, 97, 98, 100, 101, 102, 103, 104, 105, 106, 108, 109, 110, 111, 112, 115, 116, 117, 118, 119, 120, 124, 125, 126, 127, 128, 129, 135, 136, 138, 139, 140, 141, 142, 143, 151, 161, 188, 209, 214, 225, 229, 234
improvements, 11, 36, 63, 65, 66, 76, 77, 85, 89, 90, 94, 99, 102, 111, 136, 153, 157, 166, 176, 197, 227, 234, 238, 239, 249
income, 19, 77, 78, 79, 176, 182, 191, 192, 193, 210, 211, 216, 218, 219, 220, 223, 224, 225, 232, 242, 249
Indians, 136
indigenous peoples, 231
indirect effect, 73, 163
individual differences, 93
individuals, 1, 18, 19, 20, 21, 22, 25, 30, 32, 33, 46, 57, 59, 63, 67, 69, 77, 81, 97, 102, 106, 109, 123, 124, 136, 138, 177, 191, 193, 194, 208, 219
industrial wastes, 132
industry, 96, 98, 100, 115, 126, 134, 171, 183, 209
inferences, 61, 104, 108, 109, 174
information exchange, 144
informed consent, 57
infrastructure, 115, 116, 125, 144, 180, 228, 232
ingredients, 227, 230
initiation, 226
injuries, 135, 140, 168
injury, iv, 60, 74, 136, 138, 139, 140
insects, 139, 142
inspections, 214
institutions, 191, 192, 208, 239, 240
integration, 47, 138, 174
integrity, 50, 71, 118, 163, 164
interface, 125
internal reduction, 208
Internet, 237
intervention, 185, 187, 240
intrinsic value, 17, 121, 136, 175
invertebrates, 12, 117, 135
investment, 121, 181, 196, 202, 216
investments, 5, 38, 74, 111, 117, 146, 181
Iowa, 208, 216, 227, 231, 245, 248
Ireland, 244
iron, 133
irrigation, 100, 176, 189, 209
issues, 14, 18, 22, 23, 24, 25, 50, 53, 55, 76, 87, 89, 91, 95, 102, 111, 114, 116, 117, 122, 139, 154, 159, 162, 167, 173, 178, 183, 184, 208, 209, 220, 229, 232, 233, 234
iteration, 40, 91

J

juries, 59, 67, 68, 69, 71, 123, 155, 157, 158
justification, 109

K

Kentucky, 125, 216
kill, 212

L

labeling, 181, 184, 186, 226, 227, 229, 232, 233
lakes, 21, 28, 79, 100, 103, 118, 133, 164, 188
landfills, 133, 208
landings, 134
landscape, 31, 47, 60, 61, 62, 70, 71, 114, 118, 124, 125, 134, 143, 159, 166, 168, 169, 177, 190, 192, 237, 242
landscapes, 114, 174
laws, 12, 35, 96
lead, 12, 19, 31, 34, 64, 81, 88, 89, 90, 104, 107, 115, 129, 130, 133, 138, 141, 143, 167, 185, 210, 220, 225, 226
leadership, 208
leakage, 210, 239
legislation, 68, 96, 166, 216
lending, 239
liability insurance, 221
livestock, 100, 189, 211
living environment, 12, 15
loans, 239
local community, 117, 129, 131
local conditions, 139
local government, 9, 24, 126, 134, 151, 185, 212, 214, 235
logging, 30
Louisiana, 216

M

machinery, 99
magnesium, 133
magnitude, 8, 20, 30, 31, 33, 37, 38, 49, 51, 57, 58, 72, 86, 87, 90, 94, 98, 99, 106, 109, 139, 150, 177, 188, 213
Maine, 63, 103, 174, 228
majority, 42, 67, 99
mammals, 12, 117, 190
manganese, 133
manipulation, 22
manufacturing, 134, 176
manure, 27, 28, 100, 176, 177, 189, 209, 211
mapping, 22, 31, 36, 38, 41, 43, 98, 111, 118
market failure, 191, 192
market supply curve, 193
marketing, 213, 219, 225, 226, 228, 232
Maryland, 2, 216
materials, 15, 41, 71, 72, 100, 119, 153, 154, 157, 177, 189, 227
matter, iv, 35, 38, 43, 47, 100, 105, 121, 130, 131, 138, 143, 156, 160, 187, 189, 207, 227
measure of value, 72
measurement, 4, 11, 64, 66, 81, 85, 90, 94, 97, 146, 153, 162, 175, 208, 209
measurements, 52, 110
media, 131
median, 67, 224
mental model, 30, 46, 62, 86, 137, 158, 167
mercury, 103, 132, 133
messages, 92
meta-analysis, 6, 49, 51, 54, 66, 68, 101, 148, 158, 164, 167, 170
metals, 98, 99, 100, 101, 102, 132, 133, 135, 139, 177, 227
metamorphosis, 130
methodology, 62, 134, 140, 158, 169
Mexico, 207
Miami, 200
microorganism, 12, 15
migration, 44, 190
military, 134
mining, 135, 140
miscommunication, 18
mission, vii, 3, 12, 13, 25, 36, 37, 128, 145, 213
missions, 12, 25, 103, 197, 207, 244
Missouri, 216, 246
modeling, 10, 11, 27, 29, 38, 40, 43, 44, 46, 49, 50, 51, 54, 63, 84, 86, 106, 137, 139, 152, 153, 155, 164, 167, 173, 174, 178, 234
modelling, 69, 157
modification, 221
monitoring, 11, 51, 54, 137, 138, 140, 152, 164, 194, 214, 216, 234
Montana, 2
Monte Carlo method, 88
morbidity, 65, 176
mortality, 12, 65, 100, 173, 176
mortality risk, 173
motivation, 67, 162, 187
multimedia, 61, 171
multiple factors, 123
multivariate statistics, 51

N

narratives, 50, 59, 61, 62, 93, 155
National Aeronautics and Space Administration, 179

National Ambient Air Quality Standards, 112
national income, 176
national income accounts, 176
National Park Service, 72
national policy, 161
National Pollutant Discharge Elimination System, 171
National Priorities List, 132, 134, 157
national product, 226
National Research Council, 14, 36, 46, 74, 94, 157, 167, 178
National Science Foundation, 52, 54, 153, 157, 179, 180
National Survey, 179, 180, 222, 248
native population, 107
native species, 99
natural disturbance, 177
natural resource management, 63, 175
natural resources, vii, 66, 125, 160, 163, 181, 182
natural science, 115
natural sciences, 115
New England, 114
NGOs, 217
nickel, 135, 139
nitrogen, 12, 43, 52, 99, 100, 101, 103, 104, 189, 196, 201, 202, 203, 204
nitrogen gas, 43, 189
nitrous oxide, 189
noise, 61, 65, 166, 170
North America, 162, 243, 245, 249
Northern Ireland, 244
NPL, 75, 132, 136, 157, 173
nuisance, 189
nutrient, 16, 21, 39, 41, 69, 100, 103, 114, 187, 188, 189, 195, 203, 205, 209, 227, 234, 235, 237, 238
nutrients, 15, 43, 44, 99, 100, 101, 102, 186, 188, 189, 203, 227
nutrition, 226

O

objectivity, 166
observed behavior, 64
obstacles, 200, 228, 232
oceans, 51
Office of Management and Budget, 4, 18, 24, 25, 55, 85, 96, 157, 173, 176
officials, 15, 124
OFPA, 226
OH, 200
oil, 50, 208
oilseed, 227
Oklahoma, 216, 246
omission, 138
open spaces, 168, 190
operations, 27, 79, 96, 100, 171, 176, 189, 194, 197, 208, 211, 216, 221, 227, 228, 230, 231, 232, 249
opportunities, 4, 8, 24, 59, 69, 79, 84, 89, 112, 114, 115, 116, 121, 122, 124, 130, 146, 150, 154, 190, 203, 210, 213, 218, 221, 225, 232, 238, 241
Opportunities, 129, 243, 246, 248
organic compounds, 132, 189
Organic Foods Production Act, 226
organic matter, 43, 100, 187, 189, 227
organism, 48
organize, 44
organizing, 52
outreach, 93, 202, 233, 237, 240
outreach programs, 203
overlap, 34, 91
overlay, 72
oversight, 4, 7, 25, 26, 37, 80, 83, 97, 146, 148
ownership, 190, 194, 219
oxygen, 46, 99, 102
ozone, 12, 103, 124

P

Pacific, 225, 229
parallel, 10, 126
parameter estimates, 32
participants, 2, 4, 22, 58, 62, 68, 106, 117, 123, 146, 183, 193, 194, 195, 196, 197, 198, 210, 214, 217, 237, 239, 240
pasture, 188, 221, 228, 230
pathogens, 99, 100, 101, 188
pathways, 28, 43, 129, 135
peer review, 26, 29
penalties, 197, 227
percolation, 119
performance, 25, 87, 98, 130, 131, 143, 182, 183, 195, 200, 201, 202, 233, 234, 235, 236, 238
permission, iv, 197, 220
permit, 27, 44, 75, 83, 119, 176, 197, 221, 232, 237
personal benefit, 67
personal values, 21, 194
pesticide, 12, 189, 226, 227, 229
pesticides, 98, 99, 100, 133, 186, 188, 189, 229
pests, 3, 12, 145
petroleum, 161
pH, 103
pharmaceuticals, 177
phosphorus, 99, 100, 198, 200, 203, 204
photographs, 61, 92, 95
photosynthesis, 103
physical characteristics, 118, 177
physical environment, 15
physical well-being, 15

plant growth, 44
plants, 12, 43, 44, 98, 101, 134, 187, 197, 207, 248
playing, 201, 236, 238
policy choice, 69, 85, 118, 120, 163
policy initiative, 232
policy instruments, 185, 186
policy makers, 3, 13, 29, 33, 35, 48, 71, 88, 90, 104, 107, 109, 145, 178
policy making, 39, 149, 154, 177
policy options, 22, 28, 30, 34, 35, 44, 59, 71, 85, 86, 90, 94
politics, 68, 161, 162, 163, 169
pollination, 3, 12, 15, 16, 42, 145, 163, 165
pollinators, 42
pollutants, 27, 28, 96, 100, 101, 103, 105, 119, 139, 172, 176, 177, 188, 189, 197, 201
pollution, 12, 21, 39, 46, 74, 75, 103, 114, 158, 177, 183, 184, 188, 190, 196, 197, 198, 199, 203, 204, 205, 227, 232, 234, 238, 240
ponds, 135, 187
population, 8, 11, 30, 31, 48, 57, 58, 60, 62, 67, 72, 78, 98, 99, 100, 102, 107, 108, 111, 113, 114, 124, 133, 136, 137, 139, 140, 150, 153, 162, 167, 168, 177, 185, 219, 241, 250
population growth, 62, 162
positive externalities, 188
potato, 229
power plants, 197, 207
precipitation, 119, 135
preservation, 20, 78, 115, 116, 120, 164, 182
President, 249
prevention, 50
price changes, 103
principles, 14, 21, 24, 33, 35, 36, 47, 72, 92, 95, 107, 157, 159, 161, 171, 175, 176, 178, 185
prior knowledge, 91, 92
prioritizing, 5, 28
private benefits, 221
private good, 183, 192, 197, 198, 213, 218, 226
private investment, 181
probability, 62, 86, 87, 89, 128, 218, 241, 242
probability distribution, 86, 87, 89
producers, 103, 177, 183, 185, 188, 189, 191, 192, 193, 195, 196, 197, 201, 202, 203, 205, 206, 211, 213, 215, 216, 217, 218, 220, 221, 229, 232, 234, 235, 236, 237, 238, 239, 240, 249
product market, 193
production function, 6, 10, 31, 38, 39, 41, 42, 43, 45, 46, 49, 51, 53, 54, 66, 98, 111, 118, 120, 125, 127, 137, 138, 139, 141, 144, 147, 152, 170
production possibility frontier, 193
production technology, 197
productivity, 15, 16, 31, 42, 43, 44, 49, 51, 52, 131, 186, 227
profit, 183, 192, 197, 198, 217
program outcomes, 176
project, 13, 28, 44, 69, 82, 86, 124, 125, 126, 127, 137, 161, 175, 200, 208, 209, 210, 211, 235, 249
proliferation, 220, 229, 232
proposed regulations, 4, 29, 146
proposition, 241
protection, iv, vii, 2, 3, 5, 7, 12, 13, 14, 16, 17, 20, 21, 24, 25, 26, 27, 28, 30, 32, 33, 36, 37, 55, 60, 64, 74, 88, 91, 96, 108, 109, 115, 116, 117, 118, 120, 121, 123, 125, 126, 127, 128, 131, 138, 144, 145, 148, 154, 156, 162, 164, 169, 175, 177, 190, 206, 217, 226, 229
psychological value, 60
psychology, 13, 17, 22, 62, 84, 159, 162, 175
public access, 169, 221
public concern, 5, 6, 35, 36, 54, 90, 95, 106, 147, 150
public concerns, 5, 6, 35, 36, 54, 90, 95, 106, 147, 150
public discourse, 30
public education, 4, 146
public goods, 68, 166, 169, 192, 197, 212, 229, 232
public health, 227
public interest, 128
public investment, 181
public opinion, 158
public parks, 214
public policy, 21, 88, 178
public support, 21, 165
publishing, 27
Puerto Rico, 72, 163
pumps, 189
purification, vii, 3, 16, 122, 138, 145

Q

quality improvement, 77, 101, 122, 124, 158, 162, 178, 197, 202, 239
quality of life, 115, 116, 120, 176
quality of service, 182, 195, 212
quality standards, 25, 96, 209
quantification, 4, 27, 31, 33, 71, 103, 109, 136, 146

R

Radiation, 1, 2, 176
radon, 62, 158
ranchers, vii, 181, 182, 185, 187, 197, 234, 240
rangeland, 190, 208, 209
rating agencies, 208
rating scale, 92, 179

raw materials, 41
reading, 154
real estate, 163
real time, 131
realism, 159, 169
reality, 124
reasoning, 143
recognition, vii, 12, 16, 28, 37, 130, 133, 208
recommendations, iv, vii, 2, 3, 5, 8, 9, 10, 15, 22, 24, 26, 38, 45, 94, 96, 112, 118, 120, 126, 127, 133, 144, 146, 150, 151, 152, 154, 156, 171
recovery, 118, 122, 140, 160, 211, 216
recovery plan, 160
recreation, 3, 29, 31, 39, 63, 65, 69, 71, 77, 78, 101, 118, 122, 124, 125, 134, 138, 142, 143, 145, 158, 162, 170, 174, 175, 176, 178, 179, 190, 218, 219, 220, 250
recreational, 21, 27, 30, 65, 75, 79, 80, 81, 84, 99, 101, 102, 103, 104, 108, 121, 122, 124, 133, 135, 136, 137, 138, 139, 142, 143, 162, 166, 173, 177, 179, 190, 221
recycling, 16
redevelopment, 9, 10, 24, 128, 129, 130, 131, 133, 134, 137, 138, 139, 142, 143, 144, 151, 152, 174
regional policy, 15
regionalism, 164
regionalization, 52
Registry, 210, 233, 238, 249
regulations, 3, 4, 5, 12, 27, 29, 77, 96, 112, 135, 140, 145, 146, 171, 176, 178, 188, 191, 197, 212, 217, 227, 232, 239
regulatory agencies, 236
regulatory impact analyses (RiAs), 4
regulatory requirements, 32
rehabilitation, 118
relevance, 4, 9, 30, 31, 35, 39, 57, 91, 145
reliability, 52, 53, 88, 177, 242
remedial actions, 130, 131, 141
remediation, 3, 9, 10, 24, 128, 129, 130, 131, 133, 136, 137, 138, 139, 140, 141, 142, 143, 144, 151, 152, 161, 162
removals, 210
renewable energy, 208
replacement, 74, 75, 101, 122, 136, 142
reproduction, 12
requirements, 4, 24, 25, 26, 27, 29, 32, 53, 57, 75, 97, 99, 100, 115, 146, 176, 177, 197, 199, 211, 227, 236, 237, 239
researchers, 19, 20, 50, 51, 52, 57, 63, 176
residential neighborhood, 134
residues, 227
resilience, 164
resistance, 89
resolution, 163
resource allocation, 144
resource management, 63, 175
response, vii, 30, 38, 41, 44, 45, 68, 93, 100, 125, 135, 140, 164, 169, 172, 177, 178, 181, 182, 184, 223, 229
restoration, 15, 24, 71, 74, 75, 114, 117, 118, 120, 121, 129, 131, 134, 136, 139, 140, 141, 143, 159, 161, 162, 173, 195, 214, 216, 218, 229, 230, 238, 239, 240
restrictions, 9, 151, 213
retail, 208, 232, 236, 237
retirement, 186, 210, 221, 247
revenue, 203, 219
rights, iv, 18, 64, 165, 197, 213, 219, 231, 238
risk, 10, 35, 36, 62, 65, 70, 88, 89, 90, 91, 95, 121, 129, 131, 137, 138, 139, 142, 144, 149, 152, 157, 159, 163, 164, 167, 170, 171, 172, 173, 176, 195, 216, 218, 235, 237
risk assessment, 10, 35, 36, 88, 129, 131, 138, 139, 144, 152, 164, 167, 171, 172
risk perception, 164
risks, 12, 65, 93, 100, 106, 125, 128, 130, 133, 164, 171, 177, 208, 228, 232
roots, 42
rotations, 189, 209, 227, 228
rules, 12, 24, 25, 32, 52, 78, 96, 97, 112, 123, 141, 158, 176, 178, 184, 195, 207, 223, 233, 239
runoff, 100, 101, 119, 135, 138, 142, 187, 188, 196, 198, 201, 235

S

SACE, 214
safety, 96, 133, 157, 226, 229
salmon, 79, 124, 229, 230
salts, 100, 101, 177, 188
savings, 74, 197, 198
scaling, 53
scarce resources, 56, 104
scarcity, 70
school, 134
science, vii, 3, 6, 7, 8, 11, 13, 14, 17, 21, 22, 46, 48, 50, 53, 54, 57, 59, 61, 69, 80, 84, 85, 94, 106, 107, 112, 113, 118, 123, 124, 126, 127, 131, 136, 143, 145, 147, 148, 149, 150, 152, 155, 164, 167, 169, 172, 173, 174, 175, 177, 178
Science Advisory Board (SAB), vii, 2, 13, 172, 173
scientific knowledge, 45, 124, 126
scientific understanding, 4, 146
scope, 26, 32, 100, 101, 103, 110, 140, 174, 234
screening, 30, 80, 135, 176, 203, 204
Secretary of Agriculture, 235
security, 68, 132, 158

sediment, 50, 139, 174, 187, 188, 189, 235
sedimentation, 21, 186
sediments, 21, 100, 135, 139, 161
self-interest, 19, 176
seller, 194
sellers, 183, 191, 192, 198, 201, 207, 221, 237, 238
semi-structured interviews, 177
sensing, 236
sensitivity, 7, 45, 87, 88, 89, 92, 94, 149, 177, 178
service provider, 32, 236
sewage, 98, 197, 237
shape, 87, 193
shareholders, 208
shellfish, 101, 177
shortfall, 117
showing, 40, 104, 126
signals, vii, 108, 181, 182, 185
simulation, 44, 63, 162, 163, 174, 200, 221, 222, 238
simulations, 93
site use, 15
soccer, 134, 138
social consequences, 5, 21, 32, 34, 39, 60, 147
social costs, 176
social justice, 229
social sciences, 6, 40, 52, 54, 127, 147, 177
social welfare, 193
society, vii, 3, 4, 12, 13, 16, 19, 21, 30, 36, 39, 60, 74, 142, 145, 167, 181, 182, 185, 190, 192, 194, 229
sociology, 84
software, 87
soil erosion, 221, 227
soil type, 53, 119, 214
solution, 93, 202, 226
South Dakota, 216, 222, 246, 250
specialty crop, 228
species, 12, 20, 23, 31, 36, 39, 41, 42, 44, 45, 47, 50, 51, 52, 79, 81, 84, 89, 99, 100, 103, 117, 118, 119, 120, 121, 123, 124, 126, 136, 137, 138, 139, 142, 163, 166, 169, 176, 187, 190, 191, 218, 227, 232
species richness, 20, 118
specifications, 79, 86
speculation, 207
Spring, 242
stakeholder groups, 125
stakeholders, 125
standardization, 47
state, vii, 3, 9, 12, 13, 14, 24, 25, 27, 29, 30, 36, 44, 84, 99, 106, 112, 114, 121, 126, 128, 132, 135, 140, 151, 163, 174, 192, 222
states, 25, 50, 60, 86, 128, 176, 179, 207
statistics, 51
statutes, 3, 20, 24, 115, 150, 164
storage, 15, 43, 122, 162, 187, 189, 211, 235
stormwater, 27, 176
strategic planning, 25
streams, 21, 28, 40, 44, 50, 51, 100, 118, 119, 135, 218
stress, 16, 31, 44, 138, 167
stressors, 10, 12, 31, 32, 38, 39, 40, 47, 51, 54, 98, 99, 107, 111, 114, 115, 118, 139, 152
structural changes, 100
structure, 6, 13, 26, 32, 35, 37, 41, 43, 51, 100, 147, 177
subsistence, 135, 176
substitutes, 59, 228
substitution, 73, 165
subsurface flow, 119
success rate, 103
sulfur, 12, 124, 135, 197
sulfur dioxide, 12, 124, 197
sulfuric acid, 135
Superfund, 24, 109, 128, 129, 133, 134, 135, 144, 161, 172, 173, 174
supplier, 203, 236
suppliers, 182, 185, 216, 221, 236, 237
supply curve, 193
surplus, 101, 103
surrogates, 72, 139, 170
survey, 4, 14, 21, 22, 23, 30, 56, 58, 59, 60, 62, 64, 66, 68, 77, 79, 95, 102, 108, 110, 112, 122, 123, 143, 146, 154, 156, 168, 169, 174, 176, 177, 178, 219, 220
survey design, 66, 156
survival, 12, 48
sustainability, 17, 71, 160, 176, 178, 235
Sweden, 245
Switzerland, 167

T

target, 103, 131, 221
taxa, 48
taxpayers, 195
teams, 125, 127
technical assistance, 185, 186
technical change, 73, 165
techniques, 25, 49, 55, 82, 99, 113, 144, 154, 178
technology, 27, 72, 98, 176, 193, 197, 236, 237
telephone, 61
temperature, 124
tenants, 138
tenure, 211
territory, 135
test procedure, 167
testing, 4, 23, 61, 115, 132, 143, 146, 167, 168

tetrahydrofuran, 132
textbook, 197
thoughts, 17
threats, 100, 118, 134, 178, 237
tissue, 131, 139
Title I, 98
Title II, 98
toluene, 133
tourism, 84, 143
toxic contamination, 166
toxic substances, 138
toxicity, 139, 161
trade, 23, 70, 75, 76, 163, 183, 195, 198, 200, 201, 207, 208, 210, 231, 238
trade-off, 23, 70, 76, 163
trading partner, 201, 202, 237
trading partners, 201, 202, 237
training, 220
transaction costs, 187, 192, 194, 201, 202, 210, 217, 221, 240
transactions, 194, 210, 214, 233, 240
transcripts, 59, 62, 106
transformation, 15, 20
transformations, 81
translation, 27, 89, 137
transparency, 40, 45, 47, 85, 92, 95, 110, 149, 249
transport, 228
transportation, 232
treatment, 21, 22, 75, 98, 101, 173, 177, 197, 237
trial, 144
turnover, 49

U

U.S. Army Corps of Engineers, 214, 216
U.S. Department of Agriculture XE "Department of Agriculture" (USDA), 182, 185
U.S. Department of Commerce, 219
U.S. Department of the Interior, 157
U.S. economy, 73, 160
U.S. Geological Survey, 1, 157, 203, 204, 248
UK, 66, 159, 161, 163, 246
uniform, 119, 226
unit cost, 183
United, v, 1, 11, 52, 68, 99, 104, 153, 158, 163, 166, 170, 171, 175, 181, 187, 188, 190, 191, 198, 203, 205, 207, 210, 211, 218, 219, 226, 232, 242, 243, 244, 245, 247, 248, 249
United Kingdom, 171, 175
United Nations, 187, 244
United States, v, 1, 11, 52, 68, 99, 104, 153, 158, 163, 166, 170, 181, 188, 190, 191, 198, 203, 205, 207, 210, 211, 218, 219, 226, 232, 242, 243, 245, 247, 248, 249
urban, 47, 65, 122, 129, 133, 158, 163, 165, 166, 168, 170, 171, 177, 190, 214, 218, 241
urban area, 122, 241
urban areas, 122, 241
urbanization, 226, 242
USA, 63, 73, 174
USDA, 1, 182, 183, 184, 185, 188, 189, 190, 191, 195, 200, 202, 203, 204, 206, 209, 211, 212, 214, 215, 217, 219, 221, 222, 225, 226, 227, 229, 230, 231, 234, 235, 236, 237, 238, 239, 241, 242, 247, 248, 249
utilitarianism, 19

V

validation, 7, 11, 85, 90, 94, 145, 148, 153, 183
Valuation, 2, 3, 4, 21, 23, 24, 25, 29, 32, 38, 39, 48, 49, 53, 55, 59, 66, 67, 80, 85, 86, 90, 91, 92, 96, 98, 100, 111, 114, 116, 120, 126, 128, 129, 131, 137, 142, 144, 145, 146, 148, 149, 150, 153, 154, 156, 157, 158, 160, 161, 164, 165, 170, 175, 190, 244
valuation surveys, 112
variables, 44, 45, 49, 50, 51, 58, 87, 88, 94, 114, 178, 241, 242
variations, 76, 87, 211
vegetables, 228
vegetation, 119, 134, 135, 139, 140, 142, 192
vehicles, 143
vision, 208
visions, 118
visualization, 63, 93, 143, 162
VOCs, 132
volatile organic compounds, 132, 189
vomiting, 77
voters, 67, 68, 116, 118, 166
voting, 67, 68, 122, 161, 164, 169, 174

W

Washington, 15, 64, 66, 68, 74, 75, 134, 156, 157, 158, 159, 160, 161, 162, 163, 164, 165, 166, 167, 170, 171, 172, 173, 174, 207, 216, 220, 227, 243, 245, 246, 248, 249
Washington, George, 134
waste, 3, 21, 22, 24, 39, 43, 74, 106, 128, 132, 135, 150, 187, 211
waste disposal, 39
waste treatment, 21, 22
wastewater, 27, 98, 176
water quality standards, 25
water quality trading, 182, 184, 195, 196, 197, 198, 200, 201, 202, 203, 211, 213, 235, 236, 238, 239, 240

water resources, 122, 135, 189
water supplies, 101, 177
watershed, 29, 70, 74, 114, 116, 117, 118, 119, 120, 121, 124, 140, 197, 201, 202, 214, 229, 235
waterways, 52, 177, 197, 221
web, 92, 178
web sites, 92
welfare, 4, 24, 26, 27, 28, 33, 35, 64, 71, 97, 103, 123, 146, 166, 176, 193
welfare economics, 4, 24, 97, 146, 176
well-being, 3, 4, 5, 9, 13, 15, 16, 17, 18, 19, 21, 24, 30, 31, 36, 52, 56, 60, 67, 70, 71, 84, 101, 106, 124, 126, 128, 130, 131, 141, 142, 143, 144, 145, 148, 151, 166, 176, 193
wells, 132, 133, 177, 220
West Virginia, 216
wetland mitigation, 215, 218, 234, 238, 239, 242
wetland restoration, 24, 195, 218, 238, 239
wetlands, 12, 47, 65, 69, 74, 108, 117, 119, 121, 122, 132, 133, 138, 157, 166, 169, 182, 183, 187, 189, 190, 195, 209, 212, 213, 214, 216, 218, 235, 236, 239, 240, 241, 242, 248, 249
wilderness, 47, 79
wildland, 63, 175
wildlife, vii, 120, 127, 134, 135, 137, 138, 140, 142, 143, 167, 171, 176, 181, 182, 185, 188, 189, 190, 196, 213, 214, 218, 219, 220, 221, 222, 223, 224, 225, 226, 229, 232, 233, 234, 235, 236, 237, 239, 249
Wisconsin, 114, 115, 198, 216, 227, 229, 230, 245, 249
woodland, 188
workers, 229, 231
working groups, 117
World Bank, 2, 167, 170
worldwide, 207

Y

yes/no, 242
yield, 16, 23, 28, 33, 56, 57, 58, 65, 78, 81, 97, 103, 225

Z

zinc, 135, 139